FLORA OF CONNEMARA
AND
THE BURREN

FLORA OF CONNEMARA
AND
THE BURREN

D. A. WEBB

Honorary Professor of Systematic Botany in the University of Dublin

and
MARY J. P. SCANNELL

Head of Herbarium, National Botanic Gardens, Glasnevin

with contributions by M. D. Guiry, M. E. Mitchell, A. R. Perry and W. A. Watts

ROYAL DUBLIN SOCIETY
and
CAMBRIDGE UNIVERSITY PRESS
Cambridge
London   New York   New Rochelle
Melbourne   Sydney

Published by the Royal Dublin Society and the Press Syndicate of the
University of Cambridge
The Pitt Building, Trumpington Street, Cambridge CB2 1RP
32 East 57th Street, New York, NY 10022, USA
296 Beaconsfield Parade, Middle Park, Melbourne 3206, Australia

First published 1983

Printed in Great Britain
at the University Press, Cambridge

Library of Congress catalogue card number: 82–4425

British Library Cataloguing in Publication Data
Webb, D. A.
Flora of Connemara and the Burren.
1. Flowers—Ireland—Connemara, Galway (County)
2. Flowers—Ireland—Burren, The, Clare (County)
I. Title   II. Scannell, M. J. P.
582.13′09417′4   QK307
ISBN 0 521 23395 X

CONTENTS

List of Illustrations   *page* vi

Acknowledgements   viii

Introduction   ix

The Flora region and its eight districts   xii

Geology and soils   xx

Climate   xxv

Character of the flora   xxviii

Notes on some of the habitats   xxxi

The vegetation since the last glaciation   xxxvi

The history of our knowledge of the flora   xli

FLOWERING PLANTS, CONIFERS AND PTERIDOPHYTES   1

Notes on the text   3

Dicotyledons   5

Monocotyledons   195

Gymnosperms   260

Pteridophytes   263

Mosses and liverworts   279

Lichens   284

Marine algae   287

Freshwater algae   292

Fungi   294

Bibliography   295

MAPS, INDEXES   303

Map 1. Connemara (comprising districts 5, 6, 7 and 8,   304–5
with part of 4)

Map 2. The Burren (comprising districts 1, 2, and 3, with   306–7
parts of 4, 5 and 6)

Topographical index   308

Index of English and Irish names   315

Index of Latin names   319

ILLUSTRATIONS

Colour Plates

Between pp. 114 and 115
1 The Twelve Pins (*D. C. Cabot*)
2 (*a*) *Daboecia cantabrica* (*V. Gordon*). (*b*) *Saxifraga spathularis*. (*c*) *Erica mackaiana* (*M. C. F. Proctor*)

Between pp. 146 and 147
3 (*a*) *Spiranthes romanzoffiana*. (*b*) *Neotinea maculata* (*D. C. Cabot*). (*c*) *Orobanche alba* (*D. C. Cabot*). (*d*) *Potentilla fruticosa*
4 (*a*) The Cliffs of Moher (*D. C. Cabot*). (*b*) *Rhodiola rosea* at the Cliffs of Moher (*D. C. Cabot*). (*c*) *Adiantum capillus-veneris* at Poulsallagh (*D. C. Cabot*)

Figures

 1 Aerial view of Inishmore (Aran Islands), towards the western end
   (*Air Corps, courtesy of the Stationery Office, Dublin*)   *page* xiv
 2 Aerial view of Inishmore (Aran Islands), south-west of Kilrónan (*Air
   Corps, courtesy of the Stationery Office, Dublin*)   xv
 3 Glaciated granite landscape on Gorumna Island   xviii
 4 Shattered and smooth types of limestone pavement in the Burren
   (*D. C. Cabot*)   xxii
 5 Wind-shorn hawthorn at Gleninagh (*R. Welch, courtesy of the
   National Botanic Gardens, Glasnevin*)   xxvi
 6 *Dryas octopetala* and *Neotinea maculata* near Black Head (*R. Welch,
   courtesy of the Ulster Museum*)   xxx
 7 Wooded lake-island near Maam Cross   xxxiii
 8 The Hill of Doon (*Bord Fáilte Eireann*)   xxxiv
 9 Evidence of continuous human occupation of the Burren:
   gallery-grave, stone ring-fort, high cross, tower-house   xxxviii
10 Leamanegh Castle   xxxix
11 Holly-tree on limestone pavement near Mullaghmore   49
12 *Sedum anglicum* at Dog's Bay   79
13 *Gunnera tinctoria* naturalized near Leenane (*D. C. Cabot*)   83
14 Comparison of *Erica tetralix* with *E. mackaiana*   131
15 *Cuscuta epithymum* on the dunes at Fanore (*D. C. Cabot*)   148
16 *Ajuga pyramidalis* at Poulsallagh (*R. Welch, courtesy of the Ulster
   Museum*)   172
17 *Baldellia ranunculoides* near Carraroe   216
18 *Eriocaulon aquaticum* at Maam Cross   225
19 Yew-trees in the Burren   261
20 Limestone pavement near the sea at Black Head   267
21 *Ceterach officinarum* in its natural habitat   269
22 Hazel-scrub in the Burren, with rich growth of mosses   282
23 Maerl (fragmented calcareous algae) constituting one of the 'coral
   strands' near Ballyconneely   288

24 A small-holding in southern Connemara (*Bord Fáilte Eireann*)   290
25 Solution-hollows in limestone pavement near Poulsallagh, filled with
     Nostoc (*D. C. Cabot*)   293

Figures or plates for which no acknowledgement is given were taken by the authors.

ACKNOWLEDGEMENTS

Apart from those who helped us in the collection of field records (p. xi), we should like to thank the following for assistance in the production of this book:

A. O. Chater, R. Czapik, the late J. E. Dandy, L. Farrell, P. Hackney, A. C. Jermy, B. Jonsell, C. Page, R. J. Pankhurst, F. H. Perring, C. Preston, A. J. Richards and D. M. Synnott for identification of specimens or information about herbaria or literature.

Mr M. Gardiner of An Foras Talúntais, Mr T. McEvoy of the Department of Forestry, and the staff of the Meteorological Office, for information about soils, afforestation and climatic data respectively.

Dr A. Phillips and Mr C. McDermott for guidance on the geology of the region.

Dr M. de Valéra for advice on Irish names of plants.

Mrs C. MacDaeid for help with the bibliography and with the search for first records.

Mrs S. Reynolds for preparing the maps, checking the species-lists, and helping with proof-reading and the preparation of the index.

The persons and institutions named in the List of Illustrations for permission to reproduce the photographs taken by them or in their possession.

We should also like to thank the Department of Forestry and Fisheries and Bord Fáilte Eireann for grants towards the cost of colour plates; the Royal Irish Academy for a grant from the Praeger Fund towards the expenses of the earlier part of the field-work; and the National Botanic Gardens and Trinity College, Dublin for various facilities given to the authors.

The maps are based on the half-inch sheets of the Ordnance Survey of Ireland by permission (permit no. 3892).

INTRODUCTION

Connemara and the Burren are two of the most interesting regions of Ireland for the botanist; Kerry alone can rival them. But whereas Kerry has had since 1916 a county Flora which, though badly in need of updating, is still very useful and informative, no systematic account of the plants of Connemara and the Burren has been attempted on a larger scale than the summaries given by Praeger in his *A Tourist's Flora of the West of Ireland* (1909), reprinted with only minor alterations in *The Botanist in Ireland* (1934). Stimulating and useful though they are, these accounts deal for the most part only with rare species, and indicate the distribution even of these only in broad outline. When, therefore, the Irish committee of the Botanical Society of the British Isles decided in 1963 to give a focus to its activities by encouraging the production of a local Flora, it was agreed after a short discussion that Connemara and the Burren would form a suitable region.

Before going further we should explain the sense in which we use these terms. Connemara has never been an administrative unit. Originally the name was used to denote the westernmost part of Co. Galway – west, roughly, of a line running from Costelloe through Maam Cross to Leenane – but by the middle of last century it was being used more and more in a wider sense, and it gradually came to indicate all that part of the county which lies to the west of Galway city and Lough Corrib. It is in this sense (corresponding exactly to the biological vice-county H16) that we use it here. It forms a natural unit, apart from an arbitrary delimitation from Co. Mayo in the north-east; and in this connection it should be pointed out that here the boundary follows not the present administrative boundary, but that which existed before 1898 (for details see Webb, 1980*b*).

Burren, on the other hand, is one of the baronies into which Co. Clare is divided, forming its north-west corner. But a realization of the unique character of the landscape and flora which were shared by the barony and the country to the east and south-east of it led some time ago to the use of the term 'the Burren' to include all the karstic limestones of north Clare and the adjacent parts of south-east Galway. In this sense it too is a natural region, though it lacks a clear eastern boundary. We have adopted as the most convenient the main road from Galway to Ennis, as this approximates to the line which divides the mainly bare rock on the west from the mainly drift-covered area on the east.

Connemara and the Burren, so defined, are not widely separated, but they are not actually contiguous, and it might have been thought wiser to treat them in two separate Floras. But as most botanical tourists who visit the west of Ireland sample both Connemara and the Burren we thought it better to include both in one volume. Although sharply contrasted in their land-forms and their soils they share an extreme Atlantic climate which gives to their floras a partial unity.

Two smaller areas are included in our survey which belong neither to Connemara nor to the Burren. One is the narrow strip of drift-covered limestone at the head of Galway Bay, which the tourist must traverse in driving from Connemara to the Burren or vice versa; and as he will often spend a night in Galway we have added a small area to the north and north-east of the city. The other is a zone of shale lying to the south of the Burren, running from near Ennis to the coast south of Lehinch, and extending northwards in a large salient around Lisdoonvarna and Slieve Elva. The visitor to the Burren is likely to penetrate into this area, if only to see the Cliffs of

Moher, and we have therefore thought it worth while to draw his attention to the sharp contrast between the flora of the limestone and that of the shales.

We cover, therefore, in this volume the whole of the vice-county H16, a substantial part of H9 and small parts of H15 and H17. The precise limits of the region are set out on p. xii. So delimited, it occupies an area more than 1½ times that of an average vice-county, and there have been times when we wondered whether our project was too ambitious. To write a satisfactory regional Flora two conditions should be fulfilled; the author should live, or at least spend a large part of the year in or near his region, and he should have no other major botanical preoccupation. Neither of these conditions has been fulfilled for us. For most of the year we were both living in Dublin, separated from the Flora region by almost the entire breadth of Ireland, and our administrative commitments and editorial duties relating to other works meant that for long periods work on this Flora had to be set aside. Nor were many of our helpers more favourably situated, and although their contributions were invaluable they were made for the most part under difficulties – in flying visits, mostly in midsummer, and often hampered by the problems of organizing student field-trips or parties of very varied botanical expertise. Nevertheless we decided in 1978 that the work, imperfect though it was, should be hurried on towards publication, and we hope that we may be found to have provided at least a useful foundation (in earlier days we might have termed it a *Prodromus* or a *Tentamen* rather than a definitive Flora), which our successors can in due course extend or amend. Like Praeger (*Irish Topographical Botany*, 1901, p. viii), we believe that 'the chance of the remaining blanks... being filled up is much increased by their being herein pilloried and exposed to public notice'. For this reason we have tended in the case of most questionable records to err on the side of scepticism, in the belief that the omission of a true record is easily rectified later, while the insertion of a false one is likely to mislead for half a century or more.

We have as a general rule excluded casual aliens, by which we mean plants which do not reproduce themselves within the region and show no prospect of doing so; if they persist it is merely from constant re-introduction. This category includes not only many annuals of cultivated or waste ground, but also such plants as the vine and the fig, of which single plants have survived for several years in the city of Galway. Some species which we believe to have been merely casual, but which appear several times in the literature, are briefly mentioned, but casuals of which the latest record is more than 100 years old have been excluded. We have also gone to some trouble to distinguish between shrubs and trees which are naturalized (i.e. self-sown) and those which are merely planted – a distinction in which earlier writers have shown curiously little interest.

This book, like most local Floras, is meant to be used in the field in conjunction with an identificatory Flora which contains descriptions and keys. In practice, however, we realize that many users of it will not have such a book at hand, or may not be accustomed to its use. We have, therefore, within the limits of space available, included a little help towards identification of species within some of the genera, by indicating differential characters which can be described in simple language and which do not, in most cases, need the use of a lens for their recognition. There are, unfortunately, many genera such as *Callitriche*, *Potamogeton* and *Carex* for which this is impossible, but even for *Carex* we hope that our grouping of species in accordance with a few easily recognized characters may be of some help. We have also tried to give some assistance at generic level with those traditional bugbears of the amateur – umbellifers and yellow-flowered composites.

All records from 1959 onwards are treated as 'recent' rather than 'historical'. This date was chosen because, although it was not until 1962 that we began to accumulate records ourselves, 1959 was the year in which Ivimey-Cook and Proctor made most of the observations on which their masterly paper on the communities of the Burren (Ivimey-Cook & Proctor, 1966) was based. Our collection of records for the rare species was mostly done by *ad hoc* methods: search in the literature and in suitable herbaria, search in the field in recorded localities and in likely habitats, and the noting of chance finds. The data for the commoner species were accumulated, not by the now fashionable technique of searching 'tetrads' (squares of 2 × 2 km), since the large size of our region would not have permitted this with the man-power available, but by choosing a few (usually four or five) points in each 10 × 10 km square and recording all the plants to be seen within a radius of 400 m of the point. The points were chosen so as to give the maximum of ecological diversity; some were on mountain-tops, others on lake-shores or river-banks, some in bogs or marshes, some by the coast, and some in towns or villages. We believe that by this means we gained a fairly accurate estimate of the frequency of all the commoner species in each district of our region. The data were recorded on special cards, with coded information as to habitat and frequency. Time did not permit a search of all the points which were originally chosen, nor was there time to analyse some of the cards which came in rather late but, for each of our eight districts, lists from at least 20 points (and in several districts over 30) were analysed and the results entered in loose-leaf notebooks with a page for each species. From these, to which were added later data from the herbaria, the literature or other sources, were compiled the accounts which constitute the main text of this book. Only exceptionally are more than four localities cited for any one district; 5–7 occurrences are summarized as 'occasional', 8–11 as 'frequent'; 12–15 as 'very frequent'; and over 15 as 'abundant'.

In the collection of field data we were fortunate in receiving the help of a large number of botanists, both from Ireland and Britain, for without their help the total number of records would not have been much more than half of what it finally amounted to. In this respect, therefore, the Flora has a somewhat collegiate quality, and to emphasize this we have for recent records not cited the name of the recorder, except in a small number of cases where the rarity or the critical nature of the species makes such citation desirable. Similarly we have not given dates later than 1958 except for very rare species or those which seem to be showing a significant increase or decline.

It is perhaps invidious to pick out for special mention any names from the long list of our helpers, but we feel that we must mention the very special value of the work of Professor J. J. Moore (mainly in the Burren and Aran) and of Dr G. Halliday (mainly in district 8, and including the great majority of the mountain records). Our other helpers in the field included (and we say 'included', as we fear that our inadvertence or their modesty has led to the omission of some names) the following:

G. Argent, J. R. Akeroyd, H. J. B. Birks, E. M. Booth, C. Breen, Lady Anne Brewis, A. O. Chater, A. Conolly, T. Curtis, J. Cross, E. S. Edees, I. K. Ferguson, A. C. M. Folan, R. M. Goodwillie, V. Gordon, J. Hodgson, M. Keane, M. P. H. Kertland, E. N. Kirby, D. Lambert, M. McCallum Webster, D. C. McClintock, G. Mackie, M. E. Mitchell, H. O'Donnell, A. M. O'Sullivan, F. H. Perring, E. M. Rix, T. Robinson, C. Roden, J. Ryan, M. S. Skeffington, J. Shackleton, M. Telford, S. M. Walters, W. A. Watts, V. Westhoff, J. White, A. Willmot.

In conclusion we should add that we have borne in mind not only those who will use this book in the field, but also readers who will never visit the region but will use it for reference or for armchair reading. With this in view we have tried to present our data in as readable a form as possible. We have preferred place-names to austere lists of grid-squares, and we have included as much information on ecology, plant-geography, history and infraspecific variation as space permits.

THE FLORA REGION AND ITS EIGHT DISTRICTS

The region covered by this Flora has been defined in general terms on p. x, but a precise delimitation is desirable and is given here. On the west it is bounded by the Atlantic Ocean, but includes the Aran Islands, Inishbofin and Inishshark (though not Inishturk), as well as smaller islands nearer the shore. In the north, its boundary runs up the centre of Killary harbour to meet the land at the road-junction which lies less than 1 km south of the mouth of the Erriff R. It follows the county boundary as far as a point just east of the highest point 682 m (2239 ft) of the Maamtrasna plateau, but there diverges from it to follow the pre-1898 boundary, passing along the saddle south of Glenawough L. and north of the small lakes east of it to run first eastwards and then south-eastwards to reach the shore of L. Mask at a point 2½ km south-south-west of Toormakeady.

The eastern boundary of the region runs southwards from this point across L. Mask to meet the present county boundary east of Ferry Br., and then follows this boundary past Clonbur and Cong into L. Corrib. In L. Corrib it runs approximately down the middle (following the baronial boundary) to near the south end, where it cuts across to the east shore of the Menlough marble quarries. From here it follows by-roads through Ballindooly, Twomileditch and Doughiska to join the Galway–Limerick road just east of Rosshill Ho. The boundary then follows this main road as far as the major road-junction on the northern outskirts of Ennis.

From this road-junction the southern boundary follows the main road to Inagh and continues westwards along it for a further 4 km, but then diverges northwards to follow by-roads past White's Br. and Glenville Ho. to meet the sea just south of Moy Ho.

The total area of the region so delimited is approximately 3100 sq. km, including islands and all inland waters except those parts of L. Corrib and L. Mask which lie within the region.

It extends from 52° 51′ to 53° 38′ north latitude, and from 8° 48′ to 10° 18′ west longitude. It is covered by sheets 10, 11, 14 and 17 of the Ordnance Survey maps on a scale of ½ in. to a mile, and all the names mentioned here are to be found on these maps though in a few cases we have adopted a different spelling as more widely used.

We have divided it for floristic purposes into eight districts, as follows.

District 1. The Clare shales

Area, about 295 sq. km. Greatest height 348 m (1134 ft).

The southern and western boundaries are those of the region. On the north it is separated from district 2 by a line following by-roads from the coast at Fisherstreet past Ballynalackan Ho., Balliny and Ballyelly, thence round the north flank of Slieve Elva to Lismorahaun and Doonyvardan, passing later west of Noughaval to join the Ennis–Lisdoonvarna road 2 km east of Kilfenora and then following it eastwards to Roughan Ho. On the east it is separated from district 3 by this same road from Roughan Ho. to the point where it meets the Ennis–Inagh road 4 km north-west of Ennis.

The district coincides almost exactly with that part of the region composed of

Carboniferous shales and sandstones. The limestone is exposed in some areas around Kilfenora and north-west of Lisdoonvarna, but here it is mostly covered by siliceous downwash from the shales, and harbours only a few calcicole species. Elsewhere the vegetation is mainly poor, wet grassland, with some high-level moorland and a few patches of bog and acid fen. On the northern and eastern margins slight deviations from the geological boundary lead to the inclusion in the district of small patches of better grassland over partly calcareous drift, while in a few places, notably near O'Dea's Castle and Elmvale Ho. there are outcrops of limestone which bring in a fair number of Burren species. There are a few lakes, of which Inchiquin L. is the largest. Apart from a few scraps, as by L. Goller, there is no natural woodland, and most of the landscape is remarkably treeless, but demesne woods at Ennistymon and Lisdoonvarna bring in some woodland species. There is a fine sandy beach at Lehinch; otherwise the coast is rocky, culminating in the Cliffs of Moher, which, unlike most other cliffs described as 'sheer', are actually vertical; one can drop a pebble into the sea from a height of 160 m (530 ft) (Plate 4). Besides these cliffs and the Lehinch dunes the chief places of floristic interest are Inchiquin L., L. Raha, L. Goller and the valley of the Aille R.

Recorded species total 519,* of which 33 have not been seen recently. The species known only from this district are *Vaccinium oxycoccos*, *Butomus umbellatus* and the hybrid *Potamogeton × nerviger*.

District 2. The Burren hills

Total area (including the Aran Islands) about 345 sq. km. Greatest height 329 m (1073 ft).

The western and northern boundaries are formed by the sea; the southern has been defined for district 1. On the east it is separated from district 3 by the 200 ft contour, running from Roughan Ho. northwards to where it meets the road on the east side of Abbey Hill; thence eastwards along the road to the head of Aughinish Bay and thence out to sea, passing south of Aughinish.

A karstic peneplain, consisting entirely of limestone, pierced on the north by two fairly broad dry valleys leading down to Ballyvaughan and Bealaclugga respectively, and on the west by the much narrower valley of the Caher R., the only river in the district. In the west and east the limestone is largely bare; in the centre considerable areas bear a nearly continuous cover of thin soil supporting calcareous grassland. Hazel-scrub is scanty in the west, but fairly plentiful in the east, where there are also a few fragments of natural woodland, notably at Poulavallan, near Clooncoose and on the side of Mullaghmore. There are no lakes, except for some brackish lakes along the north coast, but there are turloughs (p. xxxii) near Turlough village, in the Carran basin and at L. Aleenaun (the last, unfortunately, recently drained). The coast consists mainly of low cliffs and rocky shores, but there are good sand-dunes at Fanore and north-east of Ballyvaughan, while at Rine Point, north-west of Ballyvaughan, there is an interesting spit. Apart from the woods and turloughs mentioned above, the chief areas of floristic interest are Poulsallagh, Fanore and the Caher R. valley, Black Head and Cappanawalla.

The Aran Islands resemble the western part of the mainland, differing mainly in the sparser vegetation and the considerably higher cliffs of their coastline.

Recorded species total 635, of which 35 have not been seen recently. The species known only from this district are *Astragalus danicus*, *Pyrola media*, *Limonium*

---

* This and the corresponding figure for other districts refer to vascular plants only, and exclude microspecies of *Rubus*, *Taraxacum* and *Hieracium*.

Fig. 1. Aerial view of Inishmore (Aran Islands), towards the western end. On the right is the beach of Portmurvy, with the village of Kilmurvy above it. Towards the bottom left is the village of Gortnagapple. The ground rises in successive terraces from Portmurvy to the cliffs of the south-west coast, which here vary from 25 to 40 m (80 to 125 ft) in height. The area around Portmurvy is drift-covered and furnishes reasonably good farmland. (Scale 7 cm = 1 km; 4½ ins = 1 mile.)

transwallianum, *Cuscuta epithymum*, *Carex strigosa*, the hybrid *Potamogeton × lanceo-latus*, and also *Arenaria norvegica* and *Atriplex littoralis*, both of which have been seen once but cannot now be re-found.

District 3. The Burren lowlands

Total area about 420 sq. km. Greatest height 67 m (219 ft).

Separated on the west from districts 1 and 2 by boundaries defined above, and bounded on the east by the Ennis–Galway road. The northern boundary is formed

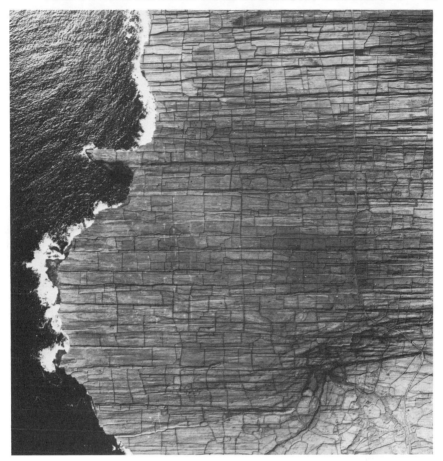

Fig. 2. Aerial view of part of Inishmore (Aran Islands), S.W. of Kilronan. The cliffs towards the top of the picture are *c*. 65 m (over 200 ft) high. The terrain is composed entirely of bare limestone pavement, and is uninhabited; the flora, however, found mainly at the base of walls and in the crevices of the pavement, is richer than the photograph would suggest. (Scale 7 cm = 1 km; $4\frac{1}{2}$ ins = 1 mile.)

by the estuary of the Dunkellin R., and from its mouth runs into Galway Bay between Mweenish I. and Eddy I.

A very flat plain, with a few hillocks in the south-east; more than half the district, however, has an elevation of less than 30 m (100 ft). Bare limestone pavements prevail in the west, but in the east and extreme south the drift cover is plentiful. Lakes are very numerous, especially in the south, and the larger ones are fringed by extensive fens and reed-beds. There are a few surviving fragments of raised bog in the south-centre. The only river is the Fergus, which enters the region at Corofin and leaves it at Ennis. Turloughs (p. xxxii) are numerous. There is a good deal of hazel-scrub, mainly in the west, and a fair-sized wood at Garryland and Coole which, though it has been subject to much management, still preserves some of its natural features. The coastline is sheltered and comparatively short; the shores are muddy or stony, with considerable areas of salt-marsh. Floristic interest centres mainly on the turloughs, the lakes and their fens (especially L. Bunny), Garryland wood, and the country between Dromore and Ennis.

Recorded species total 607, of which 20 have not been seen recently. The species known only from this district are *Ceratophyllum demersum, Sium latifolium, Teucrium scordium, Carex spicata* and the hybrid *Potamogeton × salicifolius*; also *Elymus caninus*, only once collected in 1855, but quite possibly still surviving.

District 4. The northern limestones

Total area 143 sq. km. Greatest height 72 m (234 ft).

The northern boundary runs along the road from Oughterard to Oughterard quay and thence due east to meet the baronial boundary in the middle of L. Corrib. On the east the boundary is that of the Flora region; on the south it is separated from district 3 by the Dunkellin estuary. On the west it is separated from district 5 by the main road from Oughterard to Galway as far as the Cathedral bridge, and thence down the R. Corrib to the sea. Most of the city of Galway is in this district, but the Cathedral and the south-western suburbs are in district 5.

A district without a marked character, and perhaps the least natural of the eight. It represents an extension northwards of the limestone plain of district 3, running round the head of Galway Bay and then up the west shore of L. Corrib. Apart from some hillocks north-east of Galway it is very flat. On these hillocks, and around and north of Ross L., there are some exposures of limestone pavement, but elsewhere there is a mainly gravelly drift cover. On the western edge there are some very small areas of Connemara granite, and there are some bogs overlying the pavements north of Ross L. The shore of L. Corrib is stony and exposed, and most of the islands in the lake belonging to this district are grazed, but a few bear natural scrub or woodland. Apart from this there is no woodland in the district, but there is some hazel-scrub around Menlough. The coast is similar to that of district 3. The most interesting areas floristically are around Menlough and Ballindooly (N. of Galway), Ballycuirke L., where the floras of the limestone and granite are juxtaposed, and the country north and east of Ross L.

Recorded species total 613, of which 35 have not been seen recently. The species known only from this district are *Carex riparia, Puccinellia distans*, and the somewhat doubtful *Carex flava*.

District 5. Iar-connaught (south-east Connemara)

Total area about 540 sq. km. Greatest height 348 m (1138 ft).

Separated on the west from district 6 by the road from Maam Cross to Costelloe, and thence down the middle of Cashla Bay to Galway Bay; on the north from district 8 by the road from Maam Cross to Oughterard; on the east from district 4 by the line described above. On the south it is bounded by Galway Bay.

The largest, most compact, and most homogeneous of the districts. It is composed almost entirely of granite, low-lying in the south but rising in the north-west to a conspicuous escarpment. Most of the granite is covered with blanket-bog, but there is a coastal strip in which a fairly plentiful drift cover gives mineral soils, and along this strip there is intensive settlement. At one place 5 km west of Galway, marked on maps as Seaweed Point, but known universally in the literature and locally as Gentian

Hill, there is a complex of highly calcareous gravelly drift on which many of the rare Burren plants find an outlying station. The coastline is elsewhere very uniform and unindented, mainly of low rocky shores, but with several sandy beaches; these, however, do not give rise to dune-systems of any size. Small lakes are numerous, but few of them have been investigated. The flora of much of the blanket-bog is monotonous: in one circle of 400 m radius we were unable to find more than 27 species of higher plants. For this reason, and because of the scarcity of roads, the district is less well explored than the others. There is a patch of natural woodland by the Owenboliska R. north of Spiddal and a semi-natural wood north-west of Moycullen. These, together with Gentian Hill and some fenny marshes near the boundary with district 4, are the regions of greatest floristic diversity.

Recorded species total 559, of which, however, 46 have not been seen recently. No species from the Flora region is confined to this district.

District 6. South-central Connemara

Total area about 365 sq. km, of which a considerable proportion consists of off-shore islands. Highest point 360 m (1178 ft).

Separated on the east from district 5 by the boundary described above; on the north from districts 7 and 8 by the Galway–Clifden road from Maam Cross to Canal Br.; on the west from district 7 by the road from Canal Br. to Toombeola and thence down Bertraghboy Bay to reach the ocean S. of Inishlackan; on the south bounded by the sea.

The general character of this district in its eastern and southern parts is similar to that of district 5 – fairly low granite hills covered partly by blanket-bog, but in some areas with a large proportion of bare rock. In the north and west the rock is mainly gneiss, with basic intrusion forming rather abrupt hills. The coastline is extremely complex and tortuous, with an extensive development of salt-marsh and of brackish inlets almost cut off from the sea, but in several places there are beaches of highly calcareous sand or occasionally of 'maerl' (p. 287). Lakes are numerous, though mostly fairly small; a few have wooded islands. There are semi-natural woods south of Glendollagh L. and north-north-east of Toombeola, but the latter has been largely underplanted with conifers.

Calcareous soils are confined to rather small patches of blown sand, and are less in extent than in any other district. Mainly for this reason the total flora is small, amounting to only 479 species, of which, however, only 13 have not been seen recently. The species known only from this region are *Spergularia rubra*, the alien *Juncus planifolius* and the somewhat doubtful *Ruppia cirrhosa*.

District 7. West Connemara

Total area about 495 sq. km. Greatest height 733 m (2395 ft).

The boundary separating this district from district 6 has been described above. From district 8 it is separated by the road running northwards from Weir Br. on the Galway–Clifden road, past L. Inagh to the junction with the Clifden–Leenane road, which it then follows eastwards to a point above Killary harbour due south of Bundorragha on the opposite shore, whence it runs down to join the regional boundary in Killary harbour. Elsewhere the district is bounded by the sea.

A district of very varied terrain, and including most of the Connemara localities best known to botanists. The eastern part is dominated by the mass of the Twelve Pins, composed of quartzite except for a few areas of mica-schist, which bring in the

Fig. 3. Glaciated granite landscape near the centre of Gorumna I. (S. Connemara). The bush pressed against the boulder in the right foreground is *Juniperus communis*, subsp. *nana*.

greater part of their not very abundant alpine flora. To the north is a smaller mountain mass rising to 605 m (1975 ft) between Letterfrack and L. Fee. The Twelve Pins have a considerable area of foothills on their western flank, but elsewhere drop steeply down to the plain; the only two other hills of any consequence (Errisbeg and Tully Mt.) rise abruptly from the coastal plain. There is much low-level blanket-bog, but around the coast there is some bare granite, some grassland over podsols, and considerable areas, especially on the Slyne Head peninsula, in which the large quantities of blown calcareous sand give rise to extensive sheets of machair. The coast is very diversified, rocky headlands alternating with sandy beaches and a few muddy inlets. There are two fair-sized rivers, Dawros and Owenglin, but their banks harbour few species not to be found elsewhere; more important floristically are the innumerable lakes, including the three largest in Connemara: Inagh, Derryclare and Ballynahinch. Practically all are markedly oligotrophic, though not strongly acid. There are stands of natural or semi-natural woodland west of Derryclare and Ballynahinch Loughs, at Kylemore, Letterfrack, and east and north-west of Clifden, and on many lake-islands. There are numerous areas of great floristic interest; special mention may be made of the Roundstone area, the bogland north of Errisbeg, the Slyne Head peninsula, the coast west and south-west of Cleggan, and the neighbourhood of Letterfrack and Kylemore.

Recorded species total 655, of which 36 have not been seen recently. The species known only from this district are *Elatine hexandra*, *Saussurea alpina*, *Erica ciliaris*, *Erica erigena*, *Euphrasia frigida*, *Typha angustifolia*, *Hydrilla verticillata*, *Cryptogramma crispa* and *Asplenium septentrionale*.

District 8. Joyce's country (north-east Connemara)

Area about 395 sq. km. Greatest height 706 m (2307 ft).

Although its highest point is slightly lower than that of district 7, this district contains far more high ground; it can, in fact, be described as a mountain massif, interrupted only by the Maam valley, and with a narrow fringe of low ground around Loughs Corrib and Mask and to the east of L. Inagh. The Maumturk Mts. and Leckavrea to the south of them consist largely of very bare quartzite, relieved only by a marble band at the Maumeen gap, but the mountains to the north and east are composed mostly of unaltered sediments, and are rather more hospitable to vegetation. The small hill of Lissoughter contains bands of marble and serpentine, but more important floristically is a band of magnesium-rich schist which harbours some alpines at a remarkably low level. In the east the district includes some small exposures of Carboniferous limestone north of Clonbur and west of Cong; here there are some patches of Burren-like pavement, but not many of the characteristic Burren species are to be found. Elsewhere most of the low ground is covered with level blanket-bog, except for some sandy alluvial flats around the south-west part of L. Mask. The shore-lines on L. Mask and L. Corrib are extensive; elsewhere lakes are less frequent than in the other Connemara districts and have a very restricted flora. The coast is limited to some 7 km of low stony shore, with a little salt-marsh, in the upper part of Killary harbour, and the district therefore lacks a number of common maritime species. There are some woodlands around the shores of L. Corrib, but all have been subject to a fair amount of management; that on the Hill of Doon is the most natural, the others comprising estate woods at Ashford Castle and in some of the properties north-west of Oughterard.

The montane flora is interesting, but very scattered; on the low grounds the richest areas are at Cong and Maam and around the south-west corner of L. Mask.

A total of 577 species have been recorded from the district, of which 31 have not been seen recently. The species known only from this district are *Stellaria palustris, Hypericum canadense, Vaccinium vitis-idaea, Spiranthes romanzoffiana, Potamogeton filiformis, Carex aquatilis,* and the hybrid *Potamogeton × sparganifolius.*

GEOLOGY AND SOILS

In broad outline the geology of the region is simple enough. Connemara is built almost entirely of siliceous palaeozoic rocks (some igneous, some metamorphic, some sedimentary); the Burren proper consists entirely of Carboniferous limestone; and the area south of it, corresponding to our district 1, consists of the shales, together with some sandstones and siltstones, which overlie the Carboniferous limestone and constitute the younger strata of the lower Carboniferous. The boundaries between these areas are for the most part very clear, and the land-forms and vegetation characteristic of each contrast strongly with those of the other two. Although the whole region has been thoroughly glaciated the amount of drift cover now remaining is as low as anywhere in Ireland and does little to obscure the pattern imposed by the underlying rocks.

The rocks of Connemara present endless and complex problems for the student of petrology or of earth-movements, but from the point of view of the botanist or the tourist interested in landscape they may be described in fairly simple terms. It is only in the north-east that we find unaltered sedimentary rocks: sandstones and shales of Ordovician or Silurian age. They cover a fairly small area, but this includes the mountains surrounding L. Fee, L. Nafooey and the 'Narrow Lake' (the arm of L. Mask that lies to the west of Ferry Br.). Like most rocks of their age they are hard and slow to weather. On the south they are flanked by a large area of metamorphosed sediments, originally laid down in the Cambrian or pre-Cambrian era, and metamorphosed by earth movements not very long afterwards to form quartzites, interspersed here and there with mica- or hornblende-schists or veins of marble. The marble is mixed in some of its veins with green serpentine to form the 'Connemara marble' of commerce. These metamorphic rocks extend southwards as far as the Oughterard–Clifden road, and in places for a few kilometres beyond it. Most of southern Connemara is, however, composed of a large mass of granite of Devonian age, constituting the whole of our district 5 and the southern part of district 6, and continuing along the coast to beyond Roundstone, with detached fragments up the west coast as far north as the area west of Cleggan. The granite is, however, separated from the metasedimentary rocks by a relatively thin band of gneiss, best developed around Cashel and on the Slyne Head peninsula. This band is interrupted here and there by base-rich plutonic intrusions, which form the upper parts of Errisbeg and Cashel Hill. There are also a few small exposures of Tertiary basalt, contemporary with that which covers most of Co. Antrim; of these the most conspicuous is Bunowen Hill, near Ballyconneely.

This petrological diversity gives rise to a variety of land-forms, the gentle, rolling contours of the southern granite contrasting with the steep quartzite peaks of the Twelve Pins or the Maumturk Mts. But from the point of view of soils and vegetation the whole of Connemara is remarkably uniform. It is only in a few small areas that the base-rich character of the rock modifies the generally calcifuge aspect of the vegetation; Bunowen Hill, the mica-schists of Muckanaght and Bengower, Lissoughter (which has a band of a remarkable magnesium-rich schist) and a few calcareous bluffs at the Maumeen gap are the only examples of any consequence. Here and there on the low ground a vein of marble gives rise to a streak of green grass, interrupting the brown of the blanket-bog, but such land is nearly always farmed and the flora is merely that of neutral grassland. Far more important than the basic rocks in affording conditions congenial to calcicole species is the transport inland by wind of calcareous sands from the western beaches. Throughout Connemara the sea-sand has a high content of calcium carbonate, usually from the comminuted shells of molluscs, but sometimes, as at Dog's Bay, from a great abundance of Foraminifera, or, as in the well-known 'coral strands' of Ballyconneely and Carraroe, by the virtual replacement of quartz grains by fragments of the dead calcareous skeletons of red algae (*Lithothamnion* or *Phymatolithon* spp.). We find, accordingly, that the base-demanding *Euphrasia salisburgensis*, though absent from the upper part of Errisbeg, which is composed of an ultrabasic rock, occurs on the lower (granite) slopes, because here the influence of blown sand is significant.

The eastern boundary of the siliceous rocks of Connemara lies, in the northern part of our region, beneath the waters of L. Corrib, but further south it follows very closely the main road from Oughterard to Galway. Here even the tourist quite unversed in geology can hardly fail to notice the contrast between the green pastures, broken here and there by outcrops of white limestone, on the north-eastern side of the road, with the grey granite and purple-brown blanket-bog on the south-west side. This strip of limestone, lying between the road and L. Corrib, is a sort of no-man's-land; geologically it represents a northward extension of the Burren limestones, and harbours a somewhat reduced Burren flora, but topographically it falls within the limits of Connemara. Its transitional nature is further emphasized by the fact that over fair-sized tracts the limestone is overlaid by peat, which bears a flora somewhat intermediate between that of the midland raised bogs and the Connemara blanket-bog; and in places where mounds of calcareous glacial drift break through the peat a fascinating, if confusing, medley of calcicole and calcifuge species can be seen jostling each other.

The boundary between the granite and the limestone meets the sea actually in the city of Galway, and eastwards and southwards from the city the shores of Galway Bay consist of Carboniferous limestone. This is the territory which connects Connemara to the Burren, but belongs to neither. It bears a more continuous drift cover than does either of its neighbours, and on account of this, together with its relatively low rainfall, it bears a flora which is more midland than western in character.

As we drive southwards from Kilcolgan, however, the drift cover thins out, the outcrops of bare limestone on the west side of the road become more frequent, and we soon realize that we have reached the edge of the Burren. The whole of this tract, which extends over about a quarter of Co. Clare (H9) and the north-western corner of south-east Galway (H15) consists of limestones of Lower Carboniferous (Visean) age with almost horizontal bedding. There is, however, a very slight dip towards the south, and eventually, as will be seen later, the limestone beds disappear under the shales and sandstones of slightly later (Namurian) age, while in the south-east the strata have been subjected to some folding, seen most clearly on Mullaghmore. On the whole, however, the picture is of level pavements, for the most part bare of soil and with only a scanty cover of vegetation, interrupted on the sides of the hills by small vertical cliffs, giving a terraced effect. Most of the limestone is very pure and leaves little residue after weathering; here and there some beds contain a little chert, but only the trained eye of the geologist is likely to pick them out. They are, however, of some importance, in that their weathering gives rise to a thin film of mineral soil, and although the building up of peat can take place directly over the limestone, it goes ahead more vigorously if it is insulated by a thin layer of silt or clay.

Most of the Burren is free from drift, but there are substantial patches of it here and there. Being mainly calcareous in composition, it gives rise to good farmland, and it is only when it occurs in very small patches, not worth farming, that its natural

Fig. 4. Limestone pavement in the Burren. (*a*) Shattered type, with a scanty vegetation consisting mostly of *Prunus spinosa* (blackthorn) and *Rosa pimpinellifolia* (burnet rose). (*b*) Smooth type, with rounded angles and deep crevices. The shrub in the foreground is holly; in the background near the centre are some shoots of ash.

vegetation can be seen; it differs only slightly from that of the shallow, peaty soils directly overlying the pavements. Whether the drift cover was much more extensive in the early post-glacial period is not clear, and different views on the question have been expressed. There is no doubt that some of the original drift cover has been washed down the joints in the pavements, but perhaps not on a very large scale.

The Burren falls fairly sharply into two regions, forming our districts 2 and 3. In the east it is low-lying and very flat; in the west it forms a table-land, with summits mostly around 300 m (1000 ft), to some extent dissected by valleys and enclosing a remarkable basin at Carran. This is the type of formation known to geographers as a *polje*: a basin with an entire rim, from which the drainage is entirely underground.

The level pavements of the Burren fall into two sharply contrasted types: the smooth and the shattered. Intermediates exist, but they are rare. The smooth type has very smooth, level surfaces, interrupted here and there by deep crevices which vary in width but are mostly 10–30 cm wide and often a metre or more in depth. There are no sharp angles: everything is rounded off, looking as though it has been licked or sucked. In the shattered type the surface of the pavement is irregular, but it can hardly be seen, so thickly covered is it with detached angular blocks of limestone of very varying size, strewn around apparently at random. Joints are few and shallow, being mostly blocked by rock-fragments. On the whole the smooth type predominates in the west and the shattered in the east, but there are many exceptions to this rule, and in several places extreme examples of both types can be seen within a few hundred metres of each other. There seems to be no generally agreed explanation of this variation, but it is of considerable importance to the flora; for the vegetation of the smooth type of pavement consists partly of shade-loving species in the crevices, and partly of a varied flora developing on peat found in small pockets on the surface, while the shattered pavements have a much more restricted flora, consisting mainly of *Rosa spinosissima*, *Prunus spinosa* and *Teucrium scorodonia*.

The denudation which removed the shales and sandstones originally overlying the limestone would seem to have taken place in the southern and western parts of the Burren in comparatively recent times (geologically speaking), and traces of the capping still exist. These traces consist mainly of clay soils derived from the shales, which are most extensively developed in the central and east-central part of our district 2 (on both sides of a line joining Ballyvaughan to Killinaboy), and account for the relatively continuous cover here of a grassy (and floristically rather poor) vegetation. But in a few places, notably Poulavallan, fragments of sandstone can be found. Apart from these, the only non-calcareous rocks to be seen in the Burren are occasional erratics of Connemara granite on the north shore.

To the south of the Burren limestones lies a band of Carboniferous shales, for the most part covered by a clayey drift, here and there in the form of drumlins. The shales are locally diversified by sandstones, which can be seen outcropping by L. Goller and elsewhere, while the Cliffs of Moher show a long sequence of alternating sandstones and silt-stones.

No extensive treatment of the soils of our region is called for, as its most striking feature is the very large areas of rock which carry no soil at all. This applies not only to the limestone pavements of the Burren, but also to much of the granite of southern Connemara and the quartzite summits of the mountains further north. Even where there is some cover on the rocks it is, over large areas, pure peat – scarcely a soil in the ordinary sense of the word. Where mineral soils exist they are for the most part shallow and immature. In Connemara podsols prevail; they are to be seen mainly in the coastal belt, and in places on the foothills of the mountains. Elsewhere there is only peat, shallow on the mountains, locally of some depth on the low ground. On the lowland limestones by L. Corrib and Galway Bay brown earths are developed, which are for the most part farmed. Further south and west, where the bare limestones begin, the soil, such as it is, is a sort of rendzina, but usually of a very peaty type, with the limestone particles few and mostly in the lower layers. It is only when we reach the Carboniferous shales and their associated drift that mineral soils of any depth become at all widespread, and they are almost exclusively gleys with a remarkable water-retaining power, which leads to the pastures, even on the steep slopes of drumlins, being choked with rushes.

CLIMATE

The climate of the region is, as might be expected, of a markedly Atlantic type, but it is not as extreme, at least in some parts, as is generally supposed.

The most pronounced variation from one area to another is in respect of annual rainfall, the wettest area receiving almost $2\frac{1}{2}$ times as much as the driest, although they are less than 60 km apart. The driest part of the Flora region, and indeed of the whole of the west of Ireland, is the country at the head of Galway Bay. Praeger drew attention to this long ago, and to the fact that this low rainfall, coupled with a porous soil, provides a suitable habitat for several species more characteristic of eastern and central Ireland. The annual rainfall here is only just over 1000 mm, which is about the average for Ireland, and corresponds with that of the Welsh borders or Dorset in Britain. At the other end of the scale we find Maam, with an annual average of almost 2500 mm, and if it is realized that it lies only 18 m above sea-level, this is remarkably high; the precipitation on the surrounding mountains must be very much higher. The wettest parts of Kerry are about as wet, and the wettest parts of the Scottish highlands rather wetter, but otherwise it is hard to match Maam among the lowland stations of the British Isles. It is, however, only the mountain mass of north-east Connemara, in which Maam lies embosomed, that returns these very high figures. The Twelve Pins are wet, but not quite as wet; Ballynahinch, under their shadow, receives 1700 mm and Letterfrack, close to their northern flank, 1600 mm, but at none of the remaining 14 stations in Connemara does the rainfall exceed 1500 mm, and all those on the coast (except Roundstone with 1380) receive less than 1300.

In the eastern Burren, the rainfall is scarcely higher than at the head of Galway Bay, and on the west coast and in the Aran Islands it is only 1150–1200 mm. On and around the hills of the western Burren, however, it is a good deal higher; Corofin receives 1240 mm, Kilfenora 1330 mm and Ballyvaughan 1525 mm. The effect of height may be seen from the fact that at the top of the Corkscrew Hill, only 6 km distant from Ballyvaughan but 190 m higher, the rainfall is as high as 1729 mm.

To the vegetation the figures for absolute rainfall are probably less important than those for the frequency of rain and its distribution through the year. The number of rain-days in the year shows a regional variation parallel to that for total rainfall, but the contrast is less marked: the driest stations have about 170 rain-days in the year as against some 240 days for the wettest (high-level stations excepted). Absolute droughts, with three weeks or so without rain, occur, but they are very rare, and are more frequent in spring, when the temperatures are low and the skies often clouded, than during a summer anticyclone. Moreover, although there is everywhere a well-marked annual cycle with a rainfall minimum in April and May and a maximum in December, the 'dry season' is only moderately dry. In Connemara and the Burren alike, 12 % of the year's rain falls in April and May as against 23 % in December and January, and even in the driest month rain falls on an average on at least ten days.

It is this frequent, though not always very heavy rain that characterizes the climate of western Ireland, and with it, of course, goes a high average level of humidity of the air. This too is of great importance to plants, but it must be realized that its beneficial effects are experienced only in sheltered positions, because in exposed situations they are nullified by the influence of the wind. Ireland is a windy country, and the west coast is its windiest part; Praeger has described the winter climate in the west as a series of westerly gales, with westerly winds in between. No exact

Fig. 5. Wind-shorn hawthorn at Gleninagh, near Black Head.

measurements are available for the Flora region, but conditions at Belmullet, in N.W. Mayo, are probably only slightly more severe than on the coasts of Clare and W. Connemara. At Belmullet, moderate to fresh winds (4 to 6 on the Beaufort scale) blow for 57 % of the year and gales for 6 %, leaving only 37 % for light winds or calm. At Shannon airport, to which the more sheltered parts of the Flora region approximate, the corresponding figures are 42 %, 3 % and 55 %. In the Flora region the average wind speed throughout the year varies from about 24 km per hour in the more exposed parts to 18 in the more sheltered. Even with very humid air, winds of this frequency and intensity have strong evaporating power, and we find, therefore, that the figures for potential evapotranspiration are in winter as high in the Burren and much of Connemara as anywhere in Ireland; in summer they are lower than on the south and south-east coasts, but distinctly higher than in the north-west midlands. These figures, of course, refer to more or less unsheltered stations; in the lee of a hill or even a boulder, in a hazel-copse, or in a crevice in the limestone pavement the evaporation rate will be very much less, and in these conditions woodland plants can thrive without the shelter of actual woodland. Even in more exposed positions it is probable that frequent cloud cover and frequent wetting by rain help to counteract the effects of wind.

The duration of sunshine is fairly low, the Burren receiving an average of about 3.7 hours a day and Connemara rather less (about 3.4 hours). These figures refer to lowland stations; on the mountain-tops, which are often clouded even on fine days, the figures are very much lower.

Temperatures are very equable, both the annual and diurnal variation being remarkably small. The mean temperature of the air throughout the year is just under 10 °C for most of the region – higher than in all parts of the British Isles except the southernmost fifth of Ireland, west and south Wales, and England south of the latitude of Cambridge. There is, however, only a very small difference between the coldest and the warmest month; the mean temperature for January and February is about 6 °C, that for July just under 15 °C. If we trace the January isotherm from the Burren coast we find that it cuts across southern Ireland and Devon and thence to south-west France and central Italy; Rome is no warmer in the winter than is Galway. The July isotherm, on the other hand, runs from Connemara obliquely across the Irish midlands to Co. Down, and thence along the Scottish borders to central Fennoscandia, so that Galway is no warmer in summer than subarctic Finland. There is, in fact, very little difference in temperature between a calm, sunny March day in Connemara and a wet and windy one in August.

Frost is by no means unknown, but it is seldom severe, the mean annual minimum being around -5 °C. What is perhaps more important to the vegetation is that it seldom lasts long. There are, however, three or four times in a century bouts of prolonged and severe frost, as in 1963, and these must to some extent limit the number of species of southern affinities that can establish themselves permanently in the region. Snow is of little importance; it falls only on a few days in the year and seldom lies on low ground for more than a day. And even the mountains are seldom snow-capped for more than two or three weeks throughout the winter.

To sum up then, the climate is fairly uniform throughout the region except in respect of rainfall. It is characterized by mild winters and cool summers, frequent and often strong winds, small incidence of frost or snow, and rainfall well distributed through the year, though with a distinct minimum in late spring. In the mountains of central and north-east Connemara the rainfall is very heavy; elsewhere it is, by the standards of the Atlantic coast, fairly moderate.

CHARACTER OF THE FLORA

The contrast between Connemara and the Burren is so sharp that what is true of one is seldom true of the other, and it is difficult, therefore, to make generalizations valid for the whole region. Admittedly the number of species common to the two is larger than one might conclude at first glance, for, apart from the fairly large number of tolerant species which are equally at home on calcareous and acid soils, a surprisingly large number of the Burren calcicoles crop up on the machair-like pastures of western Connemara, while several species characteristic of the Connemara blanket-bog hang on precariously in the fragments of raised bog which survive in the south-eastern part of the Burren. Nevertheless, the contrast between the two areas is undeniable. A second obstacle to generalizations about the region (which is to some extent a resultant of the first) is that its total flora is so rich as to constitute about three-quarters of the flora of Ireland, while if we exclude naturalized aliens and confine our attention to native species we find that out of 900 species for the whole island no fewer than 738 (82 %) are found in the region of this Flora. Generalizations must, therefore, start with a consideration of the species native to Ireland which are *not* to be found in Connemara or the Burren.

Of the 162 missing species, 73 may be reckoned as very rare in Ireland and 57 as rare, the criteria adopted being less than 6 and less than 25 dots respectively in the *Atlas of the Botanical Society of the British Isles*. The absence from the Flora region of these species demands no special explanation. Of the remaining 32 we find that almost half are confined to the south or the east of the country or to both. The most widespread of these is *Lepidium heterophyllum* (the only Irish native absent from Connemara and the Burren that can be considered common). Here also, though in less abundance, come *Luzula pilosa*, *Linaria vulgaris*, *Ononis repens*, *Linum bienne*, *Vulpia myuros*, *Echium vulgare* and *Scilla verna*. Essentially midland species which stop short of Galway include *Glyceria maxima*, *Cicuta virosa* and *Andromeda polifolia*. Five species, including *Prunus padus* and *Salix pentandra* (the latter freely planted in our region but not native there), are northern in their distribution. Finally, mention must be made of three species which one might reasonably expect to find in the region but do not occur there: *Cicendia filiformis*, *Euphorbia hyberna* and *Sisyrinchium bermudiana*. The first, centred on the south-west, is said to have an outlying station in Mayo, but it skips Connemara. The *Euphorbia*, centred also on the south-west, has also a few outlying stations further north, but although one is not far from the Burren and another close to Connemara they do not actually hit the target. *Sisyrinchium bermudiana*, which is almost entirely western, and which ranges from Kerry to Donegal, for some reason avoids our region altogether.

One interesting feature in the list of these absentees is the predominance of coastal (though not always strictly maritime) species, and the great scarcity of woodland species, only *Luzula pilosa*, *Euphorbia hyberna* and *Prunus padus* being assignable to this last category. This suggests that the destruction of woodland on which some authors lay so much stress, though it doubtless diminished the abundance of many species, led to the extinction of few or none. As regards the coastal species, among which we may number *Echium vulgare*, *Cynoglossum officinale*, *Myosotis ramosissima* and *Scilla verna*, it seems probable that they belong mostly to the last wave of immigrants from Britain and that they never got further than the eastern half of Ireland. The case of *Atriplex laciniata*, however, shows that some of these plants may still be on the move (see p. 176).

Turning now from the consideration of absentees to a discussion of the more striking ingredients of the Burren and Connemara floras, we may note that there are eight species which are found nowhere else in Ireland. *Asplenium septentrionale, Erica ciliaris* and *Hydrilla verticillata* have their solitary and very restricted stations in Connemara, and *Deschampsia setacea*, though more widespread, does not transgress its limits. *Helianthemum canum* and *Limonium transwallianum* are confined to the Burren and Aran, *Astragalus danicus* to Aran alone, while *Ajuga pyramidalis* is found in the Burren, Connemara and Aran, but nowhere else. More important, however, than these highly localized rarities are the 28 species which, though sparingly represented elsewhere in Ireland, have their headquarters in Connemara or the Burren. It is these species which give to the region its botanical fame, and it is to Connemara or the Burren that any botanist must go if he wishes to make an intensive study of these species. Apart from the eight confined to our region, which have been listed above, the species of which it may fairly be said that the number of individuals in Connemara and the Burren greatly outnumbers those in the rest of Ireland (and, in the case of species marked with an asterisk, in the rest of the British Isles) comprise the following:

In the Burren (including Aran)

*Adiantum capillus-veneris
Asperula cynanchica
Cerastium arvense
Cystopteris fragilis
*Dryas octopetala
Epipactis atrorubens
*Euphrasia salisburgensis
Galium sterneri
*Gentiana verna
Geranium sanguineum
Limosella aquatica

Minuartia verna
*Neotinea maculata
*Orobanche alba
Polystichum aculeatum
*Potentilla fruticosa
Rubia peregrina
Rubus saxatilis
Saxifraga hypnoides
Sesleria albicans
*Viola persicifolia

In Connemara

*Daboecia cantabrica
*Erica mackaiana
*Eriocaulon aquaticum

Eriophorum gracile
*Hypericum canadense
Lepidotis inundata

In Connemara and the Burren

Allium babingtonii

If we add to the eight species confined to the region the 28 which are exceptionally abundant there, and attempt a classification of them according to their total geographical range, the result is rather surprising. Of the Connemara species six are, predictably enough, strongly Atlantic in their European range (two of them, however, having their main centre in North America), and this is true also of *Allium babingtonii*. There are four, however, which show very different geographical affinities: *Asplenium septentrionale* and *Lepidotis* extend throughout the North Temperate zone, *Eriophorum gracile* is widespread in north and central Europe, and *Hydrilla verticillata* is subcosmopolitan, though the variety found in Connemara is known only from north Europe.

Fig. 6. *Dryas octopetala* (mountain avens) and *Neotinea maculata* (Irish orchid) growing together near Black Head. Nowhere else in the world can these two species be seen growing side by side.

Among the Burren specialities the diversity is greater still. It is often said that the peculiarity of the Burren flora consists in the juxtaposition of Mediterranean with arctic–alpine species. But *Dryas* is the only real arctic–alpine (*Ajuga pyramidalis* can perhaps be considered as boreal–montane, and *Gentiana* as a modified alpine), and *Neotinea* is the only species whose distribution is centred on the Mediterranean. *Adiantum* and *Rubia* are Mediterranean–Atlantic, while *Galium sterneri*, *Limonium transwallianum*, *Saxifraga hypnoides* and *Sesleria albicans* can be reckoned as Atlantic or subatlantic. There remain, however, fifteen species which are distributed throughout the greater part of Europe, in many cases extending to temperate Asia or North America. Why they should flourish so much more exuberantly in the Burren than elsewhere in Ireland is not at all clear. We must confess that a general explanation of the Burren flora is still to seek.

Note added in press

Glyceria maxima (pp. xxviii) grows in some quantity by the Dawros R., west of Tullywee Br. It seems to have been introduced some thirty years ago and to have increased greatly since then.

Glaucium flavum and *Trisetum flavescens* (pp. 13, 247) were seen in 1982 in their old stations on Inishmaan.

NOTES ON SOME OF THE HABITATS

Inland waters

Rivers are not a prominent feature of the landscape in either Connemara or the Burren, chiefly because they are mostly of small size. The Corrib alone has a mean discharge of more then 25 m³ per second, but its total length is only 8 km. Its eutrophic waters provide a habitat for some submerged hydrophytes, but its banks have few plants of interest. Next in size comes the Fergus; in its upper course it disconcertingly disappears underground for some stretches, but below Corofin its course is normal. In the reach just above Ennis its marshy banks harbour some interesting plants – notably *Sium latifolium*, known otherwise only from the basins of the Shannon, the Erne and (very rarely) the Boyne – and, if early records are to be trusted, these marshes were even richer 150 years ago. The Caher R., the only river in the Burren proper, is notable for its wealth of species and hybrids of *Equisetum*, and also for the abundance in it of the rare hybrid *Potamogeton × lanceolatus*.

All these are calcareous rivers; those draining the acid lands of Connemara are mostly without floristic interest; only at Maam, where there is *Carex aquatilis* in an ox-bow and *Potamogeton × sparganifolius* in the main stream, do they furnish plants which are at all uncommon.

Lakes, on the other hand, dominate the landscape in a large part of the region. L. Corrib, the second largest in Ireland, and L. Mask, which is not much smaller, both lie partly in the Flora region. Although their western shores are composed mainly of siliceous rocks, the greater part of the water draining into these lakes runs off the limestone; their waters, therefore, are strongly calcareous, with a pH usually about 8.3, although in the north-western arm of L. Corrib, which is much deeper than the rest of the lake and somewhat cut off from it, and receives drainage only from siliceous rocks, the pH sinks to near neutrality. In spite of the alkaline or neutral waters, some calcifuge species such as *Lobelia dortmanna* are to be found on the west shores of these lakes, the non-calcareous nature of the substratum being apparently more important to them than the calcareous water.

The remaining lakes are extremely numerous, and range in size from the smallest pools to lakes with an area well over 1 sq. km, such as Derryclare L. in Connemara and L. George in the Burren. The larger lakes in Connemara are all fairly deep; those on the limestone are all shallow. The latter, despite their high calcium content, are scarcely eutrophic, as they are fed almost entirely by springs and tend to be poor in nutrients such as nitrates and phosphates. Their submerged flora is consequently fairly limited, but many of them are fringed by species-rich reed-swamps or fens. Some of the larger Burren lakes, notably L. Bunny, show a considerable fluctuation in the water-level, and this tendency is further developed in the turloughs, one of the most characteristic habitats of the region. They consist of lakes in which the water is apt to rise and fall very rapidly, and in dry weather they can dry out almost completely. There are no streams entering or leaving them; the water flows in and out from holes in the limestone which function alternately as springs and sink-holes. Their vegetation is an extraordinary medley of aquatic and terrestrial plants, usually with a precise and characteristic zonation, the upper limit of frequent flooding being marked by the cessation of woody plants (for a preliminary survey see Praeger, 1932*a*). The turloughs range from Ballyvaughan, Carran and Killinaboy in the west to Gort and Kilcolgan in the east.

In Connemara such small lakes as are merely larger bog-pools contain acid water, but those lying in rocky basins have water which, although it has a low base content, usually has a pH of 6–7. This may explain the presence in many of them of *Cladium mariscus*, a species normally associated with base-rich waters. A considerable proportion of their scanty supply of basic ions comes from sea-spray, and chloride analyses of lake-waters suggest that the water of the lakes in western Connemara contains something like 0.1 % of sea-water. In lakes very near the coast, blown calcareous sand also makes a significant contribution to the base content of the water. Although many of these lakes have here and there a fringe of *Phragmites*, *Scirpus lacustris* or *Cladium*, it is usually rather thin and never builds up into anything which can be called a fen. The submerged hydrophytes, though not very abundant, include such rare species as *Elatine hexandra* and *Eriocaulon aquaticum*.

Bogs

The blanket-bog of Connemara bears most of the species characteristic of all western blanket-bog, but it is remarkable in the number of normally base-demanding species which crop up here and there, sometimes in recognizable flushes or in the neighbourhood of ultrabasic rocks, but also on level bog overlying quartzite or granite. *Schoenus nigricans*, which often reaches subdominance, is the most conspicuous, but here too belong *Juncus subnodulosus*, *Carex dioica* and *Eriophorum latifolium*. Much more chemical and autecological study is needed before a satisfactory explanation can be given. The bogs east and south-east of Oughterard and the fragments which remain in parts of the Burren lake-country are transitional in their flora between raised and blanket-bog. They are surrounded by calcareous rocks or gravels, and at their margins there is a medley of calcicole and calcifuge species, but the base-demanding plants do not grow on their unbroken surface.

Here and there on the blanket-bog of Connemara are small stands of *Phragmites* or (more rarely) *Cladium*, which indicate the sites of lakelets which have been overgrown by the bog.

Sea-shores

The rocky coasts call for little comment. Apart from the abundance of *Rhodiola rosea*, the flora of the cliff-faces is unremarkable, and there is a surprising scarcity on cliff-tops of the vegetation described by Praeger (1934*d*) as *Plantago*-sward. The considerable areas of salt-marsh have not been studied closely enough to permit of generalizations about their flora, but they do not differ conspicuously from those in other parts of Ireland. Sandy shores are widely distributed round the coast, but in only a few places are there well-developed dune-systems. Fanore is the best; others are to be found north-east of Ballyvaughan, and in various parts of W. Connemara. They have little resemblance to the dunes described in textbooks, as erosion is more conspicuous than deposition. Blow-outs are numerous and there are no recognizable fore-dunes, although, rather surprisingly, *Elymus farctus* forms a more constant ingredient of the vegetation than does *Ammophila*. Nor are there any well-developed dune-slacks: the hollows in the dunes at Fanore are occupied not by marsh plants but by *Ophrys apifera* and other species characteristic of dry ground. This is probably because, despite the heavy rainfall, the large calcium carbonate content of the sand enables water to drain away and not to persist for more than a few days.

Fig. 7. Wooded lake-island near Maam Cross, its vegetation contrasting strongly with the hillside behind, from which all trees and shrubs have been eliminated by grazing. The trees are chiefly oak and holly; a yew (darker foliage) can be seen towards the right-hand end.

But the calcareous nature of the sand has a more important effect on the vegetation: transported inland by wind it converts substantial areas of the coastal plain of western Connemara into a neutral or slightly alkaline grassland similar to the machair of the Hebrides, on which many of the Burren plants find outlying stations.

Woodland and scrub

Woodland is scarce throughout Ireland, and scarcest in the west, where the strong winds reinforce the discouragement it receives from felling and grazing. The natural climax in an unaltered state is to be found only on a few of the lake-islands – those that are large enough to allow the development of woodland as distinct from scrub, but not large enough to attract agricultural exploitation (Webb & Glanville, 1962). Here there are well-developed stands of *Quercus petraea*, with *Ilex*, *Taxus*, *Sorbus aucuparia* and *Salix atrocinerea* usually present as minor ingredients. The ground flora is characteristic of acid (or in the case of the islands in L. Corrib) near-neutral woodland, dominated most often by *Vaccinium myrtillus* or *Luzula sylvatica*. On the mainland this oakwood is always to some extent modified by human activity: by heavy grazing, as by Derryclare L., by selective felling, or by the planting of aliens. Nevertheless, in several of the woods such as those at Ballynahinch or on the Hill of Doon, where alien species, planted or naturalized, are conspicuous, the general character of the ground-flora is apparently unaffected.

In the Burren the natural climax is, presumably, ash–hazel-wood, but nowhere can this be seen in an untouched state, although a few fragments remain here and there which are exposed only to occasional felling and light grazing. These have a fairly rich herb-flora, consisting mainly of base-loving plants, though at Poulavallan the presence of some soil derived from the now vanished capping of shale and sandstone

Fig. 8. The Hill of Doon, with the narrow, north-western arm of L. Corrib in the foreground. The wood is composed mainly of native oak (*Quercus petraea*); a few planted conifers (mostly *Abies alba*) stand out above the general level of the canopy.

permits the establishment of several calcifuge species such as *Blechnum, Athyrium filix-femina* and *Vaccinium myrtillus*. Much larger areas are covered by hazel-scrub, in which there are few ash-trees or none. This is often very dense (the heavy shade presumably preventing the establishment of ash seedlings), with a ground-flora consisting chiefly of mosses, which also densely clothe the stems of the hazel-bushes (Fig. 22). For a recent study of this scrub see Kelly & Kirby (1982).

The total area under broad-leaved woodland (including estate woods, but excluding scrub) is about 12 sq. km, or 0.4 % of the total area of the region. To this must be added some 128 sq. km of coniferous woodland planted by the Department of Forestry. The largest plantations are in districts 5 and 8 (accounting for 29 % and 37 % of the total, respectively); the districts with the smallest area planted are 2 and 7. Among the species planted, *Pinus contorta* and *Picea sitchensis* greatly predominate (the former mainly on the most exposed and barren ground, the latter where shelter and nutrient status are slightly better), but here and there are small belts or stands of *Pinus sylvestris, Picea abies, Larix decidua, L. kaempferi,* and *Tsuga heterophylla.* Plantings of broad-leaved trees by the Forestry Department are confined to marginal amenity belts; these include, beside the native species, some sycamore, beech and hybrid poplars. The Department, however, owns some of the larger stands of semi-natural woods, and is preserving these as far as possible in their original condition.

Conclusion

Deforestation and afforestation are merely the two most obvious effects of human interference with the vegetation, for there are few communities, apart from those on cliffs, on small or remote islands or in large reed-swamps, which have not been affected by grazing, cutting, ploughing or draining. Nevertheless, the natural vegetation is resilient, and in places, although secondary, it must be not unlike what was there some thousands of years ago. A good example of this was seen when an island in L. Corrib, marked out in fields on the Ordnance maps of 1841 but abandoned after the famine, was found now to be covered with mature oak-wood only slightly different in its floral composition from that on a neighbouring island which was never farmed. But the recovery of nature prompted by the easing off of human pressure as a result of the decline in population between 1850 and 1950 has probably ceased: tourism, mechanization and a slight increase of population are all increasing the pressure. The outlook for the conservation of substantial areas of natural vegetation is much brighter in Connemara and the Burren than in many parts of Ireland, for much of the land is too rocky to plough, many of the bogs small enough to defy mechanical exploitation, and many of the wetlands are almost impossible to drain. But vigilance is needed all the same. Very substantial areas of hazel-scrub in the Burren have been cleared in the last decade; some of it, of course, can be spared, but it is important that some stands on drift as well as on bare rock should remain. Some of the smaller turloughs are also under threat. The dunes, especially at Fanore, may be degraded if holiday occupation increases, and Gentian Hill is under a serious threat from the growth of caravan-sites. In cases such as these one can only hope that the local authorities will accept their responsibility for conservation. In the matter of individual rare species we may congratulate ourselves that all the rarest are now protected by law – from botanists as well as from farmers and builders – and we can only hope that the law will be respected and enforced.

THE VEGETATION SINCE THE LAST GLACIATION

by

W. A. Watts

Trinity College, Dublin

Connemara and the Burren, like most of Ireland, were covered by ice during the last glaciation. About 12 500 B.P. (radio-carbon years before the present) the ice was finally melting and re-population of the land by vegetation had begun, but from the earliest stages the difference in bedrock between the two areas gave rise, as it still does today, to two sharply contrasted types of vegetation. This contrast has throughout been rendered all the sharper by the fact that in neither region was the bedrock extensively masked by deposits of till.

The data from which we can reconstruct the history of the vegetation are still limited. Jessen (1949) published data from two sites near Roundstone (distinguished as I and II), and his work has recently been supplemented and reviewed by Teunissen & Teunissen-van Oorschot (1980). In the Burren the only published results relate to the late-glacial (Watts, 1963, 1977), but a considerable volume of information on the post-glacial of the south-eastern part of the Burren now awaits publication.

Jessen's pioneer and fruitful work is, unfortunately, somewhat primitive by modern technical standards, because he could not recognize the pollen of several important genera, including *Juniperus*, *Populus* and *Taxus* as well as several herbs. He worked, also, before the days of radio-carbon dating. Nevertheless, his choice of site, especially his Roundstone II (where *Erica mackaiana* and *Eriocaulon aquaticum* grow today), and his use of macrofossils give great interest to his work.

Late-glacial deposits are present at both of his sites. They are rich in pollen and macrofossils of *Empetrum nigrum* throughout, and in the middle (warmer) part of the late-glacial there are leaves of *Juniperus communis* and fruits of *Betula pubescens*. Leaves of *Salix herbacea* occur in the silty sediments which mark the cold climates with which the late-glacial began and ended. Jessen concluded that *Empetrum*-heath may have been widespread in western Ireland at this time, and recent studies by Telford at Glenveagh (Co. Donegal), in a terrain generally similar to that of Roundstone (blanket-bog overlying granite), show that there too *Empetrum* was abundant throughout the late-glacial. Some other sites from western Ireland, however, as for example the vast blanket-bog of north-west Co. Mayo, tell a different tale, and we must conclude that, although vegetation dominated mainly by *Empetrum* prevailed in some regions of siliceous rock which now carry blanket-bog, this was not universal, even in the west. A few years ago a re-examination of the Roundstone II site showed that pollen of *Juniperus* was abundant in the warmer phase of the late glacial (Jessen's Zone II), very rare in the succeeding cold phase (Zone III), and abundant again at the beginning of the post-glacial – a common pattern in other parts of Ireland. At the beginning of the post-glacial (9000–10000 B.P.) *Empetrum* became rare, and was replaced as an undershrub largely by *Calluna* and other ericaceous species as *Betula* began its post-glacial expansion. The same period shows an expansion of *Cladium*, of which the pollen is easily recognized.

The vegetation in the Roundstone area during the earlier part of the post-glacial has no distinctive features, and appears to have been similar to that in many other parts of Ireland. A rapid expansion of *Corylus* was followed by *Pinus*, *Quercus* and *Ulmus*, and finally (at the end of the Boreal period) by *Alnus*. The plentiful representation of pollen from *Quercus* and *Ulmus* is, however, noteworthy in view of the modern vegetation of the area, and suggests a much more productive environment than that which we see today. Jessen was able to show that *Eriocaulon* was already present during the Atlantic period (5000–7000 B.P.), and that *Erica mackaiana* had made its appearance even earlier, in Boreal times. But, apart from making it clear that they were not introduced by man, this dating does nothing to settle the controversial problem of the post-glacial history of these representatives of the Hiberno-American and Hiberno-Cantabrian floras.

In the latter part of the post-glacial the most interesting feature is the decline of *Pinus* to extinction and the simultaneous expansion of *Myrica* and ericaceous plants, charting the origin and spread of the modern blanket-bog. In Donegal the pine had disappeared before 3000 B.P., and M. O'Connell tells me that a date close to 3500 B.P. may be accepted for Mayo. The date for Connemara has yet to be demonstrated, but is probably about the same; in the Burren, as we shall see, it is much later. Whether man played a part in this changeover from pine-forest to blanket-bog is not clear. Unambiguous data from Mayo show that the land was cultivated about 3300 B.P., when the pine had already disappeared, and shortly before bog-formation began. It is possible that the change, due in part to an alteration in the climate, may have been accelerated by deforestation and impoverishment of the soil by primitive agricultural practices.

In the Burren the late-glacial was marked by two cold and two relatively warm periods. The second warm and second cold correspond to Jessen's Zone II and III at Roundstone; the first two periods, which have been demonstrated in several other Irish sites, replace the not very clearly characterized Zone I of Jessen. The first warm period is marked by an abundance of *Juniperus*, but this is followed by a period of large-scale deposition of clays derived from erosion, suggesting that a deterioration of climate had greatly reduced the vegetation-cover on the hillsides. The second warm phase (less warm than the first) was dominated by grasses. Plenty of pollen of *Helianthemum* spp. is also present, but it must be remembered that this plant is a large pollen producer. In the final cold period the most characteristic pollens are those of *Artemisia*, *Armeria*, *Sedum*, *Thalictrum* and members of the Caryophyllaceae. The total pollen is scanty, and there is much mineral matter in the deposits; this suggests that the vegetation was sparse, and perhaps approximated to something between tundra and cold steppe. In the late-glacial as a whole the macrofossils found in the Burren are very different from those found at Roundstone. *Dryas octopetala* and *Salix herbacea* are the commonest; others include *Juniperus communis*, *Linum catharticum*, *Salix repens*, *Rosa pimpinellifolia* and *Arenaria ciliata*, the last-named now surviving only as a relict on the Ben Bulben range.

In the early post-glacial, there was a rapid succession similar to that in many parts of Ireland. Juniper came first, then copses of birch and willow; next pine and hazel, with oak and elm fairly close on their heels. To these we may add *Populus tremula*, the aspen, present in the early post-glacial and possibly even in the late-glacial, and *Viburnum opulus*, the guelder-rose, which was present in the birch-woods of the early post-glacial. It now appears quite certain that, for a considerable part of the post-glacial period, a good deal of the Burren was covered by open pine-forest, and that the hazel-scrub, pasture and bare pavement of today are secondary to this. It was not until near the beginning of the Christian period (*c*. 1500 B.P.) that the pine disappeared from the Burren. What caused its disappearance we cannot say. But the fact that the

Fig. 9. Evidence of continuous human occupation of the Burren. (*a*) Megalithic gallery-grave, N. of Killinaboy (*c.* 2000 B.C.). (*b*) Stone ring-fort, E. of Lisdoonvarna (?ninth century A.D.). (*c*) High cross at Kilfenora (twelfth century). (*d*) Tower-house near L. Bunny (sixteenth century).

Fig. 10. Leamanegh Castle, between Kilfenora and Killinaboy (seventeenth century, incorporating older features).

Burren, where it lasted so late, has been since late Neolithic times one of the more densely populated parts of Ireland (to judge from its wealth of monuments, ranging from cromlechs to tower-houses) suggests that deforestation by man was not, at least here, the dominant factor.

Besides the juniper and the pine, so also the yew – the only other genus among the gymnosperms native to Ireland since the last glaciation – has played a prominent part in the vegetation of the Burren. Its pollen first appears about 5100 B.P., immediately after the celebrated and much-discussed fall in the percentage of elm pollen. The proportion of yew pollen increased rapidly, and soon reached as high a figure as 20 % of the total, suggesting that in the middle part of the post-glacial there were considerable stands of yew-woodland, such as are to be seen today on the limestone at Killarney. Thereafter the proportion of yew pollen oscillates, but with a tendency to decline, until it sank to insignificance about 2000 years ago. Today the yew is thinly scattered over the pavement, but only as bushes kept low by severe grazing, and it is only on the cliffs that an occasional tree can be seen (Fig. 19). The same period that saw the decline of the pine and the yew saw a steady increase in the proportion of pollen from herbaceous species, telling a tale of progressive deforestation.

There is little in the pollen record to spell out the history of the species which give the Burren its floristic interest today. Remains of *Dryas octopetala* are abundant in the late-glacial and the beginning of the post-glacial, but thereafter they disappear, even from sediments taken from sites beside which it is now common. One must presume that it expanded relatively recently from fairly limited stands which survived on exposed slopes during the period of maximum forest-cover. Leaves of *Potentilla fruticosa* have been found in the uppermost strata from a lake within its present area of distribution, but there is no trace of it in the earliest part of the post-glacial, when it might have been expected to be abundant. Pollen of *Helianthemum* spp. occurs sporadically throughout the post-glacial; one tends to assume it is of *H. canum*, but it is possible that *H. nummularium*, now confined to a single station in Co. Donegal, might have been present too. Of *Gentiana verna* there is no trace. The plants of southern affinities are either poor pollen producers or have few and small seeds, and their absence from the fossil record indicates nothing more than that they were never present in enormous abundance. It is worth mentioning, however, that fruits of *Najas marina*, now extinct in Ireland, occur in a deposit from near Mullaghmore from 8000 to 4000 B.P. Its disappearance is probably the result of falling summer temperatures in the latter part of the post-glacial.

For further information on the relation of the late-glacial floras of the Burren and Connemara to those of other parts of north-west Europe, reference may be made to Watts (1979).

THE HISTORY OF OUR KNOWLEDGE
OF THE FLORA

The first precise record of a plant from the Burren was made by Richard Heaton, a clergyman settled in the midlands, during the distressing period of the civil wars of the mid-seventeenth century. He observed *Dryas octopetala* and *Gentiana verna* 'in the mountains betwixt Gort and Galloway' and *Juniperus communis* 'upon the rocks near Kilmacdough', and reported his finds to William How, who published them in his *Phytologia Britannica* (1650). Half a century later Edward Llwyd, the Welsh polymath, passed through Connemara in the course of his extensive travels and thence visited Aran. In Connemara he noted 'an elegant sort of heath bearing large thyme leaves, a spike of fair purple flowers like some *Campanula*, and viscous stalks' – obviously *Daboecia cantabrica* – and also *Pinguicula lusitanica*; while on Aran he saw the maidenhair fern 'in great plenty'. Some of his discoveries were published by Petiver (1702) and Ray (1703); others by himself a few years later.

In the course of the next century we can trace only one record for the region (*Filipendula vulgaris*, noted by John K'Eogh in his *Botanologia universalis Hibernica* of 1735 as growing in the Burren). But with the turn of the century the work of William Wade marks a new departure, as he was the first to botanize Connemara on a systematic basis and to record some fairly widespread species as well as the local rarities. He seems to have operated mainly from three bases, Renvyle, Ballynahinch and Oughterard, and covered most of Connemara except the south-east. Naturally he made some mistakes, but in general he was a diligent and discerning explorer, and only some half-dozen of his records have to be set aside. Most of his records were published in 1802 (a few others being added in 1804) and they included such important finds as *Eriocaulon aquaticum*, *Saxifraga spathularis* and *Oxyria digyna*, and several inconspicuous species such as *Selaginella selaginoides* and *Anagallis minima*.

Wade was the first Professor of Botany at the Dublin (later Royal Dublin) Society, and was responsible for the botanic garden at Glasnevin, which the Society then owned. Hard on his heels in Connemara came J. T. Mackay, the tough and energetic Scot just appointed to take charge of the botanic garden of Trinity College, which had been founded a few years later than that of the Dublin Society. He visited Aran as well as Connemara, and his 'Systematic catalogue of rare plants...', published in 1806, added 16 species to the known flora of the region, including *Subularia aquatica* and *Saxifraga oppositifolia*. In 1830 he gave the first report of 'a new indigenous heath found in Cunnamara'; this was the species long known as *Erica mediterranea* (now *E. erigena*).

C. C. Babington, who was later to succeed to the chair of Botany at Cambridge, seems to have been the first visitor from England to botanize extensively in Connemara, and in 1836 he published a list of about 80 plants he had seen; naturally enough he noted mainly those to which he was unaccustomed in Cambridgeshire. He worked mostly at Oughterard and Maam, but he also visited Roundstone. He assures his readers that his journey was not as adventurous as they might imagine; there was some discomfort but no danger.

Fortified, perhaps, by this reassurance, a considerable number of English and Scottish botanists visited Connemara and Aran over the next twenty years, and recorded the results of their 'rambles' in various British periodicals. Most of them recorded only those plants which appealed to them, and curiously enough none of them visited the Burren. H. Seebohm, however, took advantage of a holiday at a friend's house near Letterfrack to do a more systematic survey, and his account of the pteridophytes of the area, published in 1851, is remarkably complete. These visitors between them added considerably to the known flora of the region, which by 1854 amounted to over 220 species.

Any botanical visitor to Roundstone during this period was sure to meet William McCalla, the son of the postmaster. He was the first resident of the region to have an interest in and a knowledge of plants, and virtually the only one apart from P. B. O'Kelly of Ballyvaughan and members of the staff of Queen's (later University) College, Galway. He was almost entirely self-taught, but he soon got to know his plants and he had a remarkably keen eye for spotting novelties, several of which he brought to the notice of Mackay, Babington, Harvey and others. He devoted most of his energies to seaweeds, but it was McCalla who brought to the notice of Babington the plants which were later to be named *Erica mackaiana* and *Allium babingtonii*. Unfortunately he fell a victim to the cholera epidemic of 1849. Some details of his life are given by Eager & Scannell (1981) and Nelson (1981*b*).

The period from 1854 to 1883 was to be dominated mainly by the work of More and Hart, though the discovery on Aran of *Ajuga pyramidalis* by David Moore in 1854, Foot's investigations in the Burren and Wright's visit to Aran in 1866 also deserve mention. Foot was a geologist, working for the Geological Survey, but he had also a keen interest in botany, and in 1864 he published the first general review of the Burren flora, based on observations made during his survey work. He lists 114 species, with perceptive distributional or ecological notes on many of them. Wright published what purported to be a full list of the plants of Inishmore, but although he mentions for the first time numerous common species which nobody had hitherto troubled to record, there are curious lacunae in his list, especially among the monocotyledons, and it was soon to be superseded by Hart's.

Hart was a man of almost demonic energy, and in a whirlwind visit to Aran in 1869 he covered the ground of all three islands far more completely than had any of his predecessors, overcoming the obstacle of the numerous high drystone walls by pushing them down and getting boys to build them up again. He made a few mistakes and overlooked a few species, but on the whole his list has stood the test of time well. Thirteen years later he traversed the principal mountain ranges of Connemara in the course of a survey of the mountain flora of Ireland, sponsored by the Royal Irish Academy. Hart was a skilled, intrepid and tireless mountaineer, who thought a day among the peaks wasted if he did not traverse at least 20 miles, and though accuracy is here and there sacrificed to speed his account of the Connemara mountain flora, published in 1883, added considerably to our knowledge, especially in showing how rich in species was the schistose cliff on Muckanaght, compared with the barren quartzite peaks all round it.

A. G. More was an Englishman who first became interested in the botany of the west in the course of holidays spent with a friend who lived at Castle Taylor, not far from Ardrahan and just beyond the eastern boundary of the Flora region. His first finds for the region were from Coole and Garryland, published in 1855, but he soon began to range further west and in 1860 added over 20 species to the known flora of the region. In 1864 he came to live in Ireland, chiefly in order to assist David Moore in the compilation of a *Cybele Hibernica*. This appeared in 1866, and included, in addition to further discoveries by More, a number of unpublished observations made near Galway by A. G. Melville, the first Professor of Natural History at Queen's College. A supplement to the *Cybele*, published by More in 1872, contained many new records for both Connemara and the Burren, and in 1876 he concluded his work in the west with a fairly complete list of the flora of Inishbofin.

The fact that complete catalogues had been attempted for Aran and Inishbofin but for nowhere on the mainland gives rise to the odd situation that for nearly all really common species the first published record is for one of these islands. But thanks largely to these lists the known flora of the region had by 1883 risen to 575 species.

The next 12 years showed a slackening of activity in the region, but not a total cessation. H. C. Levinge made some interesting discoveries in the Burren (including the first record for Ireland for *Limosella aquatica*) and Nowers and Wells, two English amateurs who visited Aran in 1890, showed that there was some gleaning to be done after Hart's survey. In 1895 a renewal of activity was signalized by a joint meeting at Galway of the Irish field clubs; the meeting was organized largely by Praeger, and his first record for the region in that year (a new station for *Pilularia*) was the forerunner of many others to come later. In 1896 another pair of English amateurs, Marshall and Shoolbred, broke fresh ground in the north-east, mainly in the neighbourhood of Clonbur and Cong, and showed that there was much of interest there. Two years later Colgan and Scully published a new edition of the *Cybele*, based on the papers of More, who had died a few years earlier, but including also some recent records by others and adding from one source or another 17 new species to the regional flora. The book also deserves notice as being, as far as we can see, the first Flora in English to list systematically the preference of plants for calcareous or non-calcareous soils – an early indication of the swing from taxonomy and distribution to ecology, which was to dominate the first few decades of the twentieth century. Meanwhile, in 1896, Praeger had begun his five-year tramp through Ireland to collect the data for his *Irish Topographical Botany*. His records were arranged on a vice-county basis, and since the records for West Galway and Clare were already fairly numerous he concentrated on the unexplored counties and spent only a little time in Connemara and virtually none in the Burren. He paid, however, a hasty visit to western Connemara in 1896 and a more extensive one to the Galway–Oughterard area in 1899, and his annotated copies of the *London Catalogue*, in which he noted, with a code-letter for each district, such plants as he thought worth recording, provide many new records for West Galway, and a few for those parts of north-east and south-east Galway that lie within our region. These are all the more valuable, as for most of the species fairly common throughout Ireland (though not in all cases in Connemara) *Irish Topographical Botany* itself gives only the information, without precise locality, that they have been seen in H16.

For a decade after the publication of his *magnum opus* Praeger maintained an interest in both Clare and Galway, and his papers of 1905, 1906 and 1911 (this last devoted to a survey of Inishbofin) all added to our knowledge of the region. After 1911 his interests turned elsewhere, but in 1932 he made a welcome return with his paper on the turloughs, and in his deservedly popular *The Botanist in Ireland* (1934) he gave for the first time to a wide public a succinct summary of the floristic peculiarities of Connemara and the Burren. His later papers (especially that of 1934), in which he up-dated vice-county records for Ireland, also contained some novelties for the region.

P. B. O'Kelly, a self-taught native of Ballyvaughan, was active mainly during the period 1890–1910, and was in contact with several experienced botanists during these years, especially H. C. Levinge. He knew the Burren intimately, had a keen eye for plants, and was responsible for several interesting discoveries which have stood the test of time. A number of his specimens, however, come from localities where the species has never been seen since and are subject to some suspicion – a suspicion deepened by the nature of the catalogue of plants from his 'nursery' which he offered for sale. All were allegedly from the Burren, but some of his species do not grow there, and the list also contains many common weeds for which fairly high prices were asked.

After 1911 there was a lapse of twenty years in which we can trace no new observations on the flora of our region. It was, however, the period in which ecology was establishing itself as a recognized part of botany, and Tansley spent nearly a week in the west of Ireland in 1908 with the British Vegetation Committee. It was then that he made the observations on Garryland wood and the Connemara bogs which were later incorporated in his *The British Islands and their Vegetation*. He returned in 1911 with the first International Phytogeographical Excursion (I.P.E.), in the course of which some floristic observations were made by Druce and Ostenfeld; most of these, however, related to hybrids or infraspecific variants. Further studies of the Connemara bogs were made in the nineteen-thirties, when the rise of palynology brought them into prominence, and the publications of Pearsall and Lind and of Osvald contained interesting comparisons of these bogs with others in Britain and Scandinavia. Further ecological investigations of the bogs were carried out later by Boatman, while, much earlier, Connolly had attempted a more general survey of the blanket-bog country.

Pearsall and Lind made a fairly extensive study of the Connemara lakes in 1939; although their paper is primarily concerned with phytoplankton, many of their observations are of general ecological interest. Further studies of the chemistry of the fresh waters of Connemara, made by Webb (1947) and Gorham (1957), draw attention to the importance of estimates of blown dust and sand and of sea-salt in the rain in explaining the pH and nutrient status of the waters. Recently these problems have been further investigated by students from Holland, but their work is not yet published as we go to press.

The next landmark was the visit to Ireland of the ninth I.P.E. in 1949. The party spent only 2½ days in the region of this Flora, which could scarcely have been expected to produce any solid results had it not been for the energy of Braun-Blanquet and Tüxen, who published in their celebrated *Irische Pflanzengesellschaften* (1952) over 80 *relevés* taken in the region, and subjected the vegetation of western Ireland for the first time to the strict discipline of the Zürich–Montpellier school of phytosociology. Several new associations were described from the region. Some of their classification has had to undergo revision, but their work remains the foundation for the steadily growing mass of phytosociological data which is being accumulated for Ireland, and not least for Connemara and the Burren. A further international party, primarily of phytosociologists, paid a short visit to the region again in 1981, and the publication which is to follow on their visit will doubtless provide further documentation for this aspect of the study of the flora.

In 1953 work began on the accumulation of data on the basis of lists for each 10 km square in preparation for the B.S.B.I. *Atlas*, which was published in 1962. Connemara and the Burren were reasonably well covered, only 4 mainland squares out of 46 achieving a total of less than 150 species. There was a natural, though unfortunate tendency for the floristically rich squares, such as those containing Roundstone and L. Bunny, to be raked over again and again, while allegedly dull or less accessible squares were neglected; nevertheless, the work yielded a number of new records for the region and a far better understanding of the frequency and distribution of species already recorded.

At about the same time the British Ecological Society agreed to sponsor an ecological survey of the Burren, on the grounds that in spite of much floristic work this aspect had been neglected. It proved difficult to induce visitors from Britain to pay visits long enough to produce detailed or extensive studies, but to this generalization there was one shining exception. Ivimey-Cook and Proctor spent many weeks in the Burren in 1959, and their observations were published in a very comprehensive paper in 1966, which is a valuable storehouse of ecological data for the area. Another product of the survey was Webb's 'Catalogue raisonné' of the Burren flora (1962), in which Foot's work of a century earlier was updated and expanded by the provision of a list of some 170 species with notes on their distribution and ecology.

Work subsequent to 1960 has been mainly directed towards the compilation of this Flora, and its history may be read in our main text. Since 1959, over 90 new records have come in. A number of these related to aliens which were well known to be present, but which nobody had recorded formally, or for which no distinction had been made between planted and naturalized stands. Many of them, however, are new records of native plants, and of these the two most spectacular, *Eriophorum gracile* and *Asplenium septentrionale*, both new to Ireland, were found by visitors from Britain who were not collecting material for this Flora.

Much still remains to be done, and we hope that the uncertainties or contradictions revealed in our text will stimulate future workers to concentrate on looking out for species for whose presence the evidence is unsatisfactory, to attempt the confirmation of the numerous old records which we have left unconfirmed, to solve the taxonomic problems which are apparent in several groups, and to investigate some of the ecological paradoxes which abound in the region.

FLOWERING PLANTS, CONIFERS AND PTERIDOPHYTES

NOTES ON THE TEXT

The Latin names of species which have been seen in the Flora region in or after 1959, and which are believed to be still extant there, are printed in **bold** type. The names used are those adopted in the sixth edition (1977) of *An Irish Flora* by D. A. Webb, and are, with a very few exceptions, identical to those used in *Flora Europaea*. Alternative names used in recent Floras or popular works, which are likely to be more familiar to the reader (or, in a very few cases, names used to replace a *Flora Europaea* name which is generally agreed to be incorrect) have been added in *italics*, enclosed in parentheses. These names are in some cases true synonyms, but in many others they are names of non-Irish plants incorrectly applied. The distinction between the two is often a technical and controversial matter, and for this reason no authorities are cited for the alternative names printed in italics, for we believe that such arcane formulae as 'auct. hibern., vix (J. & C. Presl) Reichenb. fil.' are out of place in a book of this kind.

For species which have been reliably reported from the region but have not been seen there since 1958, but for which there are no real grounds for supposing them extinct, the correct name is printed in *italics* and enclosed in parentheses. For species which have never been more than casual in the region, or have been recorded in error, or can with some confidence be presumed to be extinct, the name is set in ordinary roman type and enclosed in square brackets; others in this category are mentioned in a paragraph in smaller type appended to the species to which they are most closely related.

English and Irish names, if reasonably well established in actual speech and fairly precise in their application are added after the Latin name, the English in Roman type, the Irish in *italics*. We have, however, omitted names which are merely copied from one book to another and are rarely or never on the lips of a botanist, as well as those names, all too common in Irish, which are applied indiscriminately to two or more very different plants.

The names of each species are followed by a line which details the districts from which it has been recorded; for definitions and descriptions of these districts see pp. xii–xix. As with the Latin names, **bold** type for a district means that it has been seen there since 1958; *italic* type that it was reliably recorded earlier and may still be there but has not been seen recently; roman type in square brackets that it has been recorded in error, or is extinct, or has never been more than casual or planted. In addition to the figures for the eight districts, the letters A and B are used to indicate occurrence on the Aran Islands or on Inishbofin (including Inishshark), respectively. It should be noted, however, that the Aran Islands form part of district 2 and Inishbofin and Inishshark part of district 7, so that for a plant known only from Aran both 2 and A are given on this line.

There follows next an indication of the plant's most usual habitat in the region, and of its rarity or abundance.

In the main paragraph, localities are detailed for the various districts, unless they are so numerous as to justify a generalized description such as 'occasional' or 'frequent, mainly in the north'. As with the district numbers, a locality printed in *italic* type indicates that there are no records later than 1958; roman type that there is a recent record. Authorities are given for pre-1959 records, but, for reasons given in the introduction, not for those of later date except in the case of a few rare or critical species.

3

Finally, a note of the first record for the Flora region is given; this is the earliest *published* record which we have been able to trace which indicates unambiguously the presence of the species in our region. Herbarium sheets or unpublished manuscripts of earlier date are mentioned only if the information they provide is of special interest.

The signs *, ‡, † are used before a Latin name in their traditional sense, to indicate certainly, probably and possibly introduced by man; species which are preceded by none of these signs are presumed to be native to the region. The signs, it must be emphasized, relate to the plant's status in the Flora region, not in Ireland as a whole, for many species which are native in the east are aliens in the west. In the assessment of status much personal judgement and, indeed, guess-work is involved, but we have done our best to weigh the evidence fairly, being guided partly by the frequency of occurrence in natural or semi-natural habitats, and partly by the history of increase, stability or decline.

Horizontal distances are expressed in metric units; vertical distances are normally given in both metres and feet, but in feet only when there is direct or indirect reference to a contour-line or triangulation point on a map. In the identificatory notes, dimensions are expressed in centimetres and millimetres.

The following abbreviations are used in the text:

Herbaria
BEL   Ulster Museum, Belfast
BM   British Museum (Natural History), London
CGE   University Herbarium, Cambridge
DBN   National Botanic Gardens, Dublin
K   Royal Botanic Gardens, Kew .
NMW   National Museum of Wales, Cardiff
OXF   University Herbarium, Oxford
TCD   Trinity College, Dublin

Publications
CH1   *Cybele Hibernica*, edn 1 (Moore & More, 1866)
CH2   *Cybele Hibernica*, edn 2 (Colgan & Scully, 1898)
FWI   *A tourist's Flora of the west of Ireland* (Praeger, 1909c)
ICP   *The plant communities of the Burren* (Ivimey-Cook & Proctor, 1966)
ITB   *Irish topographical botany* (Praeger, 1901b)
P. cat.   London Catalogues, preserved in **DBN**, annotated by Praeger in his accumulation of data for *Irish Topographical Botany*
Atlas   *Atlas of the British Flora* (Perring & Walters, 1962)
Crit. Suppl. *Critical Supplement to the Atlas of the British Flora* (Perring & Sell, 1968)
Fern Atlas   *Atlas of Ferns of the British Isles* (Jermy et al. 1978)
Census   *Census Catalogue of the Flora of Ireland* (Scannell & Synnott, 1972)
 Catalogue

Collectors of field records, mostly unpublished, from 1959 onwards
GH   G. Halliday
JJM   J. J. Moore
MS   M. J. P. Scannell
DW   D. A. Webb
JW   J. White

DICOTYLEDONS

RANUNCULACEAE

Clematis

*Clematis vitalba L.　　Traveller's joy

1 2 3 *(4)* **5** . . **8** . .

Hedges and walls; rarely on limestone pavement. Occasional.

1. At the bridge over the R. Fergus below Inchiquin L. Hedge by Moy Ho. **2.** On limestone pavement 2 km E. of Burren village. Hedges near Gragan Castle. **3.** Abundant in hedges and on walls S. of Ardrahan. *(4). E. of Moycullen* (Atlas). **5.** Covering a ruin 2 km E. of Spiddal. **8.** Hedge near Clonbur. Frequent immediately W. of Cong.

First record: Atlas (1962).

In most cases at no great distance from the garden from which it may be presumed to have escaped, and never in the abundance in which it is often seen in the east and south.

Thalictrum

Thalictrum alpinum L.

. . . . . . **7 8** . .

Cliffs, ledges and rocky outcrops on mountains; rare.

7. Cliffs on Muckanaght and Bengower. *Near the summit of Benlettery* (Hart, 1883). Frequent on the N.E. side of Doughruagh; also at the head of the gully between Benchoona and Garraun. **8.** In the north-facing corrie on Benbeg. Maumturk Mts.: at Maumeen and near the summit of Letterbreckaun. Sparingly on Benwee at *c.* 500 m.

First record: Wade (1802). 'Lettery Mountain, Ballynahinch, at the south-east part.'

T. minus L.

. **2 3 4 5** . **7 8** **A** .

Limestone pavement, rocky lake-shores and mountain cliffs; occasional.

2. Pavement near Black Head. Wood-margin by the Caher R., and abundant in the valley on the S.W. side of Gleninagh Mt. Inishmore. **3.** Occasional on the pavement, as at the Ballyogan Loughs, near Funshinmore, and S.W. of Mullaghmore. **4.** On an island in L. Corrib E. of Ross L. **5.** Near the river-mouth at Spiddal. **7.** *By the Dawros R.,* 1832 (**DBN**). Frequent on the E. side of Garraun, from 300 m upwards. **8.** Occasional on the shore of L. Corrib, north-westwards from Oughterard. Cliffs on Benbeg and above L. Nadirkmore. *By a bog-pool at Recess* (Hurst, 1902).

First record: Ball (1839). 'On rocks in the bed of the river above L. Inagh.'

Less abundant now in many of its stations than the earlier records would suggest.

T. flavum L.   Meadow-rue

(*1*) **2 3** . . . .

Rocky ground beside turloughs; rare.

(*1*). *Ballycullinan L.* (Praeger, 1905). **2**. By the turlough W. of Turlough village. **3**. Among rocks at the S.E. corner of Coole L. *Dromore L.* (*E. end*) (Praeger, 1905).

First record: More (1855). He records it specifically only for the neighbourhood of Castle Taylor, which lies outside our region, but internal evidence makes it clear that he saw it also at Coole, to which several other records given in the same paper refer.

The habitat is rather unusual for this species; elsewhere in Ireland it grows mainly in water-meadows or on river-banks.

Anemone

Anemone nemorosa L.   Wood anemone   *Nead coille*

. **2 3 4 5 6 7 8** . .

Woods and scrub; sometimes in open grassland or on limestone pavement. Frequent in the Burren, and locally elsewhere, but rare in the more exposed areas.

2. Abundant in the wood at Poulavallan. Widespread in smaller quantity throughout the district, both in hazel-scrub and in crevices in bare pavement. At 170 m (550 ft) on the hill N. of Mullaghmore, and at 300 m (*c*. 1000 ft) on Carnsefin. **3**. Abundant in scrub 1 km E. of Killinaboy. Grassland at Carrowgarriff. W. of Coole L. **4**. Near Burnthouse and Carrowmoreknock; not seen in S. part. **5**. Abundant in the wood N.W. of Moycullen. **6**. In scrub near Flannery Br. On an island in Nahasleam L. **7**. Wood W. of Ballynahinch L. Scrub E. and N.W. of Clifden. On small cliffs N. of L. Fee. Frequent on an island in L. Inagh. **8**. Frequent at Glann and elsewhere on the N.W. shore of L. Corrib, and on several of the islands. Woods at Ashford Castle and the Hill of Doon. Island in L. Shindilla. By the effluent stream from L. Nacarrigeen.

First record: Foot (1864). 'Grows in Burren in the most exposed and dry localities imaginable, up to a height of upwards of 1000 ft.'

Except perhaps for occurrences in the crevices of limestone pavement, the presence of this species in open habitats is a sure indication of the former existence of woodland or scrub. It possesses, however, great powers of persistence after clearance, and in the last stages of degeneration of hazel-scrub each moribund bush can be seen surrounded by a circle of anemones. See also Webb (1955).

Ranunculus   Buttercup

The yellow-flowered buttercups present no very serious difficulties in their identification, but the white-flowered aquatic species (section *Batrachium*) are difficult. *R. hederaceus*, with very small flowers and the leaves only slightly lobed, is distinctive, but the others vary greatly in accordance with season and depth and movement of water, and are best identified by specialists. The only help that can be given here is to say that if the plant has very long, limp, submerged leaves and large flowers and grows in flowing water it is likely to be *R. penicillatus*; if it grows in shallow water or on wet mud and has small flowers and shortish, bushy, fairly stiff dissected leaves it is likely to be *R. trichophyllus*; and that if it grows in brackish water or very close to the sea there is a good chance that it may be *R. baudotii*.

(*Ranunculus circinatus* Sibth.)

(*1*) . (*3*) . . . .

Lakes; very rare and not seen recently.

(*1*). *Inchiquin* L., 1892 (CH2). (*3*) *L. Atedaun and Dromore L.* (Praeger, 1905). *Near Kinvara* (Phillips, 1924).

First record: CH2 (1898). Inchiquin L.

R. trichophyllus Chaix

1 2 3 4 . . 7 8 A B

Shallow water and wet mud. Occasional to frequent on the limestone and in the extreme west of Connemara; very rare elsewhere.

1. Marsh E. of Inchiquin L. **2**. Frequent around turloughs and brackish lakes. Inishmore. **3**. Very frequent. **4**. Shore of L. Corrib E. of Moycullen. Ditch by Aughnanure Castle. **7**. By the lake at Bunowen, and by other lakelets on the Slyne Head peninsula. By the smaller lake on Omey I. In a stream E. of Ballinaboy. Inishbofin. **8**. On the county boundary at Cong.

First record: Wright (1871). Inishmore: 'in some quantity near Bungowla'.

R. aquatilis L.

. [2] **3** . . . . **8** [A] .

Lake-margins, pools and ditches. Occasional in the Burren; rare elsewhere.

3. Abundant at the S. end of L. Bunny. Pond 3 km N.E. of Ruan. N. shore of L. Atedaun. S. of Garryland. **8**. L. Mask, near Ferry Br.; *also N. of Clonbur* (Marshall & Shoolbred, 1896).

First record: Marshall & Shoolbred (1896), as *R. heterophyllus*. 'In a ditch between Clonbur and L. Mask.'

The old records for Aran are probably errors for *R. baudotii*.

R. baudotii Godr.

. **2** . **4** (*5*) . **7** . **A B**

Lake-shore and marshes, usually near the sea. Occasional, but very local.

2. Aran. L. Aleenaun. *L. Luick* (Druce, 1909). *New Quay*, 1892 (**DBN**). **4**. Marshes E. of Galway city. (*5*). *W. of Galway* (CH1). **7**. Near the N. end of Ballyconneely Bay. Aughrusbeg L. Inishbofin.

First record: CH1 (1866). 'Near Galway, to the west of the town; *Prof. Melville*.'

[R. peltatus Schrank]

A specimen from the W. end of Ballynakill L. has been tentatively assigned to this species, but, as it is not typical in all characters and as there are no firm records for elsewhere in the Flora region, it is best held over pending confirmation.

R. penicillatus (Dum.) Bab. (*R. pseudofluitans*)

1 . **3 4** (*5*) . . . . .

Streams and rivers; rarely in lakes. Occasional in the south; rare elsewhere.

1. Drumcullaun L. In the river at Ennistymon. Abundant in the R. Fergus above Corofin. **3**. In the R. Fergus below Corofin and above Ennis. Stream by L. Cleggan.

Kiltartan R. **4**. In a tributary of the R. Corrib near Terryland. (*5*). *Barna* (Praeger, 1903*a*).

First record: Praeger (1903*a*). 'Barna and Corrib River, '02–Phillips.'

R. hederaceus L.   Ivy-leaved crowfoot

1 [2] 3 4 5 . 7 8   [A] **B**

Ditches, marshes, streamsides and small pools. Fairly frequent on the Clare shales; occasional in N. and W. Connemara; rare elsewhere.

1. Frequent. **[2]**. Recorded by Wright and Hart for all three Aran Islands, but not seen since; apparently extinct. **3**. Ditch by L. Bunny. **4**. Marsh near the sea E. of Galway city. **5**. Streamside near Gentian Hill. 2½ km E.S.E. of Costelloe. **7**. Belleek. Doonloughan. Glenbrickeen L. Renvyle. Inishbofin. **8**. Drains at W. end of L. Nafooey. Teernakill Br. Flooded lane at Glann, and abundant in a small stream between Glann and Oughterard. Leenane.

First record: Wright (1871). Aran Islands.

R. sceleratus L.   *Torachas biadhain*

1 2 . 4 5 . 7 .   . .

Marshes; rather rare.

1. Near Ballynalackan Ho. **2**. Poulsallagh. **4**. Saline meadows at Oranmore. By the N.E. corner of L. Atalia. On the N. side of the Kilcolgan estuary. **5**. Marsh at the Claddagh. Marshy field by Gentian Hill. W. of Spiddal. *Inveran*, 1895 (ITB). **7**. By Doonloughan and other lakelets on the Slyne Head peninsula. S. of Ballyconneely.

First record: ITB (1901). Inveran.

R. flammula L.   Lesser spearwort

1 2 3 4 5 6 7 8   **A B**

Marshes, drains, lake-shores and other wet places. Abundant throughout Connemara; slightly less so on the limestone, but not absent from any considerable area.

First record: Hart (1875). Aran Islands.

Subsp. *scoticus* (Marshall) Clapham was reported from the Flora region by Praeger (1907), but it now appears that he (following Marshall) took far too wide a view of this variant, and that it grows in Ireland (if at all) only in Co. Mayo.

R. lingua L.   Greater spearwort

. [2] 3 4 . . 7 [8]   [A] .

Marshes, fens and streamsides; local and rather rare.

3. Occasional in the extreme south. Fen at S. end of L. Bunny. Drain at S.W. end of L. Atedaun. *Kilmacduagh* (ITB). **4**. Fen in N.E. part of Galway city, 1968; probably now extinct by drainage. E. bank of R. Corrib, near Terryland. **7**. By a lake near the centre of the Slyne Head peninsula.

First record: Mackay (1806). By the R. Fergus above Ennis.

There are also old, unconfirmed records for Inishmore (Wright, 1867), Maam (Babington, 1836) and near Letterfrack (More, 1872). The first two are probably errors for robust forms of *R. flammula*; the last may well be correct, but needs confirmation.

R. auricomus L. Goldilocks

. **2 3** . . . .

In hazel scrub or mixed woodland; rare.

2. In three places E. and S.E. of Carran. In scrub W. of the Caher R., near Formoyle. **3**. Garryland and Coole woods.

First record: Stirling & Beckett (1966). Garryland wood.

Perhaps overlooked elsewhere on account of its early flowering, but it is a rare plant in W. Ireland.

R. acris L. Meadow buttercup

1 2 3 4 5 6 7 8 A B

Meadows, pastures, marshes and roadsides. Abundant throughout, except in the N. part of district 6, where it is scarce.

First record: Wright (1871). Aran Islands.

R. repens L. Creeping buttercup

1 2 3 4 5 6 7 8 A B

Roadsides, ditches, marshes, damp grassland, and as a weed of cultivation. Abundant throughout, except on the driest parts of the limestone.

First record: Wright (1871). Aran Islands.

A remarkable variant, perhaps identical with *R. reptabundus* Jordan, is found at a fairly low level in some of the turloughs, where it is submerged for a large part of the year. It is a small, neat plant, with leaves deeply dissected into linear–oblong segments, at first sight suggesting those of *Apium* × *moorei*. It remains constant in cultivation.

R. bulbosus L. Bulbous buttercup

1 2 3 4 5 6 7 8 A B

Dry pastures, sand-dunes and rocky ground. Very frequent on the limestone and locally on sand-dunes; rare elsewhere.

1. Occasional by the coast and along the N. margin. **2**. Frequent. **3–4** Very frequent. **5**. Locally frequent near the coast. Pasture by the Oughterard–Galway road 5 km N.W. of Galway. **6**. Dunes at Moyrus. **7**. Frequent to abundant on nearly all the dunes and areas of blown sand right round the coast. **8**. On a peaty roadside bank 1 km E. of Maam.

First record: Wright (1871). Aran Islands.

*R. parviflorus L.

. **2** . . . . . . **A** .

Cultivated fields; rare, and only on Aran.

2. Near the W. end of Inishmore, 1966 (JJM). Inishmaan, 1974 (Goodwillie). Inishcer, 1969 (JJM).

First record: Webb (1980a). Aran Islands.

An interesting, though precarious survival. It was formerly established in S.E. Ireland, but there are no recent records.

R. ficaria L.   Lesser celandine

1 2 3 4 5 6 7 8   A B

Damp or shady places; also in pastures on water-retentive soils or near the coast. Frequent throughout, and locally abundant.

First record: ITB (1901). H16.

Commonest in district 8 and the coastal regions of district 7; rather scarce in district 6. In W. Connemara it can be seen in abundance in exposed pastures behind the beaches. All specimens are referable to subsp. *ficaria*.

Caltha

Caltha palustris L.   King-cup, marsh marigold

1 2 3 4 5 6 7 8   . B

Ditches, marshes, streams and lake-shores; also on damp grassland. Frequent and locally abundant in the east and south, but rather rare in S. and W. Connemara.

1. Abundant. **2**. By the Caher R. By L. Aleenaun and the Carran turlough. Marsh at Formoyle. L. Luick. **3**. Frequent except in the N.W., where there is little water. **4**. Frequent. **5**. Wet grassland near the sea 4 km W. of Spiddal. Marsh by the Drimneen R., S.E. of Oughterard. **6**. Island in Athry L. **7**. By lakes N.E. of Clifden and W. of Ballyconneely. 1 km W.N.W. of Cleggan, and by Ballynakill L. Island in L. Inagh. Inishbofin. **8**. Frequent.

First record: ITB (1901). H16.

On the heavy clay soils around Lisdoonvarna this species grows in great abundance in wet, rushy pastures, in the absence of any visible standing or flowing water.

Aquilegia

‡**Aquilegia vulgaris** L.   Columbine

1 2 3 4 (*5*) . . .   **A** .

Open scrub, stony grassland and roadsides. Frequent N.E. of Galway; rare elsewhere.

1. By a lane N. of Ballynalackan Ho. **2**. Inishmore and Inishmaan. On the old road above Black Head. **3**. S.W. of Mullaghmore, far from houses. *Dromore*, 1879 (**BEL**). **4**. Frequent on limestone knolls N.E. of Galway, as at Menlough, Ballindooly and Kilroghter. Hedges S. of Carrowmoreknock. (*5*). *By a stream S. of Tonabrocky*, 1891 (CH2).

First record: O'Mahony (1860). 'At no great distance from Menlo castle.'

Perhaps native N.E. of Galway, and, less probably, on Inishmore, although pink- and white-flowered plants are found here as well as blue. An obvious garden escape in most of its other stations.

NYMPHAEACEAE

Nuphar

Nuphar lutea (L.) Sibth. & Sm.   Yellow water-lily   *Duilleóg bháite*
1 . 3 4 5 6 7 8 . B
Lakes, slow streams and bog-pools. Occasional to frequent throughout, except in district 2, where suitable habitats are very scarce.
Rather evenly distributed; commonest in districts 3, 4 and 5; scarcest in 6 and 8.
First record: More (1876). Inishbofin: Church Lake.

Nymphaea

Nymphaea alba L.   White water-lily
1 . 3 4 5 6 7 8 . .
Lakes and slow streams. Frequent in Connemara and the eastern Burren; local elsewhere.
1. L. Goller and L. Lickeen. Frequent in the south-east. **3**. Frequent. **4**. In L. Corrib, E. of Moycullen; also in rivers nearby. Ballindooly L. **5-8**. Frequent.
First record: Mackay (1806). 'In Cunnemara they use the roots of it for dyeing wool black.'

A small-flowered variant with fewer rays on the stigma was described from Connemara as *N. occidentalis* (Ostenf.) Moss, but it intergrades with typical plants too completely to merit recognition as anything more than a variety.

CERATOPHYLLACEAE

Ceratophyllum

Ceratophyllum demersum L.   Hornwort
. . 3 . . . .
Pools, streams and lakes; rare.
3. In fair quantity in Coole L. R. Fergus below the lake at Ballyallia Ho. L. Atedaun. *Pool near Corofin*, 1905 (**DBN**). *Dromore L.*, 1905 (**DBN**).
First record: Praeger (1905). 'Pool near Drummeen Castle [N.E. of Corofin].'

BERBERIDACEAE

Berberis

***Berberis vulgaris** L.   Common barberry
. . 3 . . . .
Limestone pavement; very rare.

3. Several bushes scattered among an open woodland of *Pinus sylvestris* on limestone pavement about 2 km N. of L. Bunny.

First record: Webb (1982).

Now very rare in Ireland, thanks to a campaign for its elimination as the alternate host of *Puccinia graminis*, the rust-fungus of wheat.

*B. darwinii Hooker fil.

. . . . . . 7 . . .

On rough ground among other shrubs; very rare.

7. Several bushes at Ballinaboy, and two about 1 km to the east.

First record: Scannell & McClintock (1974).

Presumably bird-sown from a garden nearby; apparently well naturalized.

PAPAVERACEAE

Papaver   Poppy   *Caithleach dearg*

*Papaver rhoeas L.

. 2 3 4 5 . . . A .

Cultivated fields and roadsides. Occasional in a limited region of the Burren; rare elsewhere.

2. Occasional to frequent in a narrow strip of country extending from New Quay and the dunes E. of Ballyvaughan southwards to Deelin Beg. Roadside near Fanore. Inishmaan. **3.** Occasional in the N. half. **4.** Roadsides S. and W. of Oranmore. *Kilbeg ferry* (P. cat.). **5.** A few plants on a roadside inland from Gentian Hill.

First record: Corry (1880). W. of Ballvaughan.

Reasonably well established in the N.E. part of the Burren; little more than casual elsewhere.

‡P. dubium L.

. 2 3 4 . . . . A .

Roadsides, cultivated fields and waste ground. Occasional on limestone to the south and east of Galway Bay; very rare elsewhere.

2. By L. Luick. E. of New Quay. Fields behind the dunes E. of Ballyvaughan. Inishmaan (not seen on the other Aran Islands since 1869). **3.** Occasional in tilled fields E. of Corofin and N. of Ennis; also E. of Kinvara. Roadside $2\frac{1}{2}$ km S.W. of Ardrahan. **4.** Abundant on the esker N. of Gortachalla L. Drift hillock at Kilcaimin. By the coast between Galway and Oranmore.

First record: Wright (1871). Aran Islands. 'On cultivated land in the vicinity of the villages.'

If this species is native anywhere in Ireland the eskers and shingle-beaches of the Galway region are as convincing a habitat as any. But there is good reason to believe that none of the annual poppies is native in N.W. Europe; they were probably introduced with seed-corn in Bronze-Age or late Neolithic times.

P. dubium is easily distinguished by the hairs lying flat against the stem (in *P. rhoeas* they stand straight out), and the white sap (yellow in *P. lecoquii*).

*P. lecoquii Lamotte

. **2 3 4** . . . .   (*A*) .

Roadsides and cultivated ground on the limestone; rare.

2. A few plants on a roadside E. of Ballyvaughan. W. of New Quay. *Aran* (Hart, 1875). **3**. Sparingly N.W. and S.W. of Kinvara. One plant by a road near Tubber. **4**. Roadsides at Kilcaimin and S. of Carrowmoreknock.

First record: Hart (1875). Inishmore and Inisheer.

Always in small quantity, and less securely established than either of the two preceding species.

*P. hybridum L. used to be fairly common around the head of Galway Bay and in parts of the Burren, but it is now a very rare casual on the verge of extinction. **P. argemone* L. has twice been recorded from the Burren, but was never more than a rare casual in the west of Ireland. **P. somniferum* L. is occasionally seen in Galway city, but is not naturalized.

Meconopsis

Meconopsis cambrica (L.) Vig.   Welsh poppy

. **2** . **4 5** . **7** .   . .

Rocky places in the mountains; also as an escape from cultivation. Rare.

2. Near an old lead-mine *c*. 1½ km W.S.W. of Turlough village, at 235 m (770 ft), 1976 (Robinson). **4**. By L. Corrib, N. of Menlough, 1968 (DW); apparently introduced with tree-stumps which had been dumped there. **5**. By the river at Oughterard, 1969; clearly a garden escape. **7**. Sparingly on the N. face of Muckanaght, 1883, 1964, 1971.

First record Hart (1883). Muckanaght.

We ignore Mackay's (1836) record for the hill above Clifden, as E. C. Nelson (*in litt.*) gives good reason for supposing that it was based on a double misunderstanding. Mackay never saw the plant there.

Glaucium

Glaucium flavum Crantz   Horned poppy

. (*2*) **3** (*4*) (*5*) . . .   (*A*) .

Shingle beaches; formerly frequent around the head of Galway Bay, but now almost extinct.

(*2*). *Inishmaan* (Hart, 1875); not seen since. **3**. A few plants behind the storm beach on Aughinish, 1966 (JJM). (*4–5*). *Formerly frequent from near Kilcolgan to W. of Barna* (ITB and P. cat.), but not seen there recently.

First record: Hart (1875). 'Inishmaan, about the middle of the north-western shore.'

A diminishing species, like many sea-shore plants, and in any case always less plentiful in the west than in the east.

*Chelidonium majus L., the greater celandine, has been recorded from a few places in districts 3, 4 and 5, but in our experience it does not stray far enough from the garden in which it was planted to be considered fully naturalized.

Fumaria   Fumitory

‡**Fumaria capreolata** L.

. **2** . (*4*) (*5*) . **7** .   A (*B*)

Cultivated ground and waste places; rare.

2. Occasional on Aran. (*4*). *Oranmore* (Praeger, 1899). *Moycullen* (Atlas). (*5*). *Spiddal and Gentian Hill* (Atlas). **7**. One plant at the edge of the 'coral beach' near Ballyconneely, 1966. Dog's Bay, 1964. Belleek, 1967. *Formerly on Inishbofin and near Ballynahinch.*

First record: Wade (1802). 'At the back of Mrs O'Lea's house, Ballinahinch.'

Rapidly diminishing, as in other parts of Ireland.

A plant in **BM** collected by Shuttleworth at Renvyle in 1831 has been identified as *F. purpurea* Pugsl., but there are no other records for our area.

‡**F. bastardii** Bor.

. **2 3 4 5** . **7 8   A B**

Cultivated ground and waste places. Rather frequent on the limestone; occasional elsewhere.

2. . N.E. of Ballyvaughan. Near Corcomroe Abbey. E. of New Quay. Aran. **3**. Occasional to frequent. **4**. Frequent. **5**. Occasional W. and N.W. of Galway. W. of Spiddal. **7**. Frequent in the extreme west; not seen elsewhere. **8**. N. of Oughterard.

First record: More (1876). Inishbofin.

This species has increased since 1900 over most of Ireland, and especially in the west, at the expense of other members of the genus, and is now much the commonest Irish fumitory. Some plants from our region, and especially around Galway, approach *F. muralis* in some characters, but it seems to us rather doubtful that a firm separation between these two species can be maintained in Ireland.

‡**F. officinalis** L.

. (*2*) **3 4** . . **7** .   (*A*) **B**

Cultivated and waste ground; rare and apparently decreasing.

(*2*). *Formerly frequent throughout Aran; recorded for Inishmore in 1956* (Atlas), but not seen since. **3**. Locally frequent in fields near the sea 5–6 km N.W. of Kinvara, 1967. *Near L. Bunny* (Atlas). **4**. Allotment on E. margin of Galway city, 1969. On esker-gravels near Gortachalla L., 1971. *Rosscahill* (Atlas). **7**. Inishbofin (rare), 1965.

First record: Wright (1871). Aran Islands.

CRUCIFERAE

Matthiola

[Matthiola sinuata R. Br.]   Sea stock

[1] [2] . . . . . .   [A] .

Sandy sea-shores; apparently extinct.

[1]. *Lehinch dunes, before 1804* (Wade, 1804); not seen since. [2]. *Straw I., off Inishmore,*

1805, 1835, 1838; *leaves said to have been found by O'Kelly in 1894*; not seen since, and searched for in 1976 without result. *Dunes E. of Ballyvaughan, 1912 (ten plants), 1933 ('very sparingly')*; not seen since.

First record: Wade (1804), as *Cheiranthus sinuatus.* 'At high water mark about the sand hills of Dough, county Clare, but sparingly.' The identity of Wade's station is not certain, but the Lehinch dunes seem the most probable.

For the early records see CH2 and Praeger, 1934*d.* Known elsewhere only from Co. Wexford, where it is probably extinct; diminishing rapidly also in Britain.

*Cheiranthus cheiri L., the wallflower, was reported from two stations in Galway city in 1900 (ITB), and was seen in the W. part of the city in 1965, but could not be found in 1980. It is seldom found in Ireland on buildings less than 250 years old, and its colonies were probably destroyed by re-pointing or demolition.

Rorippa

The plants of this genus in the turloughs present great difficulties in identification, as it would seem that prolonged immersion delays flowering, leads to imperfect fruiting, and modifies the leaf-shape. It also seems probable that hybrids are found here – certainly *R. amphibia × sylvestris*, and possibly others.

Rorippa amphibia (L.) Besser

1 2 3 . . . .

Marshes, streams, lake-shores and turloughs. Occasional in the Burren; not seen elsewhere.

1. By the E. and S. shores of Inchiquin L. **2.** In the Carran turlough at Castletown. Frequent on the margin of L. Aleenaun. **3.** Occasional throughout.

First record: ITB (1901). Inchiquin L.

R. sylvestris (L.) Besser

1 2 3 . (*5*) . . . . .

Marshes and turloughs; occasional and locally frequent in the Burren; very rare elsewhere.

1. By Inchiquin L., and by the R. Fergus above it. **2.** L. Aleenaun. **3.** Frequent in the south; occasional in the north. (*5*). *Claddagh* (ITB).

First record: ITB (1901). 'Quay wall at the Claddagh '00, Philipps.'

R. palustris (L.) Besser

1 2 3 4 . . . **8** **A** .

Lakes, ponds, turloughs and other wet places. Frequent on the limestone; very rare elsewhere.

1. By Drumcullaun L. **2–3.** Frequent. **4.** Pond near Parkmore. By L. Corrib, E. of Moycullen. **8.** By the millpond at Cong. *S. of L. Mask* (Marshall & Shoolbred, 1896).

First record: CH1 (1866). 'Garryland, near Gort; *A.G.M.*'

R. islandica (Oeder) Borbás

. **2 3** . . . **7** . **A** .

In habitats similar to those of *R. palustris*; rare.

2. By the turlough at Turlough village, 1971. Inishmore (Poulnagappul, S.E. of Kilmurvy). **3.** Tirneevin turlough, 1977, 1980. **7.** In a small, damp hollow near the shore, just E. of the Renvyle Hotel, 1831, 1964.

First record: Jonsell (1968). Renvyle.

This critical species has recently been distinguished from *R. palustris* by Jonsell (1968); he found it in the Renvyle station, where it had been collected by Shuttleworth in 1831 (**BM**). By 1971, however, the hollow had been filled in and the plant had gone. As regards the remaining stations, the plants from Inishmore and Turlough village have been determined by Jonsell. MacGowran (1979) first reported it from Tirneevin; it was confirmed by MS and DW in 1980. A record from Caherglassaun L. awaits confirmation.

It differs from *R. palustris* mainly in its more prostrate habit, smaller flowers and shorter pedicels.

Nasturtium   Watercress   *Biolar*

The two species of watercress can be distinguished only by their fruits. In *N. officinale* they are short and broad (about $15 \times 3\frac{1}{2}$ mm); in *N. microphyllum* longer and narrower (20×2 mm). The hybrid has very few fruits; what there are are small and curved, with only one or two seeds.

Nasturtium officinale R. Br. (*Rorippa nasturtium-aquaticum*)

1 2 3 4 . . 7 8   A B

Marshes, ditches, streamsides, pools and turloughs. Occasional on the limestone and in S.W. Connemara; very rare elsewhere.

1–4. Thinly and rather evenly scattered. **7.** Recorded in several places from Roundstone to the Slyne Head peninsula, but some of the records require confirmation. Inishbofin, where it occurs in a very robust form up to 120 cm high (var. *siifolium* Steudel). **8.** Beside the lake immediately N.W. of Maam Cross.

First record: Howard & Lyon (1950). Killinaboy and Dog's Bay. Records before this date are ambiguous; they may refer to *N. microphyllum* or the hybrid.

N. microphyllum (Boenn.) Reichenb. (*Rorippa microphylla*)

1 2 3 4 5 6 7 8   A B

In habitats similar to those of the last species. Occasional to frequent, except in N. Connemara, where it is rare.

Evenly distributed over districts 1–6; in district 7 known only from the extreme west, and in district 8 only from Clonbur and Cong.

First record: Howard & Lyon (1950). Carran.

N. × sterile (Airy-Shaw) Oefel. (*Rorippa sterilis*; *N. microphyllum × officinale*)

. 2 3 4 . . 7 8   A B

In habitats similar to those of the last two species; occasional.

2. Abundant near Formoyle. Inishmore. **3.** By the W. side of Dromore L. Castle L. **4.** N. shore of Ross L. Galway city. **7.** Omey I. E. of Doonloughan. Inishbofin. **8.** Roadside ditch at Glann. Ditch by Teernakill Br.

First record: Howard & Lyon (1950). Maam.

In five of the above stations the hybrid was accompanied by one parent (in one case by both). In the remaining six neither was seen close by. The hybrid can persist and spread vigorously by vegetative reproduction. It is the plant most widely grown commercially.

Cardamine

Cardamine pratensis L.   Cuckoo-flower, Lady's smock   *Biolar griagáin, Léine Mhuire*

1 2 3 4 5 6 7 8 A B

Marshes, lake-shores, streamsides and wet meadows. Very frequent to abundant throughout.

Only in the western part of the Burren, where damp habitats are scarce, is there a noticeable gap in its distribution.

First record: Hart (1875). Aran Islands.

C. hirsuta L.

1 2 3 4 5 6 7 8 A .

Walls, limestone pavement, sand-dunes and waste or cultivated ground. Very frequent in the Burren; rather rare elsewhere.

1. Occasional, mainly on the northern fringe. **2–3**. Very frequent and locally abundant. **4**. Occasional. **5**. Dunes W. of Spiddal. Roadside near Gentian Hill. **6**. At the base of a wall near Moyrus. Lake-shore N.E. of Carraroe. Island in L. Athry. **7**. Dunes at Dog's Bay and on the Slyne Head peninsula. **8**. Roadside N. of Maam Cross. Frequent on walls around Clonbur and Cong.

First record: ITB (1901). H16. Earlier records under this name may indicate *C. flexuosa*.

Best distinguished from *C. flexuosa* by having only 4 stamens instead of the 6 usual in the family. An enigmatic plant in **TCD**, collected in 1980 from the edge of the dunes W. of Cleggan, has 5 stamens in most of the flowers; its inflorescence resembles that of *C. hirsuta*, but it is unusually tall. It is perhaps a hybrid with *C. flexuosa*.

C. flexuosa With.

1 2 . 4 5 6 7 8 A B

Ditches, woods, on the lee side of walls, and in other damp or sheltered places. Very frequent in district 8; occasional elsewhere, and usually in small quantity.

1. Occasional, mainly in the north and east. **2**. Frequent by the track in the Glen of Clab. Occasional on Inishmore. Ballyvaughan. **4**. E. shore of Ross L. Scrub near Burnthouse. **5**. 3 km S.E. of Costelloe. Near the school 8 km N.E. of Spiddal. **6**. Occasional near the coast. **7**. Occasional in the north and west. **8**. Very frequent.

First record: ITB (1901). H16.

*C. impatiens L.

. 2 . . . .

Waste ground; very rare.

2. Here and there around Ballyvaughan harbour. First seen here by W. Langton in May 1966; photographed later in the same year by Mrs Higgs. Reported later on waste ground in the village, but apparently did not persist here. Seen again by the harbour (in a different station from the first) several times up to 1978.

First record: Webb (1982).

Its persistence here for at least 12 years just raises this plant above casual status. Its means of introduction are quite obscure, but it is extremely unlikely to be native. The number of aliens naturalized near Ballyvaughan harbour is considerable.

Arabis

Arabis hirsuta (L.) Scop.

1 2 3 *(4)* **. . . 8   A** .

Limestone pavement, walls and sand-dunes. Very frequent in the Burren; very rare elsewhere.

1. Sand-dunes N. of Lehinch. **2**. Frequent on pavement throughout. **3**. Frequent, mainly in the W. half, and abundant on walls at Kinvara. *(4)*. *Ballyloughan* (P. cat.) *Near Moycullen* (Atlas). **8**. Limestone bluffs and pavement W. and N.W. of Cong. *Maumeen gap* (Hart, 1883).

First record: Foot (1864). 'In Burren, go where you will, you must meet with it.'

The record for Clifden by Mackay (1806) is almost certainly an error.

A. brownii Jordan (*A. ciliata*)

. 2 . . . . 7 .   A B

Sand-dunes; rarely on limestone pavement or banks. Local and rather rare.

2. Dunes E. of Ballyvaughan. *Roadside between Ballyvaughan and Black Head* (Murray, 1887). Inishmore, on dunes, pavement and roadside banks. **7**. On most of the dunes in W. and N. Connemara, from Dog's Bay to Inishbofin and beyond Gowlaun; usually in rather small quantity. A few plants behind the 'coral strand' near Ballyconneely.

First record: Mackay (1806), as *Turritis glabra*. 'In a sandy pasture by the seaside near the house of Anthony O'Flaherty, Esq., at Rinville [Renvyle], Connemara.'

Most botanists nowadays regard this plant as a variety of *A. hirsuta*. Without necessarily disputing this conclusion, we find it more convenient to record it here as a distinct species, as it has been treated so in all the Irish literature, and in western Ireland transitional forms have not been observed.

Barbarea

‡**Barbarea vulgaris** R. Br.   Winter-cress

1 2 3 4 5 . . 8   A .

Roadsides, farmyards, waste ground and lake-shores. Very rare in Connemara; occasional elsewhere, and tending to increase.

1. Occasional. **2**. Ballyvaughan. Inishmore. *Inisheer* (Hart, 1875). **3**. N. shore of L. Atedaun. Roadside E.S.E. of Kinvara. Waste ground at Gort. **4**. Occasional from Galway northwards. **5**. Roadsides at Spiddal and W. of Inveran. **8**. One plant by the county boundary at Cong.

First record: Hart (1875). 'Inisheer; on the roadside near the lighthouse.'

*B. intermedia Bor.

. 2 . . . . . . **A** .

Occasional in and around the villages on Inishmore and Inishmaan. Probably a recent introduction; first noted in 1975.

First record: Webb (1980*a*). Inishmore and Inishmaan.

Hesperis

*Hesperis matronalis L.   Dame's violet

1 2 3 4 . . **7 8   A** .

Roadsides, river-banks and waste ground; occasional.

1. Liscannor. **2**. Carran. Ballyvaughan harbour. Above Cloghmulk. Inishmore. **3**. Lanes and roadsides E. of Corofin. **4**. Waste ground in Galway city. **7**. Roadside on E. shore of Ballyconneely Bay. **8**. By the river near Maam Br.

First record: Andrews (1845). Inishmore.

The stations listed are confined to those in which the plant is growing at some distance from the garden whence it escaped. It is frequent throughout in the immediate neighbourhood of cottage-gardens.

Sisymbrium

‡**Sisymbrium officinale** (L.) Scop.   Hedge mustard

1 2 3 4 5 6 7 8   A (*B*)

Roadsides, waste ground and cultivated fields; occasional throughout.

Fairly evenly distributed, but commonest in district 4 and the Aran Islands, and rarest in districts 1, 7 and 8. It is seen mainly as a ruderal, but in district 3 it is fairly common as a weed of fields.

First record: Wright (1867). Inishmore.

Alliaria

Alliaria petiolata (Bieb.) Cav. & Grande   Garlic mustard

1 2 3 . . . (*7*) .   **A** .

Woods, hedges and other shady places; rare.

1. Woods at Elmvale Ho. **2**. Wood near Clooncoose. Below some small cliffs on Inishmore and Inishmaan. **3**. Roadside hedge S. of L. George. (*7*). *Near Clifden* (Atlas).

First record: Hart (1875). 'Inishmaan; near the small ruined church on the south-western shore.'

A mainly eastern species in Ireland.

Erysimum

*Erysimum cheiranthoides L.

. . **3** . **5** . . **8** . .

Cultivated fields and roadsides; rare.

3. Occasional in a tilled field at Carrowgarriff, 1964. A few plants by the roadside at Ballyogan L., 1966. Waste ground on the W. side of the Galway–Ennis road near Crusheen, 1979. **5**. Roadside at Rosscahill, 1962. *Claddagh*, 1900 (ITB). **8**. A few plants at Cong, near the county boundary, 1969.

First record: ITB (1901). Claddagh '00 – Phillips.

Never a common plant in Ireland, and tending to decrease. Despite its strong eastern tendency in Britain, its Irish headquarters is in the west midlands.

Arabidopsis

†**Arabidopsis thaliana** (L.) Heynh.   Thale-cress

1 2 3 4 . . . . **A** .

Limestone walls and rocks; rather rare.

1. On a wall S. of Inchiquin L. Limestone rocks near O'Dea's Castle. **2**. On the walls of Newtown Castle. *E. of Carran* (Atlas). Aran. **3**. Pavement 1 km E. of Corofin and 3 km W. of Garryland. Walls at Gort and 3 km N. of Ennis. **4**. Railway-track E. of Galway.

First record: CH2 (1898). 'Abundant...around Ballyvaughan.'

Brassica

‡**Brassica rapa** L.   Wild turnip

1 2 3 4 5 6 7 8   A B

Cultivated fields, roadsides and around farm buildings. Occasional on the limestone; rarer elsewhere, but tending to increase.

1. O'Brien's Br. **2**. Occasional throughout, and frequent on Aran. **3**. Occasional, mainly in the east. **4**. Waste ground at Galway docks. Menlough quarries. **5**. Near Gentian Hill. S.E. of Costelloe. **6**. Carna. Abundant in a field near the bridge E. of Lettermullan. N. of Kilbrickan. **7**. Doonloughan. Inishbofin. **8**. Clonbur.

First record: Marshall (1899). S. shore of L. Mask.

Tending to increase throughout Ireland, and especially in the west, where it is to some extent supplanting *Sinapis arvensis*. It is probably by now a good deal commoner in our region than the above notes would imply.

**B. napus* L., the swede turnip, is seen occasionally around Galway in waste places or cultivated ground, but here, as elsewhere in Ireland, it occurs only as a casual or short-lived relic of cultivation. The records in the *Atlas* for Inishmore and Inishbofin are almost certainly errors for *B. rapa*.

***B. nigra** (L.) Koch

(*1*) (*2*) . . . . **7** .   (*A*) **B**

Waste ground; very rare.

(*1*). *Kilfenora*, 1900 (ITB). (*2*). *Ballyvaughan*, 1900 (ITB). *Inishmore*, 1869 (Hart, 1875). **7**. Inishbofin, 1967 (DW); 1875 (More, 1876).

First record: More (1872): 'In cultivated ground and by waysides about Killeany, Aran; *H.C.H.*'

Nowhere very firmly established in Ireland, but its persistence on Inishbofin just entitles it to a place in our flora.

Sinapis

*Sinapis arvensis L.   Charlock   *Praiseach bhuí*

1 2 3 4 5 6 7 8   A B

Among crops, and on roadsides and field-margins. Very frequent on the lowland limestones; occasional to rare elsewhere.

Common in districts 3 and 4, but rather scarce elsewhere; we have only one record each for districts 6 (Gorumna I.) and 8 (Ferry Br.). Like many weeds of intensively cultivated areas this species is relatively scarce in the non-calcareous areas of western Ireland (as also in Wales and western Scotland). It seems also to have declined somewhat in frequency throughout Ireland over the past 50 years.

First record: More (1876). Inishbofin: in cultivated ground or borders of fields.

*S. alba L.   White mustard

. 2 3 4 (*5*) **. 7 .   (***A***) B**

Cultivated ground and roadsides; rather rare.

2. Roadside E. of Ballyvaughan, 1967. In a field at New Quay, and abundant among barley at Finavarra, 1969. *Inishmore* (More, 1872). **3.** Funshinmore, 1966. **4.** Among crops between Galway and Oranmore; also at Carrowmore, 1965–6. (*5*). *W. of Barna* (P. cat.). **7.** Roadside near Ballyconneely, 1978. Inishbofin, 1967.

First record: More (1872). Killeany (Inishmore).

**Diplotaxis muralis* (L.) DC. has been recorded from Ballyvaughan and around Galway (mainly on or near the railway), but only as a casual.

Armoracia

*Armoracia rusticana Gaertn.   Horse-radish

1 2 . . 5 . . 8   A .

Roadsides, waste ground, sand-dunes and river-banks; rather rare.

1. On the dunes N. of Lehinch, and on roadsides S. of the town, 1966. **2.** Well naturalized at Kilronan, Inishmore, 1971–6. **5.** Waste ground by the harbour at Spiddal, 1970. **8.** By the river at Cong, 1967.

First record: Webb (1980*a*).

Cochlearia   Scurvy-grass   *Biolar trá*

In our treatment of this genus we have been guided by the findings of P. W. Jackson, who has recently made an intensive study of the Irish populations.

Cochlearia officinalis L.

1 2 3 4 5 6 7 8   A B

Maritime rocks, salt-marshes, gravelly sea-shores and other littoral habitats; very rarely on mountain-tops. Frequent all round the coast, except in N. Connemara, where it is relatively rare; it has, however, been seen at Letterfrack and Leenane, and also on Inishbofin and Inishshark. It also occurs around the summit cairn on Carnsefin, above Black Head, at 320 m (1045 ft), but the plants there differ in no way from those

on the coast below. Var. *alpina* is reported in CH2 to have been seen on the summit
of Benlettery by S. A. Stewart, but there is no later confirmation; nor can we explain
the dot in the *Atlas* indicating its presence on the Maumturk range.

First record: Wright (1867). Inishmore.

C. scotica Druce (*C. groenlandica*)

. **2 3 4** . **6 7** .   **A B**

Exposed maritime rocks; more rarely on shingle-beaches or in salt-marshes. Locally
frequent.

2. Frequent from Poulsallagh to New Quay. Inishmore. **3**. 1 km N.E. of Kinvara. **4**.
Occasional from Parkmore to Oranmore. **6**. Pier on Gorumna I. **7**. Dog's Bay.
Ballyconneely Bay. Inishbofin.

First record: Praeger (1911). 'Inishbofin: western shore, in its usual habitat – chinks
of exposed rocks.'

A misunderstood, and to some extent controversial plant. In its extreme form it
is distinct in its very small size, its dark green leaves, and the fact that the flowers,
when they first open, do not project above the leaves. It retains all these characters
perfectly in cultivation. But it usually occurs in close proximity to *C. officinalis* and,
when this is so, intermediates of every kind can be found. It is probably best regarded,
therefore, as a subspecies or ecotype.

[C. anglica L.]

Jackson is of the opinion that this species does not occur in its pure form anywhere
in the Flora region. The records of Praeger (1932*a*) for districts 4 (Oranmore and L.
Atalia) and 5 (Gentian Hill) have in a sense been verified, but the plants found there
are considered by Jackson to be *C.* × *hollandica* Henrard (*C. anglica* × *officinalis*). *C.
anglica*, if it ever occurred there, has been 'hybridized out of existence'. Similar plants
have been seen at Liscannor (district 1). The *Atlas* records for New Quay and
Poulsallagh seem to be based on misidentifications.

C. danica L.

(*1*) **2** . **4 5** (*6*) **7** .   **A B**

Old walls, rocks, waste ground and other dry places; rather rare.

(*1*). *Fisherstreet* (ITB). **2**. Inishmore (rare). **4**. On several ruined buildings and derelict
sites in Galway city; also on the docks. *Oranmore* (ITB). **5**. Walls in the western
suburbs of Galway. Pier at Spiddal. (*6*). *Cashel*, 1894 (**DBN**). **7**. Maritime rocks on
Inishlackan. Pier at Bunowen. On the ruined church on Inishbofin.

First record: CH1 (1866). 'Roofs at Galway; *D.M.*'

We are unable to confirm the statement in ITB that it is common in the north part
of Co. Clare. The fact that in this species the petals are often white, and in *C. officinalis*
sometimes mauve, has led to some confusion. A sample was collected by DW in 1979
from a large colony of small plants with bright lilac flowers on the edge of one of
the 'coral strands' near Ballyconneely and was confidently assumed to be *C. danica*,
but closer examination showed that in characters of leaf and fruit it belonged to *C.
officinalis*, tending somewhat towards *C. scotica*.

Capsella

†**Capsella bursa-pastoris** (L.) Medic.   Shepherd's purse

1 2 3 4 5 6 7 8   A B

Cultivated ground, roadsides, waste places and sand-dunes. Widespread, but rare over considerable areas, and never in large populations.

1. Sand-dunes N. of Lehinch. Roadside at Elmvale Ho. Cultivated field on N. side of Slievebeg. **2–3.** Frequent. **4.** Galway, and the coast eastwards from it. E. of Moycullen. By Gortachalla L. **5.** Roadside near Gentian Hill. Dunes at Derryloughaun. W. of Spiddal. **6.** Frequent. **7.** Dunes on Slyne Head peninsula. 1 km W.N.W. of Cleggan. Inishbofin. **8.** Roadside at Clonbur.

First record: Hart (1875). Aran Islands.

Thlaspi

*****Thlaspi arvense** L.

. 2 3 . . . . .   A .

Roadsides and cultivated or waste ground; very rare.

2. Waste ground at New Quay, 1970. In a few fields at Oatquarter, Inishmore, 1977. **3.** In fields and on roadsides W. and N.W. of Kinvara, 1966–7.

First record: Wright (1871). Aran Islands.

Never a common weed except around Dublin, and probably now nearing extinction.

Draba

[Draba incana L.]

More (1872) quotes Foot as finding this 'in great luxuriance on the southern shore of L. Mask', but it has not been seen there since; and as Marshall & Shoolbred collected intensively here in 1895 and make no mention of it, an error seems probable. Its lowland stations in Ireland, apart from one near Ballyshannon, are on sand-dunes, not on limestone rocks.

*****D. muralis** L. has been recorded from Ballyvaughan and from Glann, but only as a transitory casual.

Erophila

Erophila verna (L.) Chevall.   Whitlow-grass

1 2 3 4 (5) . 7 .   A .

Limestone pavement, walls and sand-dunes. Frequent in much of the Burren; rare elsewhere.

1. Sand-dunes and adjacent banks at Lehinch. Lisdoonvarna, 1934 (**TCD**). **2.** Frequent on pavement throughout, but rare on Aran. **3.** Frequent in the east, rarer in the west. **4.** Wall by the sea between Oranmore and Galway. (**5**). *Spiddal* (Atlas). **7.** Sandy grassland at Dog's Bay, near Bunowen Castle and on Omey I.

First record: Foot (1864). 'Never have I seen it in such rank abundance as in Burren.'

Probably overlooked in several other stations on account of its early flowering and small size. Attempts to assign plants to the subspecies recognized in *Flora Europaea* have been somewhat inconclusive, but it would appear that in the Burren subsp. *praecox* (Steven) Walters and subsp. *spathulata* (Lang) Walters are both common, while a few plants approximate to subsp. *verna*. Further critical work is needed.

Lepidium campestre (L.) R. Br. has been recorded two or three times from the Galway Bay area, but only as a casual, and has not been seen there for many years.

Subularia

Subularia aquatica L.

. . . . . **6 7 8** . .

Lakes; rare.

6. In a small lake between Rosmuck and Kilbrickan. **7.** Glendollagh L.: in fair quantity in the narrow strip between the road and the railway causeway on the N. shore; also in the boat-harbour on the S. shore. Pool near the W. end of Kylemore L. *Ballynahinch and Derryclare Loughs* (FWI). '*Lake near Roundstone*', 1874 (D. Moore, **DBN**). **8.** At the S.W. corner of Coolin L.

First record: Mackay (1806). 'In several small lakes near Ballynahinch.'

In Glendollagh L. it grows in 50–150 cm of water, but in Coolin L. it is in much shallower water, between the *Littorella* and the *Eriocaulon* zones, and accompanied by *Eleocharis palustris*.

Coronopus   Swine's cress

*Coronopus didymus (L.) Sm.

1 2 3 4 5 6 7 . `A B

Roadsides, waste ground, around farm buildings, on sand-dunes, and as a garden weed. Widespread, but nowhere very common.

1. Near O'Brien's Br. *Corofin* (More, 1860). **2.** Occasional, mainly in the north. Frequent on Aran. **3.** Aughinish. By Rinroe Br. *Gort* (More, 1872). **4.** Waste ground near the station in Galway, and on the E. bank of the river above the city. **5.** Gentian Hill. Between Galway and Salthill. W. of Oughterard. **6.** Roadsides near Carna and Carraroe. **7.** Roadside W. of Cleggan. Omey I. Quarry near Dawros Br. Very frequent on Inishbofin.

First record: More (1860). Corofin.

All stations are either on limestone or near the sea. Only on Inishbofin has it been seen as a garden weed.

†C. squamatus (Forsk.) Asch.

1 2 3 4 (*5*) **6 7** .   **A B**

Roadsides, seashores, and waste or cultivated ground near the sea. Occasional in W. Connemara and on the S. shore of Galway Bay; rare elsewhere.

1. Liscannor harbour. **2.** Shore near Finavarra. Inishmore. *Black Head* (ITB). **3.** Roadside on the W. side of Kinvara Bay, and thence westwards to Aughinish. **4.** Shore near Mweenish I. (*5*) *Gentian Hill* (P. cat.). **6.** Roadside at Lettermullan Br. **7.** Inishlackan. Bunowen quay. Aughrusmore. Renvyle. Inishbofin.

First record: Wright (1871). Inishmore: in great profusion about Kilronan and Killeany.

Always near the sea, and usually in rather remote localities. It appears to be diminishing, while *C. didymus* is increasing. It differs from *C. didymus* in its coarser foliage and its deeply ridged and wrinkled fruit. Both species have a strong cress-like smell.

Cakile

Cakile maritima Scop.   Sea rocket

1 2 . . 5 6 7 .   A B

Sandy shores; local.

1. Abundant near O'Brien's Br. **2.** Fanore. Below the dunes E. of Ballyvaughan. Abundant at Portmurvy, Inishmore. **5.** Derryloughaun. Between Galway and Salthill. **6.** Moyrus strand, N.W. of Carna. **7.** Here and there round the whole coast.

First record: Wright (1871). Aran Islands.

Crambe

Crambe maritima L.   Seakale

. 2 . . 5 6 . .   A .

Gravelly or sandy sea-shores; rare.

2. In fair quantity on the beach below Seven Churches, Inishmore, 1972, 1976 (DW); reported here 'in great abundance' by Praeger (1895*b*). *Recorded earlier from 2 or 3 other stations on Inishmore, but not there now.* Four plants near the landing-stage on Inishmaan, 1976 (JW). *One plant at Fanore*, 1895 (CH2). **5.** *One plant at Gentian Hill* (Praeger, 1903*a*). Several plants on the shore W. of L. Nagravin, 1980 (Robinson). **6.** A large colony on the W. shore of Cashla Bay, $2\frac{1}{2}$ km S.S.E. of Carraroe, 1981 (Robinson).

First record: O'Flaherty (1824), as 'sea-cabbage'. Aran Islands.

Raphanus

Raphanus raphanistrum L.   Wild radish

. 2 3 4 5 6 7 .   A (*B*)

Sandy, muddy or rocky shores and waste ground near the sea.

2. Frequent from Ballyvaughan to New Quay. In small quantity at Fanore. Aran; especially common on Inisheer. **3.** N.W. of Kinvara. **4.** Abundant E. of Galway, and in smaller quantity round the head of Galway Bay. **5.** Frequent and locally abundant from Galway to Spiddal. **6.** Occasional, from near Carraroe to Mweenish I. **7.** Abundant along Ballyconneely Bay, and occasional elsewhere in the south-west; not seen N. of Streamstown Bay. *Inishbofin* (More, 1876; Praeger, 1911); not seen recently.

First record: Oliver (1851). Near Roundstone.

All the above records refer to subsp. *maritimus* (Sm.) Thellung. Being a biennial it varies a good deal in abundance from year to year. Subsp. *raphanistrum* has been seen in small quantity in fields and waste ground near the railway 3–4 km E. of Galway, but in many flowers the petals had a yellowish tinge, suggesting introgression from subsp. *maritimus*. Subsp. *raphanistrum* was formerly a common weed in much of the Flora region, but is now on the verge of extinction here, as in many other parts of Ireland.

RESEDACEAE

Reseda

[*Reseda lutea L.]

. [2] . [4] . . . . [A] .

Seen on Inishmore by Hart in 1869, and (in a different locality) by Praeger in 1895, but not seen since. There is also a specimen in **DBN** collected by Praeger from L. Atalia, near Galway. This species, however, has never been more than a casual in Ireland except on the east coast, and seems now to have totally disappeared from the west.

‡**R. luteola** L.　　Weld, Yellow-weed　　*Buidhe mór*

. (2) 3 4 . . . . (A) .

Roadsides, sand-pits, quarries and gravelly soils. Occasional on the lowland limestones; unknown elsewhere.

(2). *Aran* (Hart, 1875); not seen since. 3. Sand-pit near Ballyallia Ho., N. of Ennis. Roadside 2 km N. of Ruan. 4. N.E. suburbs of Galway, and at Menlough quarries. Between Galway and Oranmore. Drift hillocks at Kilcaimin.

First record: Wright (1871). Aran Islands.

In spite of its calcicole habit, the Burren offers little attraction for this species, as it favours gravels and sands rather than limestone rock. Although it is usually rated as native, there are good reasons for supposing that it was introduced to both Britain and Ireland as a dye-plant (de Wit, cited by Godwin, 1975). In Ireland it is most thoroughly established in the midlands and the south-east.

CISTACEAE

Tuberaria

Tuberaria guttata (L.) Fourr. (*Helianthemum guttatum*)　　Annual rock-rose

. . . . . 6 7 . . B

Dry, heathy ground, on shallow peat overlying siliceous rock; rare and very local.

6. In fair quantity on the knoll marked 64 ft at the S.E. corner of Gorumna I., 1966, 1968 (DW); also behind the 'coral strand' S.W. of Carraroe (Mackie, 1970). These two stations face each other across the mouth of Greatman's Bay. 7. Inishbofin and Inishshark; seen here by all botanical visitors from 1872 to 1967. It grows on the higher parts of the islands, but not in great abundance.

First record: More (1873). 'Abundant on Inishbofin Island... (W. McMillan) S. A. Stewart.'

Elsewhere in Ireland known only from Inishturk (N. of Inishbofin, but outside the Flora region) and one or two peninsulas in W. Cork. In Britain only in N. Wales, though it is also in the Channel Islands. Widespread in W., S. and C. Europe.

Helianthemum

Helianthemum canum (L.) Baumg.   Burren rock-rose

. **2** . . (*5*) . . . **A** .

Limestone rocks; very local.

2. From Black Head southwards to beyond Poulsallagh; somewhat local, but abundant in places, and especially at Poulsallagh. Frequent on the higher ground of Mullaghmore. About 2 km S. of Bealaclugga (ICP). Widespread and locally abundant on Inishmore. (*5*). *About 1 km W. of Salthill*, 1893 (CH2). This station has probably been built over; in any case the authenticity of the record is subject to some doubt.

First record: Andrews (1845). Inishmore. Mackay (1806) recorded *H. nummularium* from Aran, but even as late as 1853 was unwilling to admit his misidentification; it can hardly, therefore, stand as a record of *H. canum.*

Found nowhere else in Ireland, and very local in Britain.

VIOLACEAE

Viola   Violet   *Sail chuach*

†**Viola odorata** L.   Sweet violet

. **2** . **4** (*5*) . . **8** . .

Woods and hedges; rare.

2. In a hedge near Turlough village, 1967. **4.** In a hedge 1 km N. of Ross L., 1967. (*5*). *Barna*, 1901 (Praeger, 1903*a*). **8.** In the grounds of Ashford Castle; clearly an escape, but well naturalized, 1969.

First record: Praeger (1903*a*). Barna.

The status of this species in many of its Irish stations is hard to assess, but in the first two stations given above it has a fair claim to be considered native.

V. hirta L.

. **2 3** . . . . . **A** .

Limestone pavement; rare and very local.

2. Fairly frequent on the E. half of Inishmore; rarer on Inishmaan. **3.** In crevices of the pavement on both sides of the road which runs between the turloughs S.W. of Mullaghmore, 1965, 1969 (DW).

First record: More (1872). 'Sloping ground above Killeany, Aran; *A.G.M.*'

In view of its relative abundance on limestone pavement at Askeaton, Co. Limerick (where there is a reduced Burren flora), and its presence on Aran, the great scarcity of this species in the Burren is surprising. Doubtless some other stations await discovery, but it cannot be common. A specimen of the plant from district 3 was cultivated for some years by DW, and appeared quite typical.

V. palustris L.

1 [2] . . 5 6 7 8 [A] **B**

Marshes, bog-margins and streamsides; frequent to abundant on acid soils, but unknown on the limestone.

Common throughout Connemara, and frequent on the Clare shales. Hart's (1875) record for Aran is dismissed as a misidentification (see Webb, 1980*a*).

First record: Wade (1802). 'Common in boggy situations, Cunnamara.'

V. riviniana Reichenb.   Common wild violet

1 2 3 4 5 6 7 8   A B

Heathy ground, banks, roadsides, hedges, woods and lake-shores. Abundant throughout.

An extremely common species with a very wide range of habitats. Only in the western fringe of Connemara does it appear to be slightly scarcer than elsewhere.

First record: Wright (1871), as *V. silvatica*. Aran Islands. (The name *V. silvatica* is ambiguous; but on Aran it can indicate only *V. riviniana*.)

V. reichenbachiana Bor.   Wood violet

. 2 3 4 . . . 8 . .

Woods and scrub on basic soil; locally frequent.

2–4. Frequent in suitable habitats throughout, but absent from the Aran Islands where there is no adequate shelter. **8.** Grounds of Ashford Castle; also in an ash-wood N.W. of Cong.

First record: Marshall & Shoolbred (1896). 'Woods on limestone in the Ashford House Demesne.'

Very precise in its requirements in our region, demanding both good shelter and a fair supply of base; usually present when these are forthcoming. It is often accompanied by *V. riviniana*, from which it is easily distinguished in flower by its narrower, more widely-spaced petals and its slender, deep violet (not stout and cream-coloured) spur. After flowering, however, they are difficult to distinguish, and it is possible, therefore, that *V. reichenbachiana* is more widespread than our records indicate.

V. canina L.

. 2 3 4 5 6 7 8   A B

Stony heaths and lake-shores, edges of turloughs and fens; often near water, but sometimes on quite dry ground. Widespread but local.

2. By the turlough at Turlough village. In the graveyard on Inishmaan. **3.** Frequent by all the turloughs, near the upper limit of flooding. In a fen N. of Kinvara. Very frequent by L. Bunny in various habitats. **4.** Shores of Ross L., and of L. Corrib near Burnthouse and near the Menlough quarries. **5.** On a peaty island by the edge of the river at Costelloe. **6.** Lake-shore E. of Carraroe. **7.** Dry, stony heath on Inishbofin. *Bunowen Hill and E. shore of Ballyconneely Bay*, 1957 (DW). *Roundstone*, 1896 (ITB). *On the mountains above Kylemore L.* (More, 1872). **8.** Occasional on the shore of L. Corrib from near Oughterard to the Dooros peninsula, and on a few of the islands.

First record: CH1 (1866). 'By the lake at Garryland; *A.G.M.*' Wade (1804) refers to

V. canina, var. *alpina* Hudson, as growing 'on large stones to the west of Letter mountain, Ballynahinch', but the identity of his plant is not at all clear.

The plants on Inishmaan, though typical in flower and leaf-shape, have an unusually tall and erect habit, and may possibly be referable to subsp. *montana* (L.) Hartman. They deserve further investigation.

For hybrids, see under *V. persicifolia* and *V. lactea* below.

V. persicifolia Schreb. (*V. stagnina*)   Turlough violet

. **2 3** . . . .

Margins of turloughs; locally abundant.

2. In several places around Turlough village. **3**. Around most of the turloughs, and in some (e.g. Coole and Caherglassaun) in enormous abundance.

First record: More (1855). Garryland L.

This species occupies a zone a metre or so below that of *V. canina*, but there is sometimes an overlap, and in the intermediate zone hybrids between the two are usually present in fair abundance. Details of the zonation are given by Praeger (1932*a*). As the upper zones of the turloughs are usually heavily grazed, *V. persicifolia* is normally very inconspicuous outside its flowering season (late May to mid-June), although in places it is subdominant in the grassy sward. When in flower it can provide one of the most spectacular sights in the Irish flora, many square yards of turf being converted into a pale china-blue sheet by the presence of literally tens of thousands of flowers, mostly very close to ground level.

This species differs from *V. canina* in its narrower leaves and much paler blue flowers, with short, obtuse petals. From *V. lactea* it differs, apart from habitat, in the pale but clear china-blue (not greyish milky-blue) colour of its flowers, and the leaves, which are cut off square at the base, not tapered into the stalk.

A rare and decreasing species in Britain, almost confined to the fens of E. England.

V. lactea Sm.

. [2] . . . **6 7 8** . .

Heathy ground; very rare.

6. On a small knoll W. of Glinsk, 1970 (Rix). Opposite a school near Costelloe, 1968 (Mackie, det. Valentine). **7**. On the W. side of Errisbeg, near the base, 1962 (DW). *By Cregduff L. and Bunowen Hill* (Hall & Simpson, 1936). **8**. On a rocky outcrop by the Clifden road 5–6 km from Oughterard, 1976 (DW).

First record: Hall & Simpson (1936).

The Glinsk specimen was cultivated for a few years and proved quite typical. The specimen from Errisbeg was not preserved; there is, however, a specimen in **DBN**, collected by Praeger in 1906 from the same locality, labelled *V. ericetorum*, which looks very like *V. canina* × *lactea*, and is presumably the basis for his attribution of this hybrid to H16 in Praeger (1951).

Phillips (1900) gives a record for Ballyvaughan, but as it was third-hand when it reached publication it is best ignored. A specimen in **BM** labelled *V. lactea* from Ballyvaughan is, in fact, *V. canina*.

V. arvensis Murray   Heartsease

(*1*) **2 3 4** (*5*) **6** (*7*) .   **A** .

Cultivated fields and other open habitats; rather rare.

2. Locally frequent on Inishmore; rarer on Inishmaan. **3.** Occasional in the Kinvara area. **4.** Esker-gravels by Gortachalla L. *Ballyloughan* (P. cat.) **6.** Near Carna and Carraroe.

First record: Praeger (1895*b*). Near Kilronan and Kilmurvy.

Field-cards filled in for the *Atlas* in 1955–8 indicated its presence then in district 1 (Ennistymon–Lisdoonvarna area), 5 (Spiddal area) and 7 (Clifden area). It is a declining species in western Ireland, but it may still persist in some of these places, though not seen recently.

V. tricolor L.

Although intermediate forms are found on the Continent, the two subspecies of this species are very distinct in Ireland, and are best treated separately.

Subsp. tricolor

. **2** . **4** (*5*) **6** . **8**   **A** .

Cultivated ground and on the margins of small lanes; rather rare.

2. Formerly widespread on Aran, but now rare. Not recorded from the mainland. **4.** S. of Carrowmoreknock, 1967. (*5*). *Barna*, 1900 (ITB). *Near Inveran*, 1958 (Atlas). **6.** Occasional near the coast. **8.** On a track at Glann, 1962; perhaps a casual introduction with gravel.

First record: Hart (1875). Aran Islands.

Never very common in western Ireland, and, like many other weeds, tending to decrease.

Subsp. curtisii (Forst.) Syme   Sandhill pansy

[1] **2** . . . **6 7** .   **A** .

Sand-dunes; locally abundant.

[1]. Recorded by ICP for stabilized dunes N. of Lehinch but, in view of the presence here of forms of *V. lutea* transitional to this subspecies, confirmation is desirable. See further under *V. lutea* (below). **2.** Abundant on Inishmore near Killeany and occasional on Inishmaan. Fanore. **6.** 3 km S.E. of Carna. **7.** Present, though in varying abundance, on most of the dunes of W. Connemara, from Gorteen Bay to Mannin Bay; also N.E. of Gowlaun.

First record: Oliver (1851). Roundstone.

V. lutea Hudson   Mountain pansy

1 . . . .

Grassy cliff-tops and on sand-dunes; frequent in one limited area.

1. Locally frequent from the Cliffs of Moher southwards, largely, and perhaps entirely, replacing *V. tricolor*, subsp. *curtisii* on the extensive dunes N. of Lehinch.

First record: Carter (1846). 'The sand-hills of Lehinch are covered with it.'

The sand-dune habitat is unusual for this species, and the plants found there, though certainly very close to *V. lutea*, are in some characters suggestive of *V. tricolor*, subsp. *curtisii*; moreover some of the plants from Gorteen Bay, hitherto recorded as *V. tricolor*, have unusually large flowers and in some other respects approach *V. lutea*. Several authors have expressed doubts as to the taxonomic status of the Lehinch plants; they seem to be similar to the plant described by J. G. Baker as *V. symei*.

Cytological examination and cultivation experiments are needed before their status can be finally decided. For discussion see Baker (1901) and Drabble (1930).

POLYGALACEAE

Polygala   Milkwort   *Glúineach*

Polygala vulgaris L.

1 2 3 4 5 6 7 8   A B

Grassland, limestone pavement, dry banks and stabilized sand-dunes. Very frequent to abundant on the limestone; occasional to rather rare elsewhere.

1. In a hedge near Moy Ho. Sparingly on a scrub-covered knoll near Elmvale Ho. **2–4.** Very frequent and locally abundant. **5.** Pastures at Gentian Hill and near Spiddal. Disused railway-track near Oughterard. **6.** Sandy pastures on Furnace I. On rocks by an inlet N. of Costelloe. **7.** Frequent in the extreme west, though rather rare on Inishbofin; very rare elsewhere. **8.** By the river at Teernakill Br. Roadside near Ferry Br.

First record: Wright (1871). Aran Islands.

Except for the two in district 8, all stations are on limestone or near the sea.

P. serpyllifolia Hose

1 . 3 4 5 6 7 8   . B

Bogs and heathy grassland. Very rare on the limestone; abundant elsewhere.

1. Very frequent. **3.** Templebannagh bog (ICP). **4.** Occasional on bogs E.S.E. of Oughterard, and on knolls of leached drift by Ballycuirke L. **5–8.** Abundant.

First record: More (1876). Inishbofin.

A smaller plant than *P. vulgaris*, with shorter racemes of flowers, which are always bright blue. The most certain distinction, however, lies in the leaves at the base of the stem; in *P. vulgaris* they are alternate, like the others; in *P. serpyllifolia* they are opposite.

CARYOPHYLLACEAE

Saponaria

*Saponaria officinalis L.   Soapwort

1 2 3 4 . . . .

Roadsides and waste ground, never very far from houses. Occasional on the limestone; not recorded elsewhere.

1. Near Killinaboy. **2.** Rather frequent around Ballyvaughan. New Quay, and elsewhere in the north-east. N.W. of Castletown. **3.** Near Crusheen station. S. of L. Cullaun. W. of Kinvara. **4.** Waste ground near Galway station. Roadside N. of Ross L.

First record: Praeger (1934c). H16.

Clumps in hedgerows immediately opposite cottages have not been recorded, as in this position they are clearly planted and form an informal extension of the cottage garden. All plants appear to belong to the double-flowered form; they are derived, therefore, from rhizomes thrown out from gardens, and there is no reproduction by seed.

Silene

Silene vulgaris (Moench) Garcke   Bladder campion

1 2 3 4 5 6 7 8 A B

Cliffs, rocky and stony shores and harbour walls; roadsides, hedges and open, gravelly habitats; rarely on mountain-tops.

Subsp. *maritima* (With.) Löve & Löve is common round most of the coast, except for the north coast of Connemara, where it is apparently very rare. It also occurs sparingly on the summit of Benbaun, the highest peak of the Twelve Pins, and on the ridge between Bencorr and Derryclare Mt. Subsp. *vulgaris* is occasional to locally frequent in the Burren (mainly in the north half). There are also a few records from the Oughterard area (in districts 4, 5 and 8) and a very isolated one from district 7 – a roadside bank opposite the lane leading down to Dog's Bay.

First record: Wright (1871). Aran Islands.

Praeger's statements in ITB and FWI that the subsp. *maritima* is frequent on the mountains of W. Galway seems to be an exaggeration. There are no records apart from those from the Twelve Pins, and although Colgan (1900) describes it as occurring here in 'sheets' it is decidedly local. Hart did not report it from any other range, and the extensive exploration of the mountains of district 8 by Halliday during the preparation of this Flora did not reveal it elsewhere.

[*Silene gallica L.] (*S. anglica*)

A few plants were seen in a potato-field 1 km N. of Carraroe by DW in 1964; the only other record is from Oughterard in 1832. This species is (or at least till recently was) well naturalized in Co. Down, and perhaps elsewhere in the east, but has never been more than a casual in the west.

S. dioica (L.) Clairv. (*Melandrium dioicum*)   Red campion

(*1*) . . . . . . 8 . .

Woods and shady places; very rare.

(*1*). *Lisdoonvarna*, 1900 (ITB). **8**. Woods near the S. shore of L. Mask, 1977 (J. Cross); probably the same station as is mentioned in CH2.

First record: CH2 (1898). 'Ross Hill, shore of Lough Mask; *Miss M. F. Jackson.*'

*S. alba (Miller) Krause, the white campion, appeared in 1979 in fair quantity in a field near Killeany on Inishmore (Robinson); it is clearly a recent introduction, and it seems unlikely that it will persist. The only other record is of a casual occurrence near Cong before 1898 (CH2).

Lychnis

Lychnis flos-cuculi L.   Ragged robin

1 . 3 4 5 (*6*) 7 8 . B

Marshes, fens and wet meadows. Abundant in W. Connemara and on the Clare shales; rather rare elsewhere.

1. Abundant. **3**. Occasional in the south. **4**. By the R. Corrib 3 km above Galway. 2 km N.W. of Moycullen. **5**. Marsh half-way between Spiddal and Moycullen. Fen by L. Naneevin, near Rosscahill. Edge of salt-marsh at Gentian Hill. (*6*). *Lettermullan, and N.E. of Kilkieran* (Atlas). **7**. Abundant, especially in the west. **8**. Marsh at Teernakill Br.

First record: More (1876). Inishbofin: near Church Lake.

The *Atlas* record for Aran is almost certainly an error.

Agrostemma

[*Agrostemma githago L.] (*Lychnis githago*)   Corncockle

As late as 1900 this species was a frequent weed, mainly of oatfields, in the Burren and on the limestones E. of Oughterard, but its decline began soon after that, and it has long since vanished from the Flora region, as from virtually the whole of Ireland.

Cerastium

Cerastium diffusum Pers. (*C. atrovirens*, *C. tetrandrum*)

1 2 . 4 5 6 7 .   A B

Widespread and usually abundant on sand-dunes; occasional also in dry waste places and on rocks and walls near the sea.

1–2. Abundant on dunes at Lehinch, Fanore, and on both sides of Ballyvaughan; more sparingly on rocky parts of the coast from Black Head southwards. Very frequent on Aran. **4**. Walls in Galway city. *Shore near Oranmore; also Rinville* (ITB). **5–7**. Present on all the sand-dunes visited, usually in abundance.

First record: More (1876). Inishbofin.

C. semidecandrum L.

(*1*) **2** . . (*5*) . **7** . . .

Sand-dunes; rare.

(*1*). *Lehinch* (Atlas). **2**. Fairly frequent at Fanore. (*5*). *W. of Galway* (CH1). **7**. Frequent on dunes and sandy grassland on the N.E. part of the Slyne Head peninsula, and perhaps on other parts of it also.

First record: CH1 (1866). 'On the west side of Galway; *Prof. Melville.*'

A species often misidentified; we therefore think it wiser to hold over, pending the production of vouchers, verbal records which we have received from Dog's Bay and Inishmore.

C. glomeratum Thuill. (*C. viscosum*)

1 2 3 4 5 6 7 8   A B

Rocks, walls, roadsides, cultivated ground and other dry, open habitats. Widespread, but seldom in abundance.

1. Rocks near O'Dea's Castle. **2**. Pavement at Black Head. Beside L. Luick. Aran. **3**. By the cross-roads E. of Funshinmore. By the pier 5 km E. of Aughinish. **4**. Waste ground by the river in Galway. **5**. As a weed in flower-beds at Salthill. E. shore of Cashla Bay. **6**. Dunes at Moyrus. **7**. Waste ground S.W. of Clifden. Quay near

Dawros R. Inishbofin. **8**. Roadside at Maam. Frequent on the stony shore of L. Corrib from the Dooros peninsula to near Cong.

First record: Wright (1871). Aran Islands.

Probably under-recorded; it is easily overlooked in late summer.

C. fontanum Baumg. (*C. holosteoides*, *C. vulgatum*)   Mouse-ear chickweed

1 2 3 4 5 6 7 8   A B

Pastures, roadsides, sand-dunes and heaths; abundant throughout.

First record: Wright (1871). Aran Islands.

A rather striking variant with large flowers and dark green, shining, almost glabrous leaves is frequent at the Cliffs of Moher. Somewhat similar plants have been seen on cliff-tops on Inishmore.

C. arvense L.

1 2 3 4 5 . . .   A .

Limestone pavement and grassland overlying it; also on sand-dunes and shingle beaches. Locally frequent in the Burren and around Galway Bay; unknown elsewhere.

1. On a limestone outcrop near Elmvale Ho. **2**. Frequent in the northern and eastern parts, and on Aran; rather rare elsewhere. **3**. Shingle beach at Aughinish. **4**. Shingle beaches at Tawin I. and E. of L. Atalia. At the base of a wall W. of Carrowmore. **5**. On shingle at Gentian Hill.

First record: Andrews (1845). Inishmore.

The two principal habitats of this species are the shingle beaches around Galway Bay, and the grassland overlying limestone in the central part of the Burren, where it occupies a broad band stretching roughly from Killinaboy to Ballyvaughan. This latter region is floristically poor compared with the country to the east and west of it, but is marked by the abundance of *Cerastium arvense* and *Saxifraga hypnoides*.

Stellaria

Stellaria media (L.) Vill.   Chickweed   *Fliodh*

1 2 3 4 5 6 7 8   A B

Cultivated ground, roadsides, farmyards, waste ground, sand-dunes and other open communities. Very frequent throughout most of the region, and locally abundant, but only occasional in the mountains of Connemara.

First record: Wright (1871). Aran Islands.

On Aran this species, like several other weeds, is commoner at the base of walls bordering lanes than on cultivated ground.

S. holostea L.   Stitchwort

1 2 3 4 (*5*) **. . .   . .**

Hedges, scrub and wood-margins. Locally frequent in the Burren; rare elsewhere. **1**. Occasional in the Ennistymon area, and to the W. and N.W. of Corofin. **2**. Fairly frequent from L. Aleenaun and Clooncoose wood through Carran and Castletown to the Glen of Clab; not seen outside this area. **3**. Frequent around Dromore, extending north-eastwards to near Tubber. **4**. Hedge S. of Ross L. (*5*). *Spiddal* (Atlas).

First record: ITB (1901). H16.

Mainly on the limestone, but with several stations on acid-looking soils on the Clare shales.

S. palustris Retz. (*S. glauca*)

. . . . **8** . .

Marshes; very rare.

8. In a marsh N.E. of Clonbur, very near the county boundary, and probably extending into H26; discovered here by Marshall & Shoolbred in 1895; confirmed by DW in 1969.

First record: Marshall & Shoolbred (1896). 'In a swamp near Clonbur, within a few yards of the Mayo boundary.' Recorded by Wade (1804) for Renvyle, but certainly in error for *S. alsine*.

A midland species, here at its western limit.

S. graminea L.

1 2 3 4 5 6 7 8   A (*B*)

Hedges, roadsides, meadows, rough grazing and ditches. Common in districts 1 and 8; occasional elsewhere.

Only one station has been noted for district 6 (at the S.E. corner of Gorumna I.); in districts 2, 3, 4, 5 and 7 it is local, but has been seen in at least four stations in each. It appears to be absent from much of S. and W. Connemara and from the low-lying limestones round the head of Galway Bay. The record of Praeger (1911) for Inishbofin has not been confirmed recently; it occurs on Aran, but is rare.

First record: ITB (1901). H16.

S. alsine Grimm (*S. uliginosa*)

1 2 3 4 5 6 7 8   A B

Marshes, small streams, ditches and wet grassland. Rare on the limestone; very frequent elsewhere.

1. Very frequent. **2**. Aran; rare. **3**. Marsh by L. Cleggan. **4**. Shore of L. Corrib, E. of Moycullen. **5–8**. Very frequent.

First record: ITB (1901). H16.

Arenaria

Arenaria serpyllifolia L.

1 2 3 4 5 6 7 .   A B

Limestone rocks, walls, shingle and sand-dunes. Locally frequent.

Commonest on Aran and in district 4; elsewhere occasional to frequent around most of the coast, though rare in N. Connemara. There are a few inland stations in the Burren, but it is rarer than might have been expected.

First record: Hart (1875). Aran Islands.

A. norvegica Gunn.

. **2** . . . .

Limestone pavement; very rare.

A colony, 'probably not a large one' of this plant was discovered in 1961 'at an altitude of *c*. 800 ft on the south slope of Gleninagh Mountain overlooking Caher Lower...growing in shallow crevices and solution hollows on an area of limestone pavement' (Heslop-Harrison, Wilkins & Greene, 1961). Repeated search by numerous botanists has failed to re-discover it. There is no doubt about the identity of the plant, for a specimen was collected, and from the herbarium sheet Halliday extracted a seed which germinated and enabled a chromosome count to be made. There is, however, some doubt about the exact locality, for Gleninagh, as ordinarily understood, does not have a southern slope, but extends southwards into a long ridge, and where this drops to 800 ft (244 m) it is a long way from Caher Lower. The word Gleninagh can, however, be used *sensu lato* to cover the more north-westerly summit of the same height, usually known as Carnsefin, and the 800 ft contour on the south side of Carnsefin does, indeed, overlook Caher Lower. The find, therefore, was most probably made at M 164 092, or not far away.

We must presume either that the plant is now extinct, or that it still exists in small quantity. Despite the diligent searches made by Halliday and others, the latter hypothesis seems slightly more probable; on this sort of terrain it would take a dozen botanists provided with surveying equipment a week's hard work to demonstrate conclusively that the plant was absent.

Minuartia

Minuartia verna (L.) Hiern

1 2 3 . . . . . **A** .

Limestone rocks; rarely on sand-dunes. Frequent in most of the Burren and Aran; not recorded elsewhere.

1. Rocky outcrops near Elmvale Ho. **2**. Frequent throughout. **3**. Frequent along the west margin, but not seen E. of Boston.

First record: Mackay (1836). 'Island of Arran; Mr R. Ball.'

Elsewhere in Ireland known only from the basalt of the north-east.

***M**. *hybrida* (Vill.) Schischk. was recorded from railways at Ardrahan and Oughterard in 1899–1900, but is long since extinct, as in most of its Irish stations.

Moehringia

Moehringia trinervia (L.) Clairv. (*Arenaria trinervia*)

. . **3** . . . **7 8** . .

Woods; rather rare.

3. Frequent in a wood by the ruined castle S.E. of Ruan. Occasional in Garryland wood. **7**. Wood at Derreen. **8**. On the wall of the ruined abbey N. of Clonbur. In woodland on a small island in L. Corrib half way between Inchagoill and Gortdrishagh Ho.

First record: Halliday, Argent & Hawksworth (1967). 'A few plants...on the limestone perimeter wall of Clonbur Abbey.'

Nowhere very common in Ireland, and rare throughout the west.

Honkenya

Honkenya peploides (L.) Ehrh. (*Arenaria peploides*)

1 2 . . 5 6 7 . A B

Sand-dunes and sandy shores; occasionally on stony or rocky shores. Frequent in the west; rather rare in Galway Bay.

1. Abundant N. of Lehinch. **2**. Frequent at Fanore and E. of Ballyvaughan. Aran. **5**. Occasional all along the coast. **6**. N. end of Furnace I. Dunes near Moyrus. **7**. Very frequent round the coast from Inishlackan to Claddaghduff; rather local on the north coast. Inishbofin.

First record: Sim (1859). 'To the west of Galway.'

Sagina

Sagina apetala Ard. (incl. *S. ciliata*)

. 2 3 4 5 . (7) . A .

Walls, rocks, cultivated ground and waste places; rather rare.

2. Occasional on Aran. **3**. Railway track at Crusheen. *Garryland* (P. cat.). **4**. Waste ground by the river in Galway. Roadside by N. bank of Dunkellin R. **5**. Outskirts of Oughterard. Near Gentian Hill. Moycullen. (7). *Bunowen Hill*, 1956 (DW).

First record: Hart (1875). Aran Islands.

Probably under-recorded, but certainly not a common plant in the region.

S. maritima Don

. 2 . (*4*) 5 6 7 8 A B

Rocks and dry soil by the sea; also on walls, and especially in chinks in the masonry of piers. Occasional.

2. Frequent on all the Aran Islands. Fisherstreet. Rocks between Black Head and Poulsallagh, and on the road at Poulsallagh. Pier at New Quay. (*4*). *Renmore* (ITB). **5**. Maritime rocks near Inveran and S. of Costelloe. **6**. Pier at N. end of Gorumna I. **7**. Piers at Bunowen and Errislannan. Inishbofin. **8**. Bank by the sea near Leenane.

First record: Andrews (1845). Inishmore.

S. subulata (Swartz) Presl

(*1*) 2 . . . 6 7 8 A B

Mainly on rocks near the sea; more rarely on bare peat, sandy roadsides or mountain rocks. Occasional in W. Connemara; rare elsewhere.

(*1*). *Cliffs of Moher*, 1891 (**DBN**). **2**. Rocks at Poulsallagh. Inishmore, 1981 (Noltie). *Black Head*, 1953 (**TCD**). **6**. Occasional on Gorumna I. 4–5 km N. of Costelloe. **7**. Fairly frequent around Roundstone and Ballyconneely, mainly near the coast. Omey I. Rocks by Aughrusbeg L. On the ridge S.W. of Bengower, 1979 (Roden). **8**. Roadside N.W. of Maam. On bare peat between rocks on the summit of Lugnabrick.

First record: Oliver (1851). Roundstone. It seems probable, however, that the record by Wade (1802) of *Spergularia saginoides* from 'Lettery mountain, Ballynahinch' refers to this species.

All plants from the west of Ireland have completely glabrous pedicels, and may be referred, therefore, to var. *glabrata* Rouy & Fouc. They do not key out in most British Floras.

S. procumbens L.   Pearlwort

1 2 3 4 5 6 7 8   A B

Rocks, roadsides, damp grassland, streamsides, and occasionally in a wide variety of other habitats. Abundant almost throughout.

First record: Wright (1871). Aran Islands.

S. nodosa (L.) Fenzl

1 2 3 4 5 6 7 8   A (*B*)

Edges of marshes, lake-shores, damp ground and small hollows in limestone pavement. Locally frequent, but unevenly distributed, and apparently absent from considerable areas, especially in the inland parts of Connemara.

1. Rocks by the stream draining Drumcullaun L. **2.** Frequent near the coast; rarer elsewhere. Occasional on Aran. **3.** Frequent, but local. **4.** Frequent in the north; rare in the south. **5.** Occasional W. and S. of Oughterard; not seen elsewhere. **6.** Frequent near the coast. **7.** Occasional near the coast, from Roundstone to Omey I.; inland on Cregg Mt. **8.** Occasional in the east.

First record: Foot (1864). 'A most characteristic Burren plant, everywhere to be met with.'

The ecological requirements of this species are rather obscure. It is more frequent on the limestone than off it, but some of its stations appear to be very poor in base.

Spergula

‡**Spergula arvensis** L.   Corn spurrey

1 . . 4 5 6 7 8   . B

Cultivated ground; occasionally in other open habitats. Locally frequent in Connemara, mainly in the south and east; very rare elsewhere.

1. On peat-debris near Drumcullaun L. **4.** Field 5 km E.S.E. of Oughterard. **5.** Occasional round the margins, as at Costelloe, near Screeb and Rosscahill. **6.** Frequent in the south-east. **7.** Occasional near the coast in the west and north. **8.** Locally frequent in the south and east.

First record: More (1876). Inishbofin.

Spergularia

The species in this genus are most easily distinguished by their flowers. In *S. rubra* and *S. rupicola* the petals are entirely purple; in *S. media* and *S. marina* they are white at the base. Within each group the flower-size differs; in *S. rubra* they are $6\frac{1}{2}$ mm across, and in *S. rupicola* at least 12 mm; similarly in *S. marina* they are slightly over $6\frac{1}{2}$ mm and in *S. media* 12 mm or more.

†**Spergularia rubra** (L.) Presl

. . . . . 6 . .   . .

Gravelly tracks and quarry-floors; very rare.

6. On a track on the mainland opposite Freaghillaun, N.W. of Carna, 1964 (Ironside-Wood). *Quarry S. of Glinsk*, 1954 (DW).

First record: Webb (1956). S. of Glinsk.

Perhaps only casual, but as the two stations are only 5 km apart it seems possible that the plant is precariously established in this area. It is rare throughout Ireland, and especially in the west.

S. media (L.) Presl (*S. marginata*)

1 2 3 4 5 6 7 . **A B**

Salt-marshes; rarely on rocky or stony shores. Frequent in S. and W. Connemara; occasional elsewhere.

1. O'Brien's Br. **2**. Occasional on the north coast, from Rine Point to Bealaclugga. **3–4**. Occasional and locally frequent. **5**. Gentian Hill. Spiddal. **6**. Very frequent. **7**. Frequent from Mannin Bay northwards to Ballynakill harbour. Inishshark. Inishlackan.

First record: ITB (1901). H16.

S. marina (L.) Griseb.

1 2 3 4 5 6 7 8 **A** (*B*)

In habitats similar to those of the last, and with much the same distribution.

Somewhat less plentiful than *S. media* in most districts, but frequent in district 6. Not seen recently on Inishbofin, and on Aran confined to Inishmore. Leenane provides a record for district 8.

First record: Oliver (1852), as *S. salina*.

S. rupicola Le Jol.

. 2 . . . 6 7 . **A B**

Maritime cliffs and rock-crevices; more rarely in salt-marshes. Very local.

2. Occasional from Poulsallagh to Fanore. Inishmore. **6**. S.E. corner of Gorumna I., and on the mainland opposite. **7**. Occasional on the W. coast, from near Ballyconneely to near Renvyle. Inishlackan, Inishbofin, and probably on other islands.

First record: More (1872). 'On many of the islands off Connemara; *A.G.M.*'

Less common than might have been expected from its strongly western tendency and its preference for exposed rocky coasts.

PORTULACACEAE

Montia

Montia fontana L.   Water chickweed

1 2 . . 5 6 7 8 . **B**

Marshes, streamsides, mountain flushes and other wet places. Frequent in Connemara; rare elsewhere.

1. Wet roadside S. of Lickeen L. **2**. Wet track near Gleninagh Castle, N.W. of Ballyvaughan. By the ruined church W. of Caherconnell. **5–6**. Frequent. **7–8**. Very frequent.

First record: More (1876). Inishbofin.

Few specimens have been determined as to subspecies, but subsp. *chondrosperma* is recorded from districts 2 and 7, subsp. *amporitana* from district 2, and subsp. *fontana* from districts 7 and 8.

ELATINACEAE

Elatine

Elatine hexandra (Lapierre) DC.

. . . . . . 7 .   . B

Shallow water in non-calcareous lakes; occasional in W. Connemara; unknown elsewhere.

7. L. Rusheenduff, near Renvyle. Courhoor L., S. of Cleggan. Ballynahinch L. In a pool by the W. end of Kylemore L. *Cregduff L., near Roundstone* (Praeger, 1897); seen here by several others, and by J. E. Raven as late as 1955, but searched for since then by him and others without success. Inishbofin (L. Nagrooaun).

First record: More (1876). Inishbofin: in three of the lakes.

GUTTIFERAE

Hypericum   *St John's wort*   *Luibh Eoin Bhaiste*

Although eight species occur in the region, it is not difficult to tell them apart. Two have prostrate or trailing stems – *H. humifusum*, growing in dry places, with oblong, hairless leaves, and *H. elodes*, growing in wet places, with almost round, usually woolly leaves. *H. androsaemum* has a bushy appearance, with leaves much bigger than in other species, usually about 5 cm long. The remainder are slender, erect plants. *H. canadense* has very small flowers, only 6 mm across. *H. pulchrum* has the sepals fringed with gland-tipped teeth and the buds tinged with red. *H. maculatum* has no or few transparent dots on the leaves, and the sepals are blunt. *H. perforatum* and *H. tetrapterum* have sharply pointed petals and leaves with many transparent dots, but the former has a cylindrical stem and black streaks on some of the petals; the latter has a 4-angled stem and no black streaks.

Hypericum androsaemum L.   Tutsan

1 2 3 4 5 6 7 8   A B

Hedges, banks, rocky ground, bushy places and wood-margins. Occasional to frequent throughout.

Commonest in districts 2 and 8; rarest in 1, 3 and 6, but recorded from at least five stations in each. Apparently indifferent to soil type.

First record: Babington (1836). Leenane.

H. perforatum L.

1 2 3 4 5 . . 8   A .

Roadsides and dry grassland. Frequent on the limestone; very rare elsewhere.

1. Limestone rocks E.N.E. of Kilnamona. **2**. By L. Aleenaun. Waste ground near Glencolumbkille Ho. Roadsides S.W. of Ballyvaughan and W. of Turlough village. Inisheer (rare). **3**. Abundant. **4**. Frequent. **5**. On the old railway track at Rosscahill and Oughterard. Roadside near Gentian Hill. *L. Bofin* (ITB). *Knocknagreena* (ITB). **8**. Streamside W. of Cornamona. Frequent W. and N.W. of Cong. On an island in L. Corrib N.E. of Oughterard.

First record: ITB (1901). H16.

H. maculatum Crantz (*H. dubium*)

1 2 3 4 . . [7] **8** A .

Roadsides, hedges, scrub and grassland. Frequent on the limestone; rare elsewhere.

1. Streamside near Derrymore Br. N. of Drumcullaun L. Hedge by L. Eenagh. **2–3**. Very frequent, but very rare on Aran. **4**. Frequent. [7]. Roadside by Ballynakill harbour, 1967, but probably a casual introduction with building materials. **8**. Frequent W. of Clonbur and W. of Cong. *Maam* (Marshall & Shoolbred, 1896).

First record: Marshall & Shoolbred (1896). 'Maam; frequent to the W. of Clonbur.'

Nearly all the stations are on limestone, although in Ireland as a whole the species exhibits only a slight calcicole tendency. *H. perforatum*, on the other hand, is markedly calcicole; yet within the Flora region it is difficult to specify any real ecological differentiation between the two species. They cover much the same ground; yet they are rarely seen growing together.

H. tetrapterum Fries

1 2 3 4 5 6 7 8 A B

Ditches, marshes and damp grassland. Widespread and locally frequent.

1. Occasional. **2–4**. Frequent. **5**. Occasional. **6**. Frequent. **7**. Very frequent. **8**. Frequent.

First record: Hart (1875). Aran Islands.

Commonest on the Atlantic fringe of Connemara and on the limestones E. of Oughterard – two areas which seem to have little in common.

H. canadense L.

. . . . **8** . .

Flushes, and by small streams or bog-pools. Confined to the area W. and S.W. of L. Mask; here frequent and locally abundant.

8. Abundant on the sandy and slightly peaty flats, and also in disturbed peaty ground S. and E. of Killateeaun, near the mouth of the Owenbrin R.; also at *c*. 170 m (550 ft) on moorland 2 km E. of L. Nadirkmore, and on low ground between the village of Maamtrasna and the western arm of L. Mask.

First record: Webb (1957).

The stations within the Flora region lie within a radius of 4 km. The only other known Irish stations lie immediately to the north, in H26, and also near Glengarriff in W. Cork. Apart from Ireland, the only known European station is in Holland; records from E. France are referable to a related species, *H. majus* (Gray) Britton. Its main centre is in eastern North America, where it ranges from Newfoundland to Alabama.

For a full account of its morphology and ecology and its claim to be considered native, see Webb & Halliday (1973). It may be added that subsequent observations have given no indication of either increase or decrease in the size of the populations.

H. humifusum L.

1 [2] . . **5 6 7 8** [A] **B**

Roadsides, banks, dry grassland and stony ground. Occasional to frequent on acid soils; probably absent from the limestone.

1. Streamside near Derrymore Br. [2]. *Reported from Inishmore by Wright* (1871) *and from Inisheer by Hart* (1875), but not seen recently on Aran. **5**. Sandpit by a small lake 5 km W.N.W. of Galway. By a ditch 3 km N.N.E. of Screeb. Roadside 6 km W. of Rosscahill. **6**. Frequent. **7**. Occasional, mainly in the west. **8**. Frequent.

First record: Babington (1836). Oughterard.

Repeated search has failed to rediscover it on Aran, and its presence on bare limestone is most improbable. It may have been a long-term casual intermittently introduced with peat from the Carna region, or the records may be based on prostrate forms of *H. pulchrum*, which are not uncommon.

H. pulchrum L.

1 2 3 4 5 6 7 8   A B

Heathy ground, scrub, roadsides, banks, rough grazing and limestone pavement. Very frequent to abundant almost throughout.

Most plentiful in districts 2, 3, 7 and 8; scarcest in 1, 4 and 6.

First record: Babington (1836). Leenane.

The only substantial area in which this plant has not been seen is the low-lying limestone at the head of Galway Bay; this is perhaps because of the heavy drift cover, as the plant seems to be intolerant of clay. On the bare limestones of Aran and the Burren it grows in great profusion, despite its undoubted calcifuge tendency elsewhere; there is evidence that some degree of ecotypic differentiation may have produced here a race more tolerant of alkaline soils.

H. elodes L.

1 . . . **5 6 7 8** . **B**

Streams, wet bogs and ditches. Absent from the limestone; very frequent in Connemara.

1. Stream W. of Kilnamona. **5–6**. Very frequent. **7**. Frequent in the north; occasional in the south. **8**. Frequent.

First record: Baily (1833). 'Cunnemara bogs.'

Plants growing in slow streams flowing through bogs may be completely submerged, except for the flowers; in such plants the leaves are often very nearly glabrous.

MALVACEAE

Althaea

‡**Althaea officinalis** L.   Marsh mallow

1 (*2*) . . . . **7** . . . .

Roadsides, waste ground and salt-marshes; rare.

1. Sandy salt-marsh near O'Brien's Br. By the harbour at Liscannor. (*2*). *Ballyvaughan*, 1884 (**DBN**). **7**. In several places near Aughrusbeg L., well naturalized. 3 km S. of Ballyconneely, but only near houses. *2 km W. of Roundstone*, 1891 (Levinge, **DBN**).

First record: More (1860). Near Lehinch. The record by Mackay (1836) for between Lehinch and Miltown Malbay probably lies just outside our region.

An obvious introduction in most of its Irish stations, but in a few places near the west coast it grows far from houses and may be native; the station at O'Brien's Br. may possibly belong to this category.

Lavatera

Lavatera arborea L.   Tree mallow

. **2** . (*4*) . . (*7*) .   **A** (*B*)

Maritime cliffs and rocks; rare. Also occasionally in waste places as an escape from cultivation.

2. Undoubtedly native on Inishmore on cliffs near Dun Aengus; *also on Rock I.* (Hart, 1875). Found also on maritime rocks near Killeany on Inishmore, but here more likely to be an escape from cultivation. It also grows near Ballyvaughan harbour as an obvious escape. There are early records of its occurrence as an escape in various other places – Inisheer, Inishbofin, near Galway, and at Parkmore (district 4), but it seems to have disappeared from all these. Black Head is given as a station in ITB without date or authority; as it is not repeated in FWI it was probably an error.

First record: Wade (1802). 'Galway Bay, common.' First record as a native: Mackay (1806). 'On the very high cliffs facing the Atlantic, south isles of Arran.'

An interesting example of a native plant taken into cultivation from the wild, and now commoner in cultivation and as an escape than as a native.

Malva   Mallow

*****Malva moschata** L.

. [2] (*3*) . [5] . **7 8** . .

Roadsides; very rare.

[2]. *Ballyvaughan*, 1884 (**DBN**); apparently extinct. (*3*). *Tulla* (Praeger, 1902*a*). [5]. *Salthill*, 1901 (ITB); apparently extinct. **7**. Near Cregduff L. **8**. Roadside between Clonbur and Ferry Br., 1970. *Ashford Castle* (ITB).

First record: ITB (1901). Salthill.

Scarcely more than casual; it is only in S.E. Ireland that this species is firmly established.

*M. sylvestris L.

(*1*) **2 3 4 5 6** (*7*) .   A [B]

Roadsides, waste ground, by ruins and by the sea. Rather rare.

(*1*). *Near Lehinch* (Atlas). **2**. W. of Burren village. Common on Aran. **3**. By the ruined castle at the N. end of L. Bunny, and at Castle L. Roadside 2½ km N. of Ruan. **4**. At the top of the gravel beach at Carrowmore, and at Roscaun Point. **5**. In the Claddagh, and by the shore at Spiddal. **6**. At the S.W. end of Lettermullan I. (*7*). *Ballyconneely* (Atlas). *Inishbofin* (Praeger, 1911); apparently extinct.

First record: Hart (1875). Aran Islands.

*M. neglecta Wallr. (*M. rotundifolia*)

. **2 3 4 5** . [7] .   A [B]

Waste ground and roadsides. Occasional around Galway Bay; unknown elsewhere.

2. W. of Ballyvaughan harbour. By a lane N.E. of Ballyvaughan. New Quay. Aran. **3**. Aughinish. **4**. Parkmore strand. *Kilbeg ferry*, 1899 (ITB). **5**. Spiddal. *Inveran* (ITB). [7]. *Inishbofin* (Praeger, 1911); now extinct.

First record: Sim (1859). Near Galway.

Fairly well established on the S. shore of Galway Bay and on Aran; probably only casual elsewhere.

LINACEAE

Radiola

Radiola linoides Roth   Allseed

. . . . **5 6 7 8** . B

Roadsides, banks and patches of bare sandy or peaty soil. Occasional and locally frequent in Connemara; unknown elsewhere.

5. Rather frequent on roadsides 3 km W. of Killaguile; occasional also in the W. part of the district. **6**. Frequent; noted in 12 stations. **7**. Frequent in the north-west near the coast. Inishlackan. *Roundstone* (Linton & Linton, 1886). **8**. Frequent on sandy flats by L. Mask at Killateeaun. Occasional on the S. shore of the N.W. arm of L. Corrib.

First record: Balfour (1876). Cregduff L.

Linum

Linum catharticum L.   Fairy Flax

1 2 3 4 5 6 7 8   A B

In a wide variety of habitats; widespread and common.

1. Occasional. **2–4**. Abundant. **5–6**. Frequent. **7**. Abundant. **8**. Abundant in the south and east; occasional in the north and west.

First record: Hart (1875). Aran Islands.

Although this species is more plentiful on the limestone and near the coast than elsewhere, there is no sizeable area from which it is absent. Its most characteristic habitat is dry, semi-open grassland, but it can also be seen on walls, in quarries, on sand-dunes, limestone pavement, mountain cliffs, stony lake-shores, heaths, and even on sloping blanket-bog.

GERANIACEAE

Geranium   Cranesbill

G. sanguineum is known by its bright magenta flowers over $2\frac{1}{2}$ cm across. The flowers in the two garden escapes are nearly as big, but in *G. pratense* they are violet-blue and in *G. endressii* pink. Of the smaller-flowered species *G. lucidum* is known by its shining leaves and sharply winged calyx, and *G. robertianum* by its much-dissected leaves being divided into 3 equal primary lobes. The other small-flowered species are hard to characterize briefly, and must be identified from Floras.

Geranium sanguineum L.   Bloody cranesbill

. **2 3 4** . . . .   **A** .

Limestone pavement and rocky grassland. Abundant in the Burren and on Aran; almost unknown elsewhere.

2. Abundant. **3**. Very frequent to abundant, but thinning out eastwards. **4**. Occasional, amongst a Burren-type flora, on calcareous gravel ridges S. of Carrowmoreknock; also fairly frequent by L. Corrib W. of the village.

First record: O'Mahony (1860). ' In the neighbourhood of Bealnalack [Ballynalackan] Castle, I have seen it spreading over acres of rocky ground.'

Although it receives less publicity than *Dryas octopetala* or *Gentiana verna*, this species is in some ways the most spectacular plant of the Burren, as it flowers continuously from May to August.

G. pratense L. has been seen on waste ground in the S.W. part of Galway city and by the roadside W.S.W. of Ballyconneely, and *G. endressii* Gay by the river in Oughterard and also 4 km S.E. of the village. Neither can, however, be regarded as yet as effectively naturalized.

***G. pyrenaicum** Burm. fil.

. **2** (*3*) . . . (*7*) **8**   **A** .

Roadsides; very local.

2. Very frequent on the Aran Islands. (*3*) *Near Ballyallia Ho.* (ITB). (*7*). *Roadside near Clifden* (Phillips, 1924). **8**. By the bridge at Cong.

First record: ITB (1901). Ballyallia.

Thoroughly established in Aran; perhaps rather less securely at Cong. This is surprising, as the plant is much commoner in eastern than in western Ireland. It seems to have been introduced to Aran early in this century. It was not noted by Praeger in 1895, but there is in **DBN** a specimen of P. B. O'Kelly's, supposedly collected on Inishmore in 1894. Like many of his Aran records, however, this must be viewed with some reserve.

G. molle L.

1 2 3 4 5 6 7 8   A B

Sand-dunes, roadsides, waste places, dry grassland and limestone pavement. Frequent in most of the Burren, and in the coastal districts of S. and W. Connemara; rather rare elsewhere.

1. Sand-dunes at Lehinch. Rocks at O'Dea's Castle. **2**. Frequent. **3**. Occasional, mainly in the west. **4**. Roadside 5 km E. of Oughterard. Rocks by the shore half way between Galway and Oranmore. Around the castle at Ballindooly L. **5**. Frequent and luxuriant in a meadow by the sea E. of Inveran. Gentian Hill, and by the road 3 km north of it. *Spiddal* (Atlas). **6–7**. Very frequent near the coast; not seen far inland. **8**. On a wall at Ferry Br.

First record: Wright (1871). Aran Islands.

(†G. rotundifolium L.)

. (2) . . . .

Walls and laneways near Ballyvaughan; not seen recently.

(2). *Rocks and walls at Townross, near Ballyvaughan*, 1895 (CH2). *Wayside near Ballyvaughan*, 1907 (**DBN**). '*Extends for ½ mile along the rough road and on adjoining rocks*' (Praeger, 1934a).

First record: CH2 (1898). 'At Townross near Ballyvaughan, Co. Clare, on walls and on limestone rock, well established, 1894–5; *P. B. O'Kelly.*'

A search was made at Townross (which lies S.E. of Ballyvaughan) without success in 1979, but it would have to be more thorough before one could declare the plant extinct. Persistence for 38 years shows that it was something more than casual.

G. dissectum L.

1 2 3 4 5 6 7 8   A B

Roadsides, pastures, field-margins and stony ground; occasionally in cultivated ground. Generally frequent, but rare in a few areas.

1. Near the coast, S. of Lehinch. N.W. of Inagh. O'Dea's Castle. **2**. Occasional; frequent on Aran. **3–4**. Frequent. **5**. Roadsides W. of Spiddal and W.N.W. of Galway. On the old railway track at Oughterard. **6**. Frequent. **7**. Frequent in the extreme west. Inishlackan. **8**. Recess. Ferry Br., and by L. Mask north of it.

First record: Hart (1875). Inishmore: 'not common, and only on the roadsides'.

The considerable increase on Aran since Hart's day contrasts with the behaviour of most ruderals, and suggests that in his time it was a recent introduction.

G. columbinum L.

1 2 3 . . . (7) 8   A .

Disturbed, shallow soils over limestone; rather rare.

1. Rocks by Ballycullinan L. **2**. Inishmore; frequent around Killeany. *Ballyvaughan*, 1895 (**DBN**). **3**. Occasional on the pavement around the turloughs S.W. of Mullagh-more. *(7)*. *On rocky knolls on the N. side of the Slyne Head peninsula* (Praeger, 1907). **8**. Occasional near Cong.

First record: CH2 (1898). 'Townross near Ballyvaughan, Clare, 1894; *P. B. O'Kelly.*'

As with *G. rotundifolium*, a recent search at Townross was unsuccessful. There are records in the *Atlas* for the Poulsallagh, Slieve Elva and Black Head areas, but as they all derive from the same recorder and have not been confirmed they are best held in suspense.

G. lucidum L.

1 2 3 4 5 . [7] **8   A** .

Walls, rocks and limestone pavement. Common over most of the limestone; rare elsewhere, and usually on mortared or limestone walls.

1. Wall W. of Inagh, growing with *Umbilicus pendulinus*, and in the absence of any obvious source of lime. Occasional on limestone walls and rocks on the eastern margin of the district; also on walls in Ennistymon. **2–3**. Abundant. **4**. Occasional on walls in the N.W. part of the district; also S. of Oranmore. **5**. Limestone wall near Moycullen. On a heap of road-metal S.E. of Oughterard. On a granite wall (lightly cemented on top) 5 km W. of Barna. On the track to the pier at Spiddal. **8**. On walls and rocks around Clonbur and Cong. On sodded walls of siliceous rock N. of Cornamona.

First record: Wade (1802). 'Abundantly on the rocks around Cong.'

The records for the Clifden and Cleggan districts published in the *Atlas* require confirmation.

G. robertianum L.   Herb-Robert   *Earball rí*

1 2 3 4 5 6 7 8   A B

Woodland, scrub, hedges, walls, rocks, limestone pavement, and in many other habitats. Abundant throughout, except on the mountains and on blanket-bog.

First record: Wright (1871). Aran Islands.

A rather striking variant in which the foliage is pale green and less hairy than in normal plants, and the flowers pink rather than reddish-purple, is found on limestone rocks, usually near the sea, in the Burren and on Aran; it has also been seen (on granite) at the S.E. corner of Gorumna I. It has been distinguished as a subspecies, but is probably best named var. *celticum* (Ostenf.) Druce. A variant characteristic of shingle beaches, differing from the type mainly in its prostrate habit, has been recorded from Ballyvaughan, and perhaps also Aran and Inishbofin, but its distinctive characters are rather shadowy.

Erodium

Erodium cicutarium (L.) L'Hér.   Storksbill

1 2 3 (*4*) **5** . **7** .   **A B**

Sandy (rarely gravelly) places by the sea; local.

1. Dunes N. of Lehinch. **2**. Fanore. Rine Point. Dunes E. of Ballyvaughan. Frequent on Aran. **3**. Behind the storm-beach at Aughinish, and on a shingle-bank 4 km to the east. (*4*). *Ballyloughan* (P. cat.) *Roadside at Kilbeg ferry*, 1899 (**DBN**). **5**. Beach at Spiddal, and 4 km further west. **7**. Very frequent on the Slyne Head peninsula. Dog's Bay. On several of the dunes W. of Cleggan. Inishbofin.

First record: More (1872). 'Aran – *H. C. Hart.*'

There are very few inland stations for this species in Ireland, and most of the inland records refer to casual introductions which did not persist. The record for Kilberg ferry probably belongs here; the plant was most likely introduced with sea-sand for building work. It cannot be found there now.

*E. moschatum L. (L'Hér.)

(*1*) **2 . 4 . . (*7*) . A** .

Waste places; very rare.

(*1*). *By the Cliffs of Moher* (ITB). **2**. Inisheer, 1976 (**TCD**). *Formerly abundant on Inishmore.* **4**. By the castle at Ballindooly, 1967 (DW). *Renmore* (ITB). (*7*). *Clifden* (ITB).

First record: Graham (1840). On the shore near Galway.

Much rarer everywhere in Ireland than in 1900.

Erodium maritimum (L.) L'Hér. has been twice recorded from the region: from Aughrus, W. of Cleggan (district 7) by Wade (1802), and from Inishmore (district 2) by Nowers & Wells (1892). As there are no recent records and no specimens from any part of the west coast, it seems best to suppose that both records are erroneous, based probably on dwarfed plants of *E. cicutarium*.

OXALIDACEAE

Oxalis

Oxalis acetosella L. Wood sorrel

1 2 3 4 5 6 7 8   A B

Woods, scrub, limestone pavement, heathy ground among bracken, mountain cliffs, streamsides and other shady situations. Very frequent over most of the region, but absent from parts of W. and S. Connemara, and rarer in areas of great exposure.

1. Very frequent. **2**. Abundant, but rare on Aran. **3–5**. Frequent. **6**. Rare; noted only in the woods at Glendollagh and the scrub N.E. of Kilkieran. **7**. In most of the well-developed woods, from Derreen to Ballynahinch and Kylemore. At the base of a wall at Gowlaun. On the cliff on Muckanaght. Present on Inishbofin, but generally rare along the west coast. **8**. Occasional in the west; abundant in the east.

First record: Foot (1864). 'It grows in the most exposed spots [of the Burren], at an altitude of more than 1,000 ft.'

BALSAMINACEAE

Impatiens

*Impatiens glandulifera Royle   Balsam

· 1 2 . 4 . . 7 8   A .

Roadsides and waste ground; rather rare.

1. Field by the river near Roadford. **2**. Waste ground in Kilronan, Inishmore. **4**. Near the river in Galway, and on the roadside 4 km to the north-west. **7**. Waste ground behind houses on the south side of Clifden. **8**. Fallow field behind a cottage at Maam.

First record: Praeger (1939). 'By the canal at Galway (Mrs. Gough).'

Never very far from houses, though in several cases it cannot now be seen in any garden close by.

Fig. 11. Holly-tree on limestone pavement near Mullaghmore. Although it is 20–25 years old, the combined effects of grazing and wind keep it down to a height of about 20 cm.

AQUIFOLIACEAE

Ilex

Ilex aquifolium L.   Holly   *Cuileann*

1 2 3 4 5 6 7 8   A .

Woods, scrub, hedges, limestone pavement, open heathland, lake-islands and mountain glens. Very frequent over most of the region, being absent only from the most exposed coastal districts, where it succumbs to the combined effects of wind and grazing.

First record: Baily (1833). 'Mountains, Cunemara.'

If protected from grazing, the holly can endure a good deal of exposure, as can be seen from its constant occurrence on small lake-islands as far west as L. Fadda. In the Burren and Aran it is fairly frequent on the limestone pavement, rooted deep in the crevices and forming a low cushion only about 15 cm above the general surface of the rock. It appears, however, to be absent from the coastal area of district 1, and to be rare in Connemara (apart from lake-islands) west of a line drawn from Roundstone through Kylemore to Salruck.

CELASTRACEAE

Euonymus

Euonymus europaeus L.   **Spindle-tree**   *Feoras*
1 2 3 4 . . . **8** **A** .

Scrub, hedges and limestone pavement. Common on most of the limestone; very rare elsewhere.

1. In a wood at Elmvale Ho. **2–3.** Very frequent to abundant, though rather rare on Aran. **4.** Frequent in scrub from Ross L. northwards; not seen in the south. **8.** Lake-shore 3 km N. of Oughterard. Frequent in woods and scrub towards L. Mask, between Clonbur and Cong.

First record: Ball (1839). Kilcornan, near Galway; on level ground near the sea.

The bushes in the Oughterard station are growing on siliceous rock, but their roots are probably washed by the calcareous water of the lake.

RHAMNACEAE

Rhamnus

Rhamnus catharticus L.   Buckthorn
. **2 3 4** . . . **8** **A** .

Woods, scrub, river-banks, rocky lake-shores and the margins of turloughs. Locally frequent on the limestone; very rare elsewhere.

2. In scrub 2 km S.W. of Ballyvaughan. By the turlough W. of Turlough village. Aran (rare). **3.** Very frequent in the south; occasional by lakes and turloughs further north. **4.** Very frequent in the north, from near Oughterard to 3 km N. of Galway; not seen further south or east. **8.** Locally abundant by the S. shore of L. Mask. Sparingly on the shore of L. Corrib in the grounds of Ashford Castle, and frequent on several of the islands. Fox Hill, 2½ km N. of Ferry Br.

First record: Wade (1802). 'On many islands, Lough Corib.'

All records are from the limestone, except those from Fox Hill and some of the L. Corrib islands. The buckthorn is local in Ireland, but frequent in a few areas, mainly in the north and west, thus contrasting strangely with its predominatly south-eastern distribution in Britain.

Prostrate forms have been noted in district 3, but not nearly as frequently as in the following species.

Frangula

Frangula alnus Miller (*Rhamnus frangula*)
. . **3** . . . . **8** . .

Rocky ground, mainly on limestone pavement and usually close to lakes or turloughs; very local.

3. Occasional around the turloughs S.W. of Mullaghmore. At the N.E. corner of L. Bunny, and very sparingly between this station and the last, extending in small quantity northwards to near Castle L. and southwards to L. George. In Garryland wood, near the margin of Coole L., and by the 'race-course' turlough W. of Gort. **8.** By the S. shore of L. Mask, N. of Clonbur, in fair quantity, both on limestone and shale. *Cannaver I., L. Corrib* (More, 1860). *Lusteenmore I., L. Mask* (Praeger, 1909a).

First record: More (1855). 'Quite prostrate on limestone flags outside Garryland wood.'

Usually quite prostrate, but by the shore of L. Mask a fair number of normal, erect bushes have been seen, growing close to prostrate ones. A prostrate plant was cultivated for some years in the Botanic Garden of Trinity College, Dublin. It remained unchanged for three years, but then put up two strong, erect shoots, while others continued to grow horizontally. It appears that no single or simple explanation will suffice to account for the prostrate habit.

ACERACEAE

Acer

*Acer pseudoplatanus L.   Sycamore   *Crann bán*

1 2 3 4 5 6 7 8 . .

Woods, hedges and river-banks; occasionally on waste ground or limestone pavement. Frequent except in blanket-bog areas.

1–4. Frequent, except in the most exposed areas; planted on Aran but not naturalized. **5.** Frequent along the E. margin; also at Spiddal. **6.** In plantations near Ballynahinch. Copses at Recess. Heathy ground near Carna. **7.** Kylemore woods. Letterfrack. S.W. of Clifden. **8.** Frequent.

First record. Impossible to establish in the absence of distinction between planted and self-sown trees.

The above records refer to trees which are clearly self-sown. The sycamore is also widely planted in shelter-belts beside cottages, especially near the sea, as it stands up to salt-laden winds better than any other hardy broad-leaved tree.

In several parts of the Burren young trees may be seen on the limestone pavement, usually (though not always) fairly close to a planted parent tree. Like all other woody plants on the pavement they suffer severely from grazing, and very few grow to fruiting size. Self-sown trees have also been seen on the islands in Loughs Corrib and Mask.

HIPPOCASTANACEAE

Aesculus

[***Aesculus hippocastanum** L.]   Horse-chestnut

Noted in woods and copses in all districts except 2; most frequently in districts 3 and 8, but for the most part as obviously planted trees. It is self-sown at Letterfrack, Cong and Ennistymon, but in all cases in demesne woodlands, and we have seen no clear evidence of regeneration in competition with native vegetation.

LEGUMINOSAE

Ulex   Furze, gorse   *Aiteann*

Ulex europaeus L.

1 2 3 4 5 6 7 8   A .

Hedges, scrub, heathy ground and bog-margins. Widespread and locally common, but rare over considerable areas.

1. Frequent. **2**. Covering a small peninsula of glacial drift W. of Ballyvaughan harbour. In a few places close to the Corofin road 8–10 km S. of Ballyvaughan. A small colony S. of Ballynalackan Ho. **3**. Rare in the west; locally abundant in the east. **4**. Very frequent about Moycullen, Ross L., and the adjoining shore of L. Corrib. Abundant near Rinville, S. of Oranmore, but not seen elsewhere around the head of Galway Bay. **5**. Very frequent. **6**. Frequent in the north and east; rare in the south and west. **7**. Frequent in the extreme west; rare elsewhere. **8**. Abundant.

First record: Babington (1836). Common in Connemara, except around Clifden.

Commoner on acid soils than on limestone, but irregularly distributed on both. It is distinctly rare in S.W. and C. Connemara, and in the western part of the Burren. Only in districts 5 and 8, and on the eastern fringe of district 3 does it really look at home, and there is little doubt that over a good deal of the region it is merely naturalized, having been originally introduced as a hedging plant. In Aran it is rare and obviously introduced; on Inishbofin it is seen only as a few planted hedges and is not even naturalized.

U. gallii Planchon

. [2] . **4 5 6 7 8** [A] .

Heathy ground; locally abundant in Connemara, especially in the south and west; unknown elsewhere.

[2]. *Noted by Balfour on Inishmore in 1853* (*as* U. nanus) *and by Praeger in 1895* ('*in two places near the centre of the island*'); not to be found there now. **4**. Locally abundant by a lane leading from the main road to the N. end of Ballycuirke L. **5**. Frequent by the coast; rare elsewhere. **6**. Abundant in the south; rare in the north. **7**. Abundant in the south and west; rare in the north and east, and not on Inishbofin. **8**. A single bush by the road 3 km E. of Maam Cross. A few plants on heathy ground 7 km N.W. of Oughterard. A small colony 4 km from Leenane by the road to Maam.

First record: Babington (1836), as *U. nanus*. Near Clifden.

Although stunted bushes of *U. europaeus* in exposed situations are often mistaken for this species, it is unlikely that both records for Aran are errors, and it seems best to suppose that this strongly calcifuge species was introduced there with peat from Connemara, and continued to grow for some time on the debris. Its Irish distribution is mainly southern, and Connemara marks the northern limit of its occurrence in any quantity, except for its rather unexpected abundance in the Mourne Mts. In Britain, on the other hand, it is essentially western, but extends northwards to the Solway.

The *Atlas* record for Poulsallagh is an error.

Cytisus

Cytisus scoparius (L.) Link (*Sarothamnus scoparius*)   Broom   *Giolcach sléibhe*

. . . **4 5 6 7 8** . .

Hedges, roadsides and heathy ground. Very local and rather rare, and confined to acid soils.

4. A single plant on the S. side of the road leading from Oughterard to the quay. **5.** Occasional near the N.E. margin. **6.** A single plant in a cutting for a new road between Maam Cross and Recess. **7.** A small colony in a hedge by the N.E. corner of Kylemore L. *Near Tully Cross* (Babington, 1836). *Near Clifden* (Atlas). **8.** Widespread, but nowhere common.

First record: Babington (1836). Near Tully Cross.

Ononis

[*Ononis repens L.]   Rest-harrow   *Sreang bogha*

Although Praeger (ITB) gives native status to this plant in three stations in district 4, from Moycullen to near Clarinbridge, where he observed it in 1899, we are inclined to agree with the editors of CH2 that it is only casual in most of the west of Ireland. The only recent record is by M. E. Mitchell, who saw it at Merlin Park, E. of Galway, in 1960; it seems probable that it was introduced here with gravel for building. There is also a record from Ballyvaughan in 1895 by Colgan, who regarded it as introduced, and a specimen in **DBN** collected near Cong in 1873. In no case is there any information on habitat or abundance.

Medicago

*Medicago sativa L.   Lucerne

. **2** . . . . . . **A** .

2. Pastures and meadows on all three Aran Islands, 1972–6, mostly in small quantity, but in fair abundance between Kilronan and Killeany. Evidently introduced into cultivation there between 1869 and 1890, and noted by all visitors to Aran since the latter date. At present it seems to be 'tolerated' rather than actively cultivated, and is probably tending to diminish.

First record: Nowers & Wells (1892). Near Kilronan.

M. lupulina L.   Black medick

1 2 3 4 5 6 7 (*8*)   **A** .

Roadsides, pastures, sand-dunes and sea-cliffs. Abundant on the limestone; very rare elsewhere.

1. Abundant on dunes N. of Lehinch. Roadside by O'Dea's castle. **2–4.** Abundant. **5.** On the old railway track near Oughterard. Gentian Hill. By a small road 6–7 km N.W. of Barna. **6.** Mweenish I. **7.** Grassland near the S.W. shore of Mannin Bay. (*8*). *N.W. of Oughterard* (P. cat.).

First record: Hart (1875). Aran Islands.

Trifolium   Clover   *Seamróg*

Trifolium pratense L.   Red clover

1 2 3 4 5 6 7 8 A B

Roadsides and grassland of all kinds. Abundant in suitable habitats almost throughout, but rather scarce in central Connemara. It is impossible to tell in which districts it is native, and in which it is a relic of cultivation.

First record: Wright (1871). Aran Islands.

*T. medium L.   Zig-zag clover

. 2 3 4 . . 7 .   (A) (B)

Roadsides, waste places and rocky ground; rather rare.

2. Roadside S.W. of Ballyvaughan. Rough grazing S. of Slieve Carran. *Inishmore* (Nowers & Wells, 1892); not seen since. **3.** Locally frequent on roadsides and limestone pavement 5–7 km E. and S.E. of Slieve Carran. **4.** Roadside 5 km E. of Galway. Waste ground at Burnthouse. **7.** Rocks at Belleek. Rocky pastures on the Slyne Head peninsula near Truska L. By the mouth of the Culfin R., near Gowlaun. *Inishbofin* (Praeger, 1911); not seen by earlier or later visitors to the island.

First record: Nowers & Wells (1892). Inishmore; near the Black Fort.

In all cases apparently a relic of cultivation, and obviously transient in many of its stations; in a few, however, it seems well established.

†T. arvense L.   Hare's-foot clover

. (2) . . . 6 (7) .   (A) .

Bare, rocky ground; very rare and probably inconstant.

(2) Inisheer (Hart, 1875). *Inishmore*, 1953 (**DBN**). **6.** A few plants at the S.E. corner of Gorumna I., 1966 (DW). **(7).** *Rocky bluff near W. end of the Slyne Head peninsula* (Webb, 1947).

First record: More (1872). 'Between the lighthouse and the old fort in South Island of Aran; *H.C.H.*'

The status of this plant in the region is hard to assess. In all cases the population observed was very small, which, coupled with its apparent inconstancy, suggests that it might be only casual. All the stations, however, are remote from agricultural activity, and it is difficult to suggest a means of introduction. It may, therefore, be a native with a very precarious foothold.

It is mainly an east-coast plant in Ireland, though it is recorded also from Inishturk (Co. Mayo).

[T. striatum L.]

Recorded from Aran by Nowers & Wells (1892). As it is known from limestone rocks in N. Kerry its occurrence here is not impossible, but as there are no other records from western Ireland and no voucher specimen, and as the locality on Inishmore has several times been searched in vain it seems prudent not to admit the species to the flora until confirmation has been obtained.

T. repens L.   White clover

1 2 3 4 5 6 7 8   A B

Grassland of all kinds; more rarely on rocky ground. Abundant throughout, and especially on limestone or near the sea.

First record: Wright (1871). Aran Islands.

*T. hybridum L.

. [2] (*3*) **4** . . . .

Wet grassland; established at the head of Galway Bay.

[2]. In the forecourt of the field station at Carran, 1978 (MS), but apparently a casual brought in with building material; gone by 1980. (*3*). *Kinvara Bay* (P. cat.). **4**. Locally abundant at Carrowmore, and in smaller quantity at Rinville, 4 km to the north.

First record: Atlas (1962).

T. fragiferum L.   Strawberry clover

. . **3 4** . . . . . .

Saline meadows; very local.

3–4. In fair quantity on both sides of the estuary of the Dunkellin R., 1969 (DW & I. K. Ferguson).

First record: Webb (1982).

   Very rare in the west, but there is a fairly recent record for Clarecastle, some 45 km to the south.

T. dubium Sibth. (*T. minus*)   Yellow clover, shamrock

1 2 3 4 5 6 7 8   A B

Roadsides, semi-open grassland, sand-dunes, gravelly places and waste ground. Very frequent throughout, and locally abundant. Commonest in districts 3 and 7; scarcest in 5 and the western part of 8.

First record: Hart (1875). Aran Islands.

T. campestre Schreb. (*T. procumbens*)   Hop clover

1 2 3 4 5 . 7 8   A B

Limestone pavement, sand-dunes, semi-open grassland and rocks. Occasional to frequent on the limestone; very rare elsewhere.

1. Frequent on the dunes at Lehinch. Roadside near Moy Ho. **2**. Occasional; frequent on Aran. **3**. Fairly frequent. **4**. Occasional in the north, between Oughterard and L. Corrib. Embankment of disused railway in Galway city. **5**. Meadow 3 km E.S.E. of Costelloe. Rocks near Gentian Hill. Sandpit near Rosscahill. **7**. Rocky outcrops on the Slyne Head peninsula. Sandy pastures near Gorteen Bay. Inishbofin. **8**. Rocky outcrops near Cong.

First record: Hart (1875). Aran Islands.

[***T. micranthum** Viv. (*T. filiforme*)]

Although this species has been recorded for three stations within the region it cannot be rated as more than a casual; it is only in the south and east of Ireland that it is effectively established. There is an old record for Leenane (Babington, 1836), and

R. D. Meikle (*in litt.*) reported it on rocky grassland near L. Bunny some 30 years ago, but it has not been rediscovered in either place. In 1965 the authors saw it at Spiddal, apparently well established on sodded walls and roadside verges, but the site has now been destroyed by road-widening.

Anthyllis

Anthyllis vulneraria L.   Kidney vetch

1 2 3 4 5 6 7 8   A B

Limestone pavement, dry pastures, sand-dunes and maritime rocks and cliffs. Frequent and locally abundant on limestone and near the sea; very rare elsewhere.

1. Frequent on cliffs, and on the dunes N. of Lehinch. **2**. Fairly frequent, but local; abundant on Aran. **3**. Frequent. **4**. Rocks by the estuary near Kilcolgan. Frequent around Galway. Roadside N.W. of Moycullen. **5**. Occasional by the coast; also near Moycullen and Oughterard. **6**. Occasional near the coast. **7**. Abundant on the Slyne Head peninsula; frequent by the coast elsewhere, and on Inishbofin. **8**. Limestone pavement 2½ km N.W. of Cong, very near the county boundary.

First record: Hart (1875). Aran Islands.

All plants in the region are yellow-flowered, the majority of a deep golden-yellow, but some of a paler lemon-yellow. The vast majority seem to be referable to subsp. *vulneraria*, but subsp. *lapponica* (Hyl.) Jalas has, rather surprisingly in view of its very boreal distribution, been recorded from W. of Roundstone by Cullen (1976). It differs mainly in the fewer and unequal leaflets of the upper leaves.

Lotus

Lotus corniculatus L.   Bird's-foot trefoil

1 2 3 4 5 6 7 8   A B

Dry grassland, sand-dunes and rocky ground. Abundant throughout.

First record: Foot (1864). Burren: 'to be met with in every direction'.

A dwarf variant with fleshy leaves, sometimes distinguished as var. *crassifolius* (Pers.) Rouy, has been recorded from several of the dunes of W. Connemara.

L. uliginosus Schkuhr

1 [2] . . 5 6 7 8   [A] B

Marshes, wet grassland, river-banks and hedges. Very frequent over a large part of Connemara; very rare elsewhere.

1. Cancregga, S. of Hag's Head. **5**. Ditch 3–4 km S.E. of Costelloe. Grassland 2–3 km N.E. of Screeb. Meadows at Rosscahill. **6**. Abundant. **7**. Very frequent. **8**. Frequent in the east; rather rare in the west.

First record: ITB (1901). H16.

Despite its frequency over most of Connemara, there are some areas, as around Recess and Leenane, where it seems to be quite rare. It is totally absent from the limestone, and the record of Wright (1871) for Aran, which has never been confirmed, may be confidently dismissed as an error.

Astragalus

Astragalus danicus Retz.

. 2 . . . . . . **A** .

Stabilized sand-dunes and rough grazing; also on limestone pavement on cliff-tops. Very local.

2. On Inishmore and Inishmaan; very local, but fairly frequent over small areas.

First record: Mackay (1836). 'On the largest of the south islands of Arran; *Messrs. R. Ball and Wm. Thompson.*'

No plausible hypothesis has ever been put forward to account for the presence on Aran, but nowhere else in Ireland, of this predominantly Boreal–Continental species. In Britain it is found mainly on the east coast, but there are outlying stations in the Hebrides and the Isle of Man.

The vetches

If we omit two very rare species, the vetches (*Vicia* and *Lathyrus* spp.) can be distinguished easily enough, mainly by flower-colour. *Lathyrus pratensis* alone has yellow flowers. In *Vicia hirsuta* they are very small (less than 6½ mm) and pale blue; in *V. cracca* bright violet-blue. In the remaining three the flowers are purplish, but *Lathyrus montanus* is easily distinguished by having no tendrils on the leaves. *V. sativa* and *V. sepium* have tendrils, but in the former the flowers are bright purple and mostly solitary in each leaf-axil; in *V. sepium* they are dull purple and mostly borne in groups of 3 to 5.

Vicia

‡**Vicia hirsuta** (L.) Gray

. . **3 4** . . **7** (*8*)   . **B**

Cultivated fields, waste ground and grassy places; rare.

3. Near the ruins at Kilmacduagh. **4.** In a cultivated field near Rinville. Waste ground at Aughnanure Castle. **7.** Inishbofin. (*8*). *Near Clonbur* (Marshall & Shoolbred, 1896).

First record: More (1876). Inishbofin.

Varying in abundance from year to year, and perhaps only casual in some stations, but established at least on Inishbofin, where it has persisted from 1875 to 1967.

V. cracca L.   Tufted vetch

1 2 3 4 5 6 7 8   A B

Hedges, meadows and scrub; occasionally on limestone pavement, cultivated ground or other open habitats. Common on the limestone; occasional and locally frequent elsewhere.

1. Frequent near the coast; rare elsewhere. **2–4.** Very frequent throughout. **5.** Hedge by Gentian Hill. Rough grazing W. of Spiddal. Tilled field N.E. of Screeb. **6.** Occasional in the south and east; not seen elsewhere. **7.** Very frequent in the west; rarer elsewhere. **8.** Frequent in the east; not seen in the west.

First record: Wright (1871). Inisheer.

[V. orobus DC.]

This plant has been recorded from two stations in district 8, but we are sceptical as to its actual occurrence in them. Ogilby (1845) recorded it by a mountain stream north of Halfway House (Maam Cross), but the habitat is most unlikely. Wade (1802) reported it from 'Bilbery Island and other islands of Lough Corib', and according to CH2 its presence on Bilberry I. was confirmed by Miss Jackson in 1897. Unfortunately there are two islands in L. Corrib with this name. The southern (N.W. of Oughterard) was searched independently and thoroughly by DW in 1965 and Roden in 1974 without finding the plant. In 1980 DW and Roden together attempted a search of the other Bilberry I. (between Cornamona and the Hill of Doon). Very bad visibility resulting from driving rain prevented its recognition, but landings were made on two adjacent islands. Here there was a remarkably robust and stiffly erect form of *Lathyrus montanus* – very different from *V. orobus* in flower and leaf, but strikingly similar to it in habit, so that DW was momentarily deceived. We believe it possible that the records refer to this plant (the possibility of confusion being increased by the fact that, for most of the nineteenth century, *L. montanus* was put in the genus *Orobus*), and that *V. orobus* is best excluded from the flora until definitely confirmed. No herbarium specimens from the Flora region are forthcoming, and it is perhaps significant that whereas all known stations for the plant in Ireland are on limestone or basalt, the two supposed stations in Connemara are on siliceous rock.

[V. sylvatica L.]

The records for this species are scarcely more satisfactory than for the last. CH1 gives it for Streamstown and the Hill of Doon (before 1864), but it has not been confirmed in either station. O'Kelly is cited in CH2 as having found it 'abundant in clefts of limestone crags near Caheracloghane Fort, south of Slieve Elva' in 1895, but there are no specimens, and a search of the rather limited limestone outcrop here in 1980 was negative.

V. sepium L.   Bush vetch

1 2 3 4 5 6 7 8   A B

Hedges, scrub and wood-margins; more rarely in exposed habitats. Common on the limestone; frequent elsewhere.

Rather more frequent than *V. cracca* in most areas, but rarer in those which are lacking in shelter, such as district 6 and the western part of district 7.

First record: Hart (1875). Aran Islands.

V. sativa L. (*V. angustifolia*)

1 2 3 4 . . 7 8   A B

Rocky grassland, sand-dunes, roadsides and cultivated ground; occasional.

1. Frequent on the dunes N. of Lehinch. **2.** Frequent on Aran, but doubtfully native. Field E. of Ballyvaughan. **3.** Rough grazing by a lake E.S.E. of Killinaboy. Stony grassland W.S.W. of Kinvara. **4.** Cultivated field at Rinville. **7.** Fairly frequent on Inishbofin. **8.** Roadside W. of Ferry Br.

First record: More (1976). Inishbofin: in stubbles and along hedges or borders of fields.

Only subsp. *nigra* (L.) Ehrh. has been seen. In some of its stations it is almost certainly native, but in others it has the appearance of a relic of cultivation, although in most parts of Ireland it is subsp. *sativa* that is cultivated.

Lathyrus

Lathyrus pratensis L.   Meadow vetchling

1 2 3 4 5 6 7 8   A B

Grassland, hedges and limestone pavement. Local in Connemara, and rare over considerable areas; abundant elsewhere.

1–4. Abundant. **5.** Roadside W. of Spiddal. Hedge near Gentian Hill. **6.** Wet grassland at S.E. corner of Gorumna I. **7.** Occasional, mainly near the sea, and abundant in a few places. **8.** Dooros peninsula on L. Corrib, and on several of the islands. Hedges near Glann.

First record: Hart (1875). Aran Islands.

L. montanus Bernh.

1 2 3 4 5 6 7 8   . B

Heaths, peaty meadows, river-banks, rocky ground and limestone pavement; frequent.

1. Occasional. **2.** Very frequent, but not recorded from Aran. **3–4.** Occasional. **5.** Frequent. **6.** Annaghvaan I. S.E. corner of Gorumna I. Bunnahown. **7.** Frequent. **8.** Abundant.

First record: More (1860), as *Orobus tuberosus*. 'Upon an islet called Canova in Lough Corrib.'

Although the general distribution of this species in Ireland indicates distinct calcifuge tendencies, it is widespread in the Burren, and appears to be, with *Hypericum pulchrum*, a pioneer among heathland species in pockets on the limestone pavement where peaty soil has accumulated, later to be followed by more exacting calcifuges such as *Calluna vulgaris* and *Erica cinerea*.

ROSACEAE

Spiraea

*Spiraea salicifolia L.

. . . . 5 . 7 .   . .

Planted in hedges here and there (though apparently never on the limestone), and in a few places spreading far enough by means of suckers in boggy ground to be regarded as naturalized.

5. In the bog beside a lane near Slievaneena L., S.W. of Moycullen; also 2 km further south, by the Spiddal–Moycullen road. **7.** By the by-road S.W. of Ballynahinch Castle.

First record: Webb (1982).

Hedges of this species have been noted in districts 1, 6 and 8, but without effective naturalization.

Filipendula

Filipendula ulmaria (L.) Maxim. (*Spiraea ulmaria*)   Meadowsweet   *Airgead luachra*

1 2 3 4 5 6 7 8   A B

Marshes, lake-shores, river-banks, ditches, meadows and pastures; abundant almost throughout.

First record: Foot (1864). Burren: 'to be met with in every direction'.

Rather scarce in the central part of district 5, and even more so in the north and west of district 6. In W. Connemara, however, it makes a notable display in damp meadows with *Lythrum salicaria*.

F. vulgaris Moench (*Spiraea filipendula*)

. 2 3 . . . .

Rocky grassland and limestone pavement; very local.

2–3. Very frequent and locally abundant over a limited area W. of Gort. Southwards it extends to L. Bunny and the northern slopes of Mullaghmore; westwards to near Glencolumbkille Ho.; northwards to Funshinmore; eastwards to within 1½ km of Gort. Outside this area (which is about 11 km square) it is recorded for Ardrahan (FWI; no precise station is mentioned), which lies some 8 km to the north-east, but the station may be outside the Flora region.

First record: K'Eogh (1735). 'It grows...wild in the barony of Burrin.'

Unknown elsewhere in Ireland; the record for Portumna (Moore, 1845) has never been confirmed, and is probably best discounted.

Rubus

Rubus idaeus L.   Wild raspberry   *Subh craobh*

1 2 3 4 5 6 7 8   . .

Hedges, scrub and wood-margins. Frequent near the W. shore of L. Corrib; occasional elsewhere.

1. Locally frequent in the south. **2.** Glen of Clab. Scrub near Carran, and to the south-east of it. Wood-margin at Clooncoose, **3.** Woods at Coole. **4.** Frequent around Ross L. and in the country N. and N.E. of it. **5.** W. of Rosscahill. *Spiddal* (Atlas). **6.** Gortmore, W. of Screeb. **7.** N. of Dawros Br. **8.** Frequent from Cong westwards to Maam, and thence southwards to the lake-shore opposite the Hill of Doon.

First record: Foot (1864). 'Widely and plentifully spread all over Burren.'

R. saxatilis L.   Stone bramble

. 2 3 4 . . (7) (8)   A .

Very frequent on limestone pavement; occasional on other types of rocky ground or in stony scrub or hedges.

2–4. Very frequent throughout. (7). *Muckanaght* (ITB). (8). *Abundant on the limestone shore of L. Mask* (Marshall & Shoolbred, 1896). Sparingly near the waterfall at the Maumeen gap, 1981 (Roden).

First record: Wade (1802). 'Among the stones on the sides of the mountains,

Cunnamara, on Bilberry Island, Lough Corib.' His later record (1804) is perhaps more reliable: 'The commons of Killinaboy, barony of Burren, afford it in great profusion.'

R. caesius L.   Dewberry

1 2 3 4 . . . 8   A .

Limestone rocks, especially around lakes and turloughs; very rare elsewhere.

1. Roadside E. of L. Raha. **2**. Occasional; locally abundant on Aran. **3**. Locally abundant. **4**. On an island in the S.W. part of L. Corrib. **8**. Abundant on limestone by the S. shore of L. Mask.

First record: Ogilby (1845). Abundant on Inishmore.

R. fruticosus (agg.)   Bramble, blackberry   *Dris*

Brambles are common throughout most of the region, being scarce only in areas of uninterrupted limestone pavement or blanket bog. They have received less attention in Ireland than have other critical groups, but at our invitation Mr E. S. Edees came over in 1970 to survey the brambles of our region. He was able to name 17 species with some confidence, and added one more from Marshall's specimens in **CGE**, but he writes: 'I can find names for only two-thirds of the brambles I saw. Some of those which I was unable to name seem distinct enough to be described as new species, but it would be unwise to give them names until the *Rubus* flora of the whole of Ireland has been more thoroughly investigated.'

The only other extensive list of brambles from the region is that of Marshall & Shoolbred (1896). Several of their records have been confirmed by Edees, and some others are accepted by Watson (1958). A fair number, however, were given names which are nowadays considered to be either erroneous or ambiguous, and there seems to be little point in reprinting these. The list that follows, therefore, is restricted to names given by Edees, or accepted by Watson for H16, with two or three others recorded by Praeger which seem to have a fair probability of being correct by modern standards. All entries in roman type refer to species seen by Edees in 1970; for records before 1959, in italic type, an authority is given.

Rubus plicatus Weihe & Nees. **8**. Dooros peninsula, and near Maam Br.

R. purpureicaulis Watson. **4**. Roadside E. of Ross L. This is one of the 'Triviales' (*R. corylifolius* agg.), hybrids between *R. caesius* and other species, which are for the most part impossible to name. They are very common in the Burren and Aran, and are probably to be found in smaller quantity elsewhere. Edees says of another specimen, from Deelin Beg (district 2), 'close to *R. sublustris* Lees'.

R. selmeri Lindeb. **7**. Ballinaboy. **8**. Frequent around Cong and Clonbur, and occasional by the S.W. corner of L. Mask and the N.W. part of L. Corrib.

R. lindleianus Lees. **1**. By the Dealagh R., 2½ km below Kilshanny. (**8**). *S. of L. Mask* (Marshall & Shoolbred, 1896).

R. amplificatus Lees. **1**. Frequent between Lisdoonvarna and Lehinch. **2**. Deelin Beg. **5**. Near Moycullen, on the road to Spiddal.

R. pyramidalis Kalt. **6**. 3 km W. of Screeb. **8**. S. of Killateeaun, and between Ferry Br. and Clonbur.

R. mollisimus Rogers. (*2*). *Inishmore* (Praeger, 1895c). (*6*) *St Macdara's I.* (Praeger, 1895c). (*7*). *Inishbofin* (Praeger, 1911).

R. iricus Rogers. **1**. Frequent from Lisdoonvarna southwards to Ennistymon and Liscannor. **2**. Poulsallagh. Black Head. Deelin Beg. Turlough village. Near Glensleade castle. **3**. W. of Boston. **7**. Ballyconneely. Ballinaboy. Between Letterfrack and Kylemore. **8**. Frequent around the S.W. part of L. Mask and the N.W. part of L. Corrib, and thence westwards to Maam and Leenane.

Apparently endemic to Ireland, mainly in the west. 'One of the most beautiful of all brambles, with its ample panicle of richly coloured flowers' (Edees).

R. polyanthemus Lindeb. **1**. Frequent near the coast. **2**. Near L. Aleenaun. S.W. of Glensleade castle. **3**. Rinroe. **5**. Near Moycullen. W. of Oughterard. **7**. N. of Roundstone. Ballyconneely. Between Letterfrack and Kylemore. **8**. W. and S.E. of Ferry Br. Dooros peninsula. N.W. of Oughterard. *Maam*, 1895 (**CGE**, det. Edees).

R. cardiophyllus Muell. & Lef. **7**. S. of Ballynahinch L. Ballinaboy. **8**. Maam. S. of the 'Narrow Lake'. Dooros peninsula.

R. dumnoniensis Bab. **8**. Dooros peninsula. *Maam*, 1895 (**CGE**, det. Edees).

R. ulmifolius Schott. Present in all districts; very abundant on the limestone, and frequent near the coast elsewhere; also along many of the main roads. In some places, as around Kinvara and Gort, almost the only species present.

R. chlorothyrsus Focke. (*8*). *Maam Br*. (Simpson, 1936).

R. hesperius Rogers. **1**. Halfway between Lisdoonvarna and the Cliffs of Moher. **5**. Near Galway. S.W. of Moycullen. **6**. Near Screeb. **7**. S. of Ballynahinch L. N. of Roundstone. Ballinaboy. Between Letterfrack and Kylemore. **8**. Frequent in the east.

Endemic to Ireland. 'Not as showy as *R. iricus* nor perhaps as widely distributed, but one of the commonest brambles in Connemara' (Edees).

R. hebecaulis Sudre. **1**. W. of Lisdoonvarna. S. of Kilshanny. **7**. Ballinaboy. Between Letterfrack and Kylemore. **8**. Dooros peninsula. W. of Cornamona.

R. vestitus Weihe & Nees. **2**. Deelin Beg. **4**. Oranmore. Frequent near Galway. By Ross L. **5**. E. and N.E. of Spiddal. **6**. Screeb. **8**. Frequent around the N.W. part of L. Corrib.

R. wirralensis Newton. **2**. In two places between Kilfenora and Ballyvaughan.

R. cinerosus Rogers. (*8*). *Near Clonbur*, 1896 (ITB); accepted by Watson for H16.

R. dunensis Rogers. (*7*). Inishbofin (Praeger, 1911).

R. fuscus (agg.). **7**. S. of Ballynahinch L. (*8*). *Abundant E. of Coolin L.*, 1895 (**CGE**, det. Edees). Edees regards this as an aggregate of species which have not yet been sorted out.

R. acutipetalus Lef. & Muell. *Given by Watson for H16*.

R. pallidus Weihe & Nees. **8**. Between Ferry Br. and Clonbur.

R. scaber Weihe & Nees. **7**. S. of Ballynahinch L. **8**. Dooros peninsula. *5 km S.W. of Cong*, 1895 (**CGE**, det. Edees).

R. vectensis Wats. (*8*). *Maam*, 1895 (**CGE**, det. Edees).

R. koehleri Weihe. (*8*). *Near Clonbur* (Marshall & Shoolbred, 1896). Accepted by Watson.

R. hypochlorus (Sudre) Bouv. (*8*). *Oughterard* (Marshall & Shoolbred, 1896, as atypical *R. koehleri*). Accepted by Watson.

Dryas

Dryas octopetala L.   Mountain avens   *Leathín*

. **2 3 4 5** . . **8** . .

Limestone pavement and calcareous gravels; very rarely on mountain cliffs. Abundant in most of the Burren; very rare elsewhere (Fig. 8).

2. Widely distributed and locally very abundant (as on the slopes immediately E. of Black Head), but absent from some fair-sized areas. Not on Aran. **3**. Local; mainly in the west, but extending north-eastwards to half way between Ardrahan and L. Fingall. **4**. Abundant on a moraine S. of Carrowmoreknock. **5**. In fair quantity on Gentian Hill. **8**. Rather sparingly on magnesium-rich schist on Lissoughter at *c*. 300 m (1000 ft).

First record: How (1650). 'In the mountaines betwixt Gort and Galloway. *Mr. Heaton.*'

This species, together with *Gentiana verna*, forms the main basis of the Burren's botanical fame. It grows with a profusion unequalled in N.W. Europe, and down almost to sea-level; the nearest point at which this can be paralleled is on the north coast of Scotland, more than 5 degrees of latitude north of the Burren. Its effectiveness as a spectacle is enhanced by the fact that in addition to its main flowering season in late May it produces a second crop of flowers in August; moreover the feathery styles of the fruiting heads are almost as conspicuous as the flowers.

Its enormous abundance in some parts of the Burren makes its absence from other, apparently similar areas hard to understand. It avoids deep, clayey drift, but is equally at home on porous drift and on solid rock, on north- or south-facing slopes, at sea-level and at 300 m (1000 ft). Its absence from Aran, where almost all the other rarities of the Burren are to be found, is another fact that admits of no easy explanation.

Geum

Geum urbanum L.   Wood avens   *Machall coille*

1 2 3 4 5 6 7 8   A .

Woods, scrub, hedges and other shady places; widespread and locally frequent.

1. Occasional. **2**. L. Aleenaun. S.W. of Ballyvaughan. Abundant in Poulavallan. Frequent in Aran, mainly along the small lanes. **3–4**. Frequent. **5**. Woods at Spiddal. Frequent along the N.E. margin. **6**. In scrub at Flannery Br., and at the N.E. corner of Glendollagh L. Woods at Cashel Ho. **7**. Woods at Ballynahinch and Letterfrack. **8**. Frequent.

First record: Foot (1864). Burren: 'at all elevations and in all situations'.

G. rivale L.   Water avens   *Machall uisce*

1 2 3 4 . . **7** .   . .

Scrub, wood-margins, limestone pavement and rock-ledges. Occasional to frequent on the limestone; very rare elsewhere.

1. Very luxuriant at the edge of a forestry plantation west of the top of the Corkscrew Hill. **2**. Occasional to frequent on the pavement throughout. **3**. Wood-margins at Garryland and Coole. **4**. Frequent between Ross L. and L. Corrib. **7**. North-facing cliff on Muckanaght.

First record: Foot (1864). Burren: 'at all elevations and in the most exposed and dry places'. The hybrid with *G. urbanum* (*G.* × *intermedium* Ehrh.) was, however, recorded by More (1855) from Coole. The records cited above for *G. rivale* include those for the hybrid, as it is very difficult to draw a line between them.

Fragaria

Fragaria vesca L.   Wild strawberry   *Subh talún*

1 2 3 4 5 6 7 8   A (*B*)

Woods, scrub, roadsides and rocky ground; abundant in the Burren and near L. Corrib; occasional to frequent elsewhere.

1. Hedges and scrub S. of Lickeen L. Rocky ground W. of Killinaboy. **2–3**. Abundant. **4**. Abundant in the north; not seen around Galway Bay. **5**. Frequent. **6**. On the old railway-track at Recess. Scrub near Flannery Br. Stony lake-shore N.E. of Carraroe. **7**. Abundant in Ballynahinch woods. Kylemore. Shore of Derryclare L. **8**. Very frequent except along the western margin.

First record: Foot (1864). Burren: 'Nowhere in Britain probably does the strawberry grow in such plenty or attain to such size as it does in this rocky garden of nature.'

Easily confused, when not in fruit, with the barren strawberry (*Potentilla sterilis*). The true strawberry, however, flowers about a month later, with broader, overlapping petals, and the hairs on the underside of the leaf are silky and pressed close against the surface, whereas in the *Potentilla* they stand straight out.

A fair-sized colony of **F. × ananassa* Duch., the garden strawberry, was seen by DW in 1972 by the roadside in Inishmore, a short distance N.W. of Kilronan. It is perhaps a little too soon to say whether it has established itself as a permanent member of the flora.

Potentilla

Potentilla anserina L.   Silverweed   *Briosclán*

1 2 3 4 5 6 7 8   A B

Roadsides, sand-dunes, cultivated and waste ground, marshes, and periodically flooded grassland. Abundant throughout.

First record: Wright (1871). Aran Islands.

At the bottom of some of the shallower turloughs, and about half-way down the zonation in the deeper ones, this species dominates the sward to the almost total exclusion of other plants. In such situations the silvery hairs on the leaves are fewer than usual.

P. erecta (L.) Raüsch   Tormentil   *Nealfheartach*

1 2 3 4 5 6 7 8   A B

Heaths, bogs, pastures, meadows and limestone pavement. Abundant throughout.

First record: Babington (1836). Bogs between Clifden and Roundstone.

Like many other species with a weak calcifuge tendency, it avoids deep calcareous drift, but grows freely on the bare limestone.

P. anglica Laich (*P. procumbens*)

1 2 3 4 5 6 7 8   A B

Roadsides, banks, rough grazing and river-sides. Frequent on acid soils; rather rare on the limestone.

1. Frequent. **2.** Occasional in the east; not seen elsewhere on the mainland, but present on Inishmore. **3.** Roadside N. of Boston. **4.** Occasional, mainly in the north. **5–7.** Frequent. **8.** Occasional.

First record: More (1876). Inishbofin.

P. anglica × erecta (*P. suberecta* Zimm.) was recorded from Clonbur by Marshall & Shoolbred (1896), and probably occurs elsewhere.

P. reptans L.   Creeping cinquefoil   *Cúig-mhéarach*

1 2 3 4 5 . 7 8   A [B]

Roadsides, limestone pavement, pastures and mountain rocks. Frequent on the lowland limestones, especially around Galway Bay; rare elsewhere.

1. Pasture by Elmvale Ho. **2.** Limestone pavement near Turlough village; frequent and locally abundant on Aran. **3.** Very frequent, especially in the north. **4.** Occasional to frequent. **5.** Pastures W. of Salthill. **7.** On rocks and grassland towards the W. end of the Slyne Head peninsula. **8.** Roadside near Leenane.

First record: Wright (1871). Aran Islands.

The records in the *Atlas* for Inishbofin, and the squares centred on Lisdoonvarna, Spiddal, the Twelve Pins and Letterfrack have not been confirmed, and are probably errors.

P. fruticosa L.

. **2 3** (*4*) . . . . . .

Rocky or grassy ground subject to occasional flooding; very local (Plate 3).

2. Abundant in a depression 1½ km S.W. of Ballyvaughan. In fair quantity in a flushed valley with an intermittent stream after rain, on the W.S.W. side of Gleninagh Mt. A few bushes at the S. end of the Carran turlough. **3.** Frequent at the upper limit of flooding around lakes and turloughs from near Killinaboy north-eastwards to Castle L. and the N. end of L. Bunny. Sparingly by the N. end of L. George. (*4*). *By L. Corrib, a short distance S. of Oughterard* (Wade, 1802); often searched for here, but apparently extinct.

First record: Wade (1802). 'Amongst the rocks, on the banks by L. Corib, behind Sir John O'Flaherty's Lemonfield, Oughterard.' First confirmed record: Wade (1804). 'Abundantly in the Barony of Burren, the S.E. side, where it joins the barony of Inchiquin.'

The only stations in Ireland outside the limits of this *Flora* are on the E. shore of L. Corrib and one or two adjacent islands. In Europe the distribution of this species is remarkably discontinuous, and the Burren populations are probably the largest in W. Europe. In Britain it occurs in small quantity in the Lake District and in Upper Teesdale; otherwise its closest stations are in the Pyrenees, the Maritime Alps and S.E. Sweden.

Foot (1864) stated that it has two flowering seasons (late spring and early autumn), and that the later flowers are twice as big as the earlier. This has been repeated by several other authors, but we cannot confirm it. In our opinion it flowers fairly continuously throughout the summer, without any great variation in flower-size.

P. palustris (L.) Scop. (*Comarum palustre*)   Marsh cinquefoil

1 . 3 4 5 6 7 8   . B

Marshes, reed-swamps, fens, ditches, bogs and lake-shores. Frequent over much of the region, but rather local.

1. Frequent. **3–4.** Occasional to frequent. **5.** Occasional in the east; not seen in the west. **6.** Occasional. **7.** Frequent in the west; not seen in the mountains. **8.** Very frequent.

First record: Babington (1836). Oughterard.

A plant whose ecological requirements are not easy to assess. It is to be found both in acid and in highly calcareous waters, but is by no means ubiquitous.

P. sterilis (L.) Garcke   Barren strawberry

1 2 3 4 5 . 7 8   A .

Hedges, banks, wood-margins, scrub, rough grazing and rocks. Very frequent in the Burren; rare in S. and W. Connemara; occasional to frequent elsewhere.

1. Occasional. **2–3.** Very frequent. **4.** Frequent. **5.** Occasional. **7.** Ballynahinch woods. Roadside on N. bank of Clifden estuary. **8.** Very frequent.

First record: ITB (1901). H16.

ROSACEAE

Aphanes   Parsley-piert

Aphanes arvensis L.

1 2 3 4 5 . . 8   A .

Roadsides, walls, rocks and bare patches in pasture. Frequent on the limestone; rare (though perhaps to some extent overlooked) elsewhere.

1. Rocks by O'Dea's Castle and Elmvale Ho. **2–4**. Frequent. **5**. Roadside near Spiddal. **8**. Limestone rocks near Cong.

First record: Crit. Suppl. (1968). Earlier records are ambiguous.

A. microcarpa (Boiss. & Reut.) Rothm.

. . . . 5 6 7 8   . B

Habitats similar to those of the preceding species, but only on siliceous rocks and soils. Occasional and locally frequent.

5. Occasional, mainly in the west. **6**. Frequent around Costelloe and Screeb; rather rare elsewhere. **7**. Roadside near L. Conga. Roundstone. Renvyle. Near L. Muck. Inishbofin. **8**. Frequent near Maam; rare elsewhere.

First record: Walters (1949*b*). Roundstone.

Alchemilla   Lady's mantle

Our records for this genus are clearly incomplete, partly because we have thought it best to ignore records published before the revision by Walters (1949*a*); even when a varietal name is appended to *A. vulgaris* the identity of the plant is subject to some doubt. We hope, however, that our general picture of the distribution of the 3 species is correct, and there is little doubt that all species are more sparsely represented than might have been expected from the nature of the terrain.

Alchemilla glabra Neyg.

1 2 . . 5 . 7 8   . .

Meadows, flushes and mountain cliffs; occasional.

1. Meadow near Ballynalackan Ho. **2**. Meadow at Castletown. Meadow on Turlough Hill. S. of Ballyvaughan. **5**. Scrub-covered hillside S. of Oughterard. **7**. In the ravine by the S.W. side of L. Fee. Muckanaght. Cregg Mt. **8**. Flush on the hill W.S.W. of Leenane. Cliffs S. of the W. end of L. Nafooey, and in the valley 2 km further south.

First record: Webb (1952). Ballyvaughan and Muckanaght.

A. xanthochlora Rothm.

1 2 . 4 . . 7 (*8*)   . .

Grassland and scrub. Very frequent in parts of the Burren; rare elsewhere.

1. N.W. of Inagh. S. of Lickeen L. *Lisdoonvarna*, 1933 (**DBN**). **2**. Very frequent almost throughout, but not on Aran. **4**. By the W. shore of Ballycuirke L. **7**. Benlettery. Grassland near Recess. (*8*). *Near L. Nafooey* (Atlas).

First record: Webb (1952). Benlettery.

A. filicaulis Buser (*A. vestita*)

1 2 3 4 . . 7 8 A .

Meadows, pastures, scrub-margin, mountain cliffs and limestone pavement. Fairly frequent on the limestone; rare elsewhere, and mainly on mountains.

1. Meadow near Elmvale Ho. **2**. Frequent, but rather rare on Aran. **3**. Frequent in the south; not seen north of Gort. **4**. Frequent in the north, but not seen S. of Ballycuirke L. **7**. Cliffs on Muckanaght and Bengower. **8**. Lissoughter. Maumeen gap in Maumturk range. Valley W. of Cornamona. Near Clonbur.

First record: Marshall & Shoolbred (1896). 'In a pasture near Clonbur sparingly.'

All specimens are referable to subsp. *vestita* (Buser) Bradshaw.

Agrimonia   Agrimony

Agrimonia eupatoria L.

1 2 3 4 5 . 7 8 A .

Roadsides and rough grazing. Very frequent on the limestone; rare elsewhere.

1. Hedge by Moy Ho. **2–4**. Very frequent, but only occasional on Aran. **5**. In several places near the coast between Spiddal and Inveran. **7**. Near Dog's Bay, and near Aillebrack L. **8**. Frequent around Cong.

First record: Foot (1864). Burren: 'abundant everywhere'.

A. procera Wallr. (*A. odorata*)

(1) 2 3 4 5 6 7 8 . .

Roadsides and rough grazing. Locally frequent on acid soils; rare on the limestone.

(1). *Near the Cliffs of Moher* (Atlas). **2**. Near Carran. **3**. Pavement E.S.E. of Slieve Carran. **4**. W. of Ballycuirke L. **5**. W. of Spiddal. Frequent in hedges by Killaguile Ho. **6**. S.E. corner of Gorumna I., and on the mainland opposite. **7**. Frequent around Clifden and Ballyconneely. By Garraunbaun L., N. of Moyard. **8**. E. of Cornamona. Several places near Ferry Br.

First record: More (1872). 'Very fine near Clifden, Connemara.'

Although most specimens are quite distinct from *A. eupatoria*, some of those from district 8 are intermediate in several characters.

Sanguisorba

Sanguisorba minor Scop. (*Poterium sanguisorba*)   Salad burnet

. 2 3 (4) . . . . A .

Limestone pavement, pastures and roadsides; locally frequent.

2–3. Frequent throughout most of the Burren, and abundant on Aran. **(4)**. *Merlin Park*, 1899 (ITB).

First record: Oliver (1851). Inishmore.

Acaena

*Acaena ovalifolia Ruiz & Pavon

Naturalized in small quantity by a stream outside the gate of Killaguile Ho., S.S.E. of Oughterard (district 5), 1970–7.

First record: Census catalogue (1972). H16.

This species and another are naturalized near Aasleigh Ho., at the head of Killary harbour, just outside our region, in Co. Mayo; they are sometimes wrongly ascribed to Co. Galway on herbarium labels.

Rosa   Rose   *Rós*

Although this genus has not been the subject of critical study in our region it seems probable that the records here presented are reasonably complete. Our chief difficulty has lain in the *tomentosa*-group, but we believe that only one species is present in the region, and that it agrees better with *R. sherardii* than with *R. tomentosa*, to which all the early records were assigned.

Otherwise the species are not difficult to distinguish. *R. pimpinellifolia* has erect stems seldom 1 m (3 ft) high, bearing numerous fine prickles. *R. arvensis* is known by its long, trailing stems, neat foliage and pure white flowers. *R. rubiginosa* is distinct in the fruity (not resinous) smell of the crushed leaves; it has small pink flowers. *R. sherardii* has bluish-green, usually downy leaves with a resinous smell when crushed, and larger, bright pink flowers. *R. canina* has hairless leaves without scent, arching but not trailing stems, and large flowers, usually pale pink but sometimes white.

Rosa arvensis Huds.

. . **3 4** . . . . . .

Hedges and wood-margins; rather rare, and only on the limestone.

3. Hedges *c*. 3 km S. of Ballinderreen. Frequent in and around Garryland wood. *Hedge E. of Mullaghmore*, 1958 (DW). **4**. Hedges S. of Gortachalla L., and hence eastwards to L. Corrib. *Kilbeg ferry*, 1899 (ITB).

First record: ITB (1901). Kilbeg ferry.

R. pimpinellifolia L. (*R. spinosissima*)   Scotch rose, burnet rose

1 2 3 4 5 6 7 8   A B

Rocky, stony or sandy ground. Abundant in the Burren; occasional elsewhere.

1. Limestone rocks N.N.W. of Inchiquin L. **2–4**. Abundant. **5**. Dunes 1 km E. of Spiddal. Near Gentian Hill. **6**. Roadside bank immediately W. of Maam Cross. **7**. Frequent on the Slyne Head peninsula; also on Omey I. and the mainland opposite. Inishbofin. *Renvyle* (Atlas). **8**. Scree on the W. side of Lissoughter. Cliffs at the Maumeen gap (Maumturk range), and on the N. side of Benbeg. Frequent on the limestone near Cong.

First record: Sim (1859). Gentian Hill.

Characteristic especially of the 'shattered' type of limestone pavement in the Burren, where it is often co-dominant with *Teucrium scorodonia* and sometimes *Prunus spinosa*.

R. canina L.   Dog rose

1 2 3 4 5 6 7 8   A B

Hedges, thickets and streamsides; frequent on the limestone; occasional elsewhere.

1. Occasional by the coast; rare inland. **2**. Frequent. **3–4**. Very frequent. **5**. Hedge by Gentian Hill. Rocky ground N.N.E. of Screeb. **6**. Rocky ground 5 km N. of Costelloe. Near Cashel. **7**. Doonloughan. N. shore of Streamstown Bay. Roadside W. of Ballynahinch. Inishbofin. **8**. Occasional in the east, and frequent on the islands in L. Corrib; rare in the west.

First record: Hart (1875). Aran Islands.

‡**R. rubiginosa** L.

. . . **4** (*5*) . . **8** . .

Hedges and disturbed ground; rare.

4. Near the ruins of Menlough Castle. (*5*). *Railway bank W. of Galway*, 1933 (**DBN**). **8**. In a hedge near the Hill of Doon.

First record: Praeger (1934*c*). H16.

R. sherardii Davies

1 2 3 4 5 6 7 8 . .

Roadsides, hedges and rough grazing. Frequent in parts of Connemara; occasional elsewhere.

1. Near Moy Ho. E. end of Lickeen L. N. of Drumcullaun L. By Derrymore Br. **2**. Near the ruined church 6 km W. of Carran. **3**. By the S. shore of L. George. Hedges near Tirneevin. **4**. Hedges by Aughnanure Castle. **5**. Gentian Hill. Spiddal. **6**. Frequent by the coast. **7**. Occasional near the coast. **8**. In several places by L. Corrib, N.W. of Oughterard; also on the Dooros peninsula and on one of the islands.

First record: ITB (1901), (as *R. tomentosa*). H16.

Sorbus

Apart from *S. aucuparia*, which is quite distinct, the remaining species of this genus, constituting the whitebeams, present considerable difficulty in their identification. The commonest, *S. aria*, has ovate to elliptical leaves, very white beneath, irregularly and usually deeply toothed; the fruit is longer than wide, with lenticels evenly distributed over it. *S. hibernica* differs mainly in the fine, regular toothing of the leaves, which are at first white beneath, but by midsummer take on a somewhat greyish or greenish tint. Its fruit is at least as wide as long, with lenticels mainly in the lower half, and it ripens 3 weeks earlier than that of *S. aria*. *S. rupicola* differs from both the above in having narrowly obovate leaves, broadest well above the middle, and with the basal third almost without teeth. The underside is almost as white as in *S. aria*. The fruit is wider than long, with scattered lenticels. Finally, the alien *S. intermedia* (native of Scandinavia and the Baltic region) has slightly lobed leaves, greyish-yellow beneath, and fruit considerably longer than wide, without conspicuous lenticels.

It should be noted that descriptions of leaves refer to those on spur-shoots; those on leading or coppice-shoots are very variable in shape.

Sorbus aucuparia L.   Rowan, Mountain ash   *Caorthann*

1 2 3 4 5 6 7 8   . (*B*)

Woods, scrub, rocky ground, mountain cliffs and ravines, and on lake-islands. Occasional to frequent in Connemara; rare elsewhere.

1. Rare; seen only at the base of Clifden Hill. **2**. Occasional in the north, and frequent in Poulavallan; in the south seen only as seedlings. **3**. 3 km E. of Kinvara. 2 km N. of Boston. **4**. Occasional in the north; not seen S. of Moycullen. **5–6**. Frequent. **7**. Abundant on lake-islands; occasional elsewhere, apart from exposed coastal areas. *Inishbofin* (Praeger, 1911). **8**. Very frequent.

First record: Wade (1802). 'On the mountains and hanging from the crevices of rocks, Cunnemara, but not common.'

A variant with orange-yellow fruits was seen by Ostenfeld near Roundstone in 1911. In 1969 several trees on and S.W. of Mullaghmore were noted as having very silvery-hairy leaflets; they are probably referable to var. *lanuginosa* (Kit.) Beck.

Sorbus hibernica Warb. (*S. porrigens*)   Irish whitebeam

. . **3 4 5** . **7** (*8*)   . .

Hedges, copses and woods; rather rare.

3. Occasional around Gort and Coole, and westwards to Kilmacduagh. **4**. S. of Oranmore. Near Merlin Park. **5**. One large tree near the old railway S. of Oughterard, and another by the river near the waterfall. In a partly felled wood on the hillside 2½ km S. of Oughterard. One tree near the road S.E. of L. Bofin. **7**. *Ballynahinch* (*locus classicus*), 1938 (**BM**, Warburg). One tree in the graveyard at Roundstone, and another in a small *Salix*-copse near the tip of the Errislannan peninsula. (*8*). *Near Clonbur* (Marshall & Shoolbred, 1896; as *Pyrus aria*).

First record (as *S. porrigens*): Pugsley (1934). Garryland wood.

Endemic to Ireland. Formerly confused with *S. porrigens* Hedl., from Greece and Turkey, and later with *S. porrigentiformis* Warb., endemic to Britain.

†S. aria (L.). Crantz   Common whitebeam

1 2 3 4 5 6 . .   . .

Woods, copses and limestone pavement. Locally abundant near Galway and in the N.E. part of the Burren; rare elsewhere.

1. By O'Dea's Castle. **2**. Glasgeivnagh Hill. **3**. Frequent and locally abundant in the north, especially near Ballinderreen, where a large area of limestone pavement appears to have been recently colonized; not seen S. of L. Bunny. **4**. Occasional S. of Oranmore. Frequent and locally abundant E. of Galway. A few trees E. of Ross L. **5**. A few trees near Barna Ho. **6**. Two large trees half smothered in a conifer plantation across the road from the main gate of Ballynahinch Castle.

First record (in restricted sense): Praeger (1934*b*). 'Limestone rocks in railway cutting 3 miles E. of Galway.'

The status of this species is hard to assess. It has usually been accepted as native, but the large population E. of Galway could easily have been derived from trees planted in the grounds of Merlin Park; this is all the more likely in view of the fact that the *S. intermedia* which accompanies it almost certainly derives from this source.

The large population near Ballinderreen gives the impression of having been bird-sown fairly recently, perhaps from a few trees planted beside the road. On the other hand some of the stations of this species in Co. Sligo are remote from any obvious source of introduction. The balance of evidence seems to be slightly in favour of native status.

S. rupicola (Syme) Hedl.

. . . . . (6) 7 . . .

Rocky lake-islands; very rare.

(6). *Island in the larger Loughanillaun* (= *Gowla L.*) *5 km E.S.E. of Cashel*, 1941 (Praeger, 1939, 1946). **7**. Island near the N. end of Derryclare L., 1961 (**TCD**).

First record: Praeger (1939). Island in Loughanillaun.

*S. intermedia (Ehrh.) Pers.

. . . **4** . . . .

4. Naturalized in some quantity on roadsides and embankments *c*. 5 km E. of Galway, where the railway crosses the former main road to Oranmore.

First record: Webb (1979).

Malus   Apple

Our records of this genus are somewhat defective, on account of the confusion which persisted until lately as to the nature of the two taxa represented. The cultivated apple (which frequently becomes naturalized) was formerly treated as a subspecies of the wild crab (*M. sylvestris*); it is now agreed, however, that it is a complex hybrid, possibly including among its parentage *M. sylvestris*, but other species as well. We have excluded those records which we cannot assign confidently to one species or the other. They are not hard to distinguish: *M. sylvestris* has small, hairless leaves, usually thorny branches, and the sepals hairless outside. *M. domestica* has much larger, more or less downy leaves and no thorns; the sepals are downy outside. We have seen one tree (in district 8, 3 km S.E. of Maam) which appeared intermediate; it is perhaps a secondary hybrid.

Malus sylvestris Miller   Crab apple   *Ubhall fhiadhain*

. . **3 4** . **6** . **8** . .

In or near old-established woods; rarely in hedges. Occasional.

3. Frequent in Garryland wood. By the fen S.W. of Mullaghmore. **4**. Occasional between Ross L. and L. Corrib. **6**. A small bush in a hedge E. of Cashel. **8**. In the woods at Gortdrisha, N.W. of Oughterard, and also near L. Corrib 2 km further to the N.W. On several islands in L. Corrib.

First record: Webb (1982). Earlier records are ambiguous.

*M. domestica Borkh. (*M. sylvestris*, subsp. *mitis*)

. **2 3 4** . . . . **A** .

Hedges and rocky ground. Occasional on the limestone; not seen elsewhere.

2. Roadside S. of Bealaclugga. One small tree on Inishmaan. **3**. Hedge W. of Kinvara. **4**. By the railway 2 km E. of Galway.

First record: Praeger (1902*b*), as *Pyrus Malus*, var. *mitis*. H16.

Crataegus

Crataegus monogyna Jacq.　　Hawthorn, whitethorn　　*Sceach geal*

1 2 3 4 5 6 7 8　A .

Woods, scrub, hedges, limestone pavement and neglected pastures; abundant.

Least common in districts 1 and 6, where exposure is greatest, and absent from Inishbofin.

First record: Hart (1875). Aran Islands.

Very variable; the variation probably deserves closer study. Some of it is suggestive of slight introgression from *C. oxyacanthoides*, although this species, which is occasionally planted and perhaps naturalized in eastern Ireland, has not been observed in the Flora region.

Cotoneaster

*Cotoneaster microphyllus Lindl.

1 2 . 4 5 6 7 8 . .

Banks, walls, rocks and heathy ground; occasional to frequent, and still increasing.

1. By Drumcullaun L. **2.** On the summit of Carnsefin (about Black Head); here at least since 1949. On rocks S. of Deelin Beg. On a wall at New Quay. **4.** In great abundance on railway cuttings east of Galway. In several places around Ross L. **5.** On a wall at Spiddal. Heathy ground by the road half-way between Spiddal and Moycullen. By the road 4 km W. of Oughterard, and by the old railway-track a little further west. **6.** Frequent around Cashel, Ballynahinch and Toombeola. Roadside immediately W. of Maam Cross. **7.** In several places around Toombeola, Roundstone and Clifden. **8.** Near Clonbur. On an island in L. Corrib N. of Oughterard.

First record: Praeger (1934c). Limestone rocks near Oughterard.

*C. simonsii Baker

. . . . . 6 7 8 . .

Heathy ground and in scrub; locally frequent in Connemara.

6. Very frequent at Cashel. Roadside S.W. of Canal Br. On an island in L. Athry. **7.** Abundant around Ballinaboy. On the Errislannan peninsula. Wood-margin at Derreen. N.W. of Moyard. **8.** On several islands in L. Corrib N. of Oughterard.

First record: Druce (1911). Near Clifden.

Prunus

Prunus spinosa L.　　Blackthorn　　*Draighean*

1 2 3 4 5 6 7 8　A B

Hedges, thickets, undergrazed pastures and limestone pavement. Abundant throughout the Burren, and frequent elsewhere on the limestone and on the Clare shales. Locally frequent in Connemara, as around Screeb and Clifden and by L. Corrib; elsewhere no more than occasional.

First record: Wright (1871), as *P. communis*. Aran Islands. 'It is curious to find small trees of this species in the clefts between the limestone rocks; they sometimes grow out from between the rocks, but the heavy winds in the winter keep them on a level with the surface of the ground.'

*P. domestica L.   Bullace, Wild damson

1 . **3 4** . . . (*8*) . .

Hedges and scrub; rare.

1. Roadside below Clifden Hill. **3**. Dromore woods. **4**. Hedge W. of Ballycuirke L. Scrub near Burnthouse. (*8*). *Bilberry I., in L. Corrib* (Wade, 1802).

First record: Wade (1802). Bilberry Island, L. Corrib.

All records seem to be referable to subsp. *insititia* (L.) Schneider. We have rejected old records of '*P. communis*', as in most cases they refer to *P. spinosa*.

P. avium L.   Wild cherry   *Crann silín*

1 . (*3*) . **5 6** . **8** . .

Woods; rather rare.

1. Frequent on Clifden Hill. (*3*). *Near Gort* (Atlas). This probably refers to the woods at Coole. **5**. A few trees near the edge of the wood N.W. of Moycullen, 1968; perhaps felled now. **6**. Several trees on an island in Athry L. **8**. On two islands in L. Corrib N. of Oughterard.

First record: Praeger (1906). 'Near Moycullen, among native shrubs, not planted.'

*P. cerasus L.

1 . **3 4** . . **7 8** . .

Hedges; rare.

1. Near Ballynalackan Castle. **3**. Near Gort. **4**. 1 km S. of Rosscahill. **7**. Surviving by a ruined cottage near Clifden. Well naturalized by a stream 5 km N. of Clifden. **8**. Clonbur.

First record: Marshall & Shoolbred (1896). 'In hedges near Clonbur; doubtless introduced.'

It is rather difficult with this species to draw the line between naturalized plants and mere planted hedges. The stations cited above refer either to places where suckering has brought it to some considerable distance from the original site of planting, or to single bushes in hedges of other shrubs, which probably represent dispersal of the seed by birds.

*P. laurocerasus L.   Cherry laurel

. . **3** . **5** . . **8** . .

Woods and plantations, and on one lake-island; very local.

3. Abundant in the woods at Coole. **5**. Moycullen wood. **8**. Wood near Teernakill Br. Several small bushes on an island in L. Corrib S. of Cornamona.

First record: Atlas (1962).

The stations cited are those in which the plant is clearly naturalized and has spread beyond its intended limits in mainly native woodland. This spread is normally achieved mainly by layering, but the existence of several bushes on the island in L. Corrib (accompanied by saplings of sycamore and rhododendron) shows that spread by seed also takes place.

There is in **DBN** a specimen of *P. padus* L., the bird cherry, collected by R. D. O'Brien in 1909 from 'Moy Glen, near Lehinch'. The occurrence of this species as a native in W. Clare is just possible, but it seems much more likely that the specimen came from a planted tree in the woods of Glenville Ho. There are no other records for near the W. coast except in Co. Donegal.

SAXIFRAGACEAE

Escallonia

*Escallonia macrantha Hook. & Arn.

. . . . . . 7 . . .

Rocky ground by the sea; very rare.

7. Naturalized in two places on the shore of Ballynakill harbour; also at Ardbear Bay.

First record: Druce (1911). Near Clifden.

Ribes

*Ribes nigrum L.   Black currant

. . . **4 5** . . . . .

Hedges and fields; rare.

4. One large bush in a dense thicket of *Prunus spinosa* near the N. shore of Ross L., 1969. **5.** Naturalized in great abundance in a neglected pasture near Spiddal Ho., 1977.

First record: Webb (1982).

Both stations are within a few hundred metres of the garden from which the seed had presumably been distributed by birds.

*R. uva-crispa L.   Gooseberry

. **2 3 4 5** . 7 (8) . .

2. A few bushes by Ballyvaughan harbour. By a ruined cottage near L. Aleenaun; some young bushes were seen, as well as the surviving planted ones. **3.** In a field 3 km W. of Kinvara, and in a wall half-way between Kinvara and Ardrahan. **4.** In several places near ruins on the N. side of Ross L. In hedges near Aughnanure Castle. By the bank of the R. Corrib, 3 km above Galway. **5.** In the wood N.W. of Moycullen. **7.** One bush by a wall N.E. of Ballyconneely. (**8**). *About the Pigeon-hole, Cong* (Baily, 1833).

First record: Baily (1833). Cong.

Seldom far from gardens, but well naturalized and tending to increase.

Saxifraga

Saxifraga oppositifolia L.

. . . . . . **7 8** . .

Mountain cliffs, talus and rocky outcrops; rather rare.

7. In fair quantity on the north-facing cliff on Muckanaght at 550–575 m (1800–1900 ft); also at 360–400 m (1200–1350 ft) on the N.W. side of Bengower. *Abundant on Kylemore Mountain*, 1897 (CH2). **8.** Plentiful on a rocky ridge and the talus at its base on the W. side of Lissoughter, mostly at *c.* 270 m (900 ft), but recorded as low as 135 m (450 ft). Frequent on cliffs N. of the Maumeen gap on the Maumturk Mts.

First record: Mackay (1806). 'On the mountains in Joyce country, near Lough Corrib.'

It is remarkable that this arctic–alpine species should descend to so low an altitude on Lissoughter, its most southerly station in Ireland.

S. stellaris L.

. . . . . . **7 8** . .

Mountain cliffs, damp screes and by mountain streams; occasional.

7. On the N. face of Muckanaght, and on the talus below it. By the Gleninagh R., on the N.E. side of the Twelve Pins. **8.** On the cliffs above L. Nadirkmore; also further to the east on the cliffs above the smaller lough. On the mountain immediately S. of Leenane, and on the S.W. side of Devilsmother, near the summit. On the N. cliffs of Lugnabrick and in the north-facing corrie of Benbeg.

First record: Babington (1836). Mountains near Maam.

As was first noted by Colgan (1900), a curious variant of this species grows in fair abundance by the Gleninagh R. at an altitude of about 225 m (750 ft). It is largely viviparous, most of the flowers being replaced by leafy buds, and it retains this character in cultivation. A very similar form is known from the Austrian Alps, but it seems fairly certain that the viviparous character has arisen separately in Ireland and Austria. For a fuller discussion see Webb (1963*a*).

S. spathularis Brot. (*S. umbrosa*)   St Patrick's cabbage   *Cabáiste Phádhraic*

. . . . **5 6 7 8** . **B**

Woods, streamsides, shady rocks and banks and mountain cliffs. Widespread in Connemara, and abundant in much of the north and east; rarer in the south and west (Plate 2).

5. Rocks by the E. end of L. Formoyle, and S.E. of Lettercraffroe L. By a stream 6 km W. of Rosscahill. Rocks 3 km E.S.E. of Screeb. *Spiddal* (Atlas). **6.** Shady rocks on Cashel Mt. **7.** Frequent in the north; rather rare in the south. **8.** Very frequent to abundant throughout, except on the Cong–Clonbur limestones.

First record: Wade (1802), as *S. umbrosa*. 'Very common throughout Cunnamara.'

It will be seen from the above that Wade's statement, repeated by many later authors, requires some qualification. Although it grows in a wide variety of habitats it is by no means ubiquitous, requiring shade or some degree of shelter, or else (as on mountain-tops) frequent mist and rain. It is remarkable that it has not been recorded from any of the lake-islands, despite the presence of well-developed woodland on many of them.

The *Atlas* record for 94/21 (S.E. of Oughterard) seems to be an error.

S. spathularis is, together with *Erica mackaiana*, a Hiberno–Cantabrian species in the strict sense of the word, its distribution outside Ireland being confined to N.W. Spain and N. Portugal. It was for long confused with the Pyrenean *S. umbrosa*, which does not occur in the British Isles as a native.

S. × **polita** (Haw.) Link (*S. hirsuta* × *spathularis*)

. . . . . . 7 (*8*) . .

Shady rocks and by mountain streams; rare.

7. Fairly plentiful, with *S. spathularis*, in a ravine by the S.W. side of L. Fee. (*8*). *By a stream near Maam*, 1946 (F. Winder).

First record: Babington (1836). 'We observed *Saxifraga Geum* in only one spot in Connemara, that is, by the side of the small waterfall upon the second brook passed on the road from Maam to Leenane.'

These two colonies are of great interest as indicating the former presence in Connemara of *S. hirsuta*, which is now confined to Kerry and Cork, and corroborate the theories of Praeger on this point, based on his discovery of a similar hybrid on Clare I., Co. Mayo. Babington's record had long been forgotten, even, it would seem, by himself, until it was resurrected by Stelfox (1946, 1947) and the plant re-found by Winder. The L. Fee station was discovered by P. W. Richards in 1933 (specimen in **CGE**), and has been confirmed by DW several times recently. It is slightly closer to *S. hirsuta*, while the Maam plant is slightly closer to *S. spathularis*, but both are clearly intermediate. For a further discussion see Webb (1948, 1950*a*). A small colony of pure *S. hirsuta* has been seen under trees by the road at Derryerglinna, but it is fairly clear that it is a garden escape, not the relict of a native population.

S. × *urbium* Webb (*S. spathularis* × *umbrosa*), the London pride of gardens, which is a hybrid not known in the wild state, has been planted by a holy well on the S.E. flank of Glasgeivnagh Hill, and is reported (1980) by T. Robinson to be now naturalized in some quantity. It differs from *S. spathularis* in its leaf-margin, which is scalloped rather than toothed, its rather hairier and shorter leaf-stalk, and by seldom setting seed. '

S. rosacea Moench (*S. decipiens*, *S. sternbergii*)

. 2 . . . . 7 . A .

Locally abundant on limestone pavement; very rare on mountain cliffs.

2. Frequent at Black Head and Poulsallagh, and locally abundant between these stations, near the drift-line, where it is associated with *Armeria maritima* and *Parietaria diffusa*. Common on all the Aran Islands, on roadsides as well as on limestone pavement. **7.** Sparingly on the N. face of Muckanaght.

First record: More (1860), as *S. hirta*. Black Head.

A very variable species; see Webb (1950*b*) for details. The plant found in the Burren and on Aran is an extreme form, being less hairy, more compact, and with more distinctly mucronate leaf-segments than most plants from Kerry or the Galtees. The plant on Muckanaght is generally similar but less extreme. All forms may be distinguished from *S. hypnoides* by the absence of creeping, leafy stolons, by the erect (not nodding) flower-buds, and by the fact that the leaf-segments, though they may be mucronate, do not have the long arista found in *S. hypnoides*.

It ranges through W. and S. Ireland from Donegal to the Galtees. In Britain it is confined to one station in N. Wales, where it may be extinct. Elsewhere it ranges rather discontinuously from Iceland to the Jura and Czechoslovakia.

S. hypnoides L.

1 2 3 . . . .

Limestone pavement and rocky grassland; common in the central part of the Burren, but thinning out towards the east and west; not seen elsewhere.

1. Rocky bluffs by O'Dea's Castle and Elmvale Ho. (both stations are on the limestone, and close to the boundary with district 2). **2**. Frequent and locally abundant, especially in the east; not on Aran. **3**. Occasional in the south-west, as around L. Atedaun and Castle L. *Garryland* (More, 1855).

First record: Foot (1864). 'Covers the whole district [Burren].'

There are numerous early records for Aran, but they are all due to confusion with *S. rosacea*.

S. tridactylites L.

1 2 3 4 5 . 7 8   A .

Rocks, walls, sand-dunes and open habitats on shallow soils. Very frequent on the limestone; occasional elsewhere on calcareous sands or mortared walls.

1. Limestone rocks at O'Dea's Castle. On the parapet of the bridge 1 km W. of Corofin. **2–4**. Very frequent and locally abundant. **5**. On a wall near the waterfall at Oughterard. **7**. Sand-dunes at Dog's Bay and N.E. of Gowlaun. *Bunowen Hill and E. shore of Ballyconneely Bay*, 1957 (DW). *In rock-crevices on the N. side of the Slyne Head peninsula* (Praeger, 1907). **8**. On a wall at Clonbur.

First record: Ball (1839). On level ground near the sea at Kilcornan, near Galway.

Chrysosplenium

Chrysosplenium oppositifolium L.   Golden saxifrage

1 2 . 4 . 6 7 8   A .

Damp woods and scrub, ditches and mountain cliffs. Frequent but local.

1–2. Frequent. **4**. Shady ditch S. of Ross L. **6**. Scrub near Flannery Br. **7**. Cliff on Muckanaght. Woods at Ballynahinch, Derreen, Letterfrack and Salruck. **8**. Very frequent.

First record: Babington (1836). 'Leenane; near a small waterfall above Mr. J. Joyce's house.'

Less abundant than might have been expected from a damp-loving plant of essentially western distribution in Europe; on the other hand its frequency in the mainly arid terrain of district 2 is remarkable; it has been noted in eight distinct stations, including two on Aran.

PARNASSIACEAE

Parnassia

Parnassia palustris L.   Grass of Parnassus

1 2 3 4 . 6 7 8   . .

Lake-shores, marshes, fens, cut-away bog over limestone, flushes and small, damp hollows. Frequent, though local, on the limestone; rare elsewhere.

1. Marsh E. of Inchiquin L. **2**. Frequent in the N. half; not seen in the south, nor on Aran. **3**. Frequent in the west; also between Kinvara and Ardrahan. **4**. Frequent N. of Galway, and by Ross L. and L. Corrib. **6**. In a flush at *c*. 180 m (600 ft) on Cashel Hill. **7**. Dunes S. of Kingstown Bay. **8**. In a small moorland flush at Glann, and on two of the islands in L. Corrib nearby. Shore of L. Corrib, E. and W. of the Dooros peninsula.

First record: More (1860). Canova I. in L. Corrib.

CRASSULACEAE

Umbilicus

Umbilicus rupestris (Salisb.) Dandy   Pennywort   *Carnán caisil*

1 2 3 4 5 6 7 8 A .

Rocks and walls. Occasional in Connemara; rare on the limestone.

1. Occasional on walls near the coast. **2**. Limestone wall at Corcomroe Abbey. Farmyard wall at Glensleade Castle. In the fork of a large ash-tree S. of the Corkscrew Hill. Inishmore. **3**. Wall 3 km W.N.W. of Ennis. **4**. Walls near Moycullen. **5**. Occasional near the coast. A few plants on a wall S.W. of Rosscahill. **6**. Occasional in the south-east. **7**. On walls 1 km E. of Clifden. **8**. Rather frequent around the N.W. part of L. Corrib, and on some of the islands.

First record: Graham (1840). Walls by the roadside between Galway and Oughterard.

In view of its distribution elsewhere in Ireland, where it behaves as a fairly strict calcifuge with a southern and western tendency, its scarcity in most of Connemara is remarkable, as is also its wide, though thin, distribution in the Burren.

Rhodiola

Rhodiola rosea L. (*Sedum rosea*)   Roseroot

1 2 . . . . 7 8 A B

Mountain and sea cliffs; very local (Plate 4).

1. Very frequent on the Cliffs of Moher. **2**. Abundant towards the W. end of Inishmore, where it grows on level pavement and even on the road as well as on cliffs; occasional elsewhere on the islands. **7**. Fairly frequent on cliffs on the Twelve Pins; also on Altnagaighera. Sparingly on the north faces of Doughruagh and Benchoona. Locally abundant on sea-cliffs on Inishbofin. **8**. Frequent on north-facing cliffs on Lugnabrick and Benbeg. At the base of the cliffs above L. Nadirkmore. Frequent on Benwee.

First record: Mackay (1806). 'In the south isles of Arran.'

Sedum   Stonecrop

S. anglicum Huds.

. 2 . 4 5 6 7 8 A B

Rocks, walls, and among heathy vegetation on shallow soils. Very frequent to abundant in Connemara; very rare elsewhere.

2. In considerable quantity in a 'field' (an enclosed area of limestone pavement with a partial vegetation cover) near the middle of Inisheer, 1976 (JW). *Formerly on Inishmore and Inishmaan* (Wright, 1871; Colgan, 1893). **4**. On a dyke of igneous rock E. of Galway railway station. **5**. Very frequent. **6**. Abundant. **7**. Abundant in the south and west; occasional in the north-east. **8**. Rare in the south; occasional to frequent elsewhere.

Fig. 12. *Sedum anglicum* (pink stonecrop) at Dog's Bay.

First record: Wade (1802). 'On the rocks in many parts of Cunnamara.'

The occurrence of this species on limestone on Inisheer is very remarkable; we know of no other instance in Ireland. What is even more remarkable is that despite this fact the earlier authors mention its occurrence on Aran without any reference to this anomaly. Its disappearance from Inishmore and Inishmaan might be explained by supposing that it had been introduced with turf from Connemara, and grew on in the debris from turf-stacks (as it does very readily in siliceous districts), were it not that the station on Inisheer in which it survives is far from houses and is not at all likely to have been the site of a turf-stack. Moreover, the plant is now growing happily in pockets of very shallow and not particularly peaty soil. We can offer no explanation of this anomaly.

*S. album L.

1 2 3 4 5 . 7 8 A .

Walls, roadsides, limestone pavement and sand-dunes; occasional almost throughout, and in some places at a considerable distance from houses.

1. Frequent on walls in the central part, as around Ennistymon, Lehinch and Corofin. **2**. On walls at Ballyvaughan, New Quay and Burren village. On pavement and bare sand at Fanore. In two places on Inishmore. **3**. Occasional throughout, on walls and limestone pavement. **4**. On walls by a ruined cottage at Srue, and spreading thence onto the limestone pavement. On the pier at Oranmore. **5**. Roadside S. of Moycullen. River-wall at Oughterard. **7**. Wall 1 km E. of Clifden. **8**. Walls at Clonbur.

First record: Atlas (1962).

Commonest on the limestone, but not confined to it. It may be distinguished from

S. anglicum by its more robust habit, its smaller, but more numerous flowers, and its leaves, which are usually twice as big, and not spurred at the base, as are those of *S. anglicum*.

S. acre L.

1 2 3 4 5 6 7 .   A B

Limestone pavement, walls and calcareous sands. Very frequent in the Burren; locally frequent on dunes elsewhere.

1. Frequent on the dunes N. of Lehinch. **2–3**. Very frequent throughout. **4**. Rocks at Rinville. Wall E. of Galway city. On an island in the S.W. part of L. Corrib (Roden, 1979). **5**. Dunes on both sides of Spiddal. **6**. Dunes at Moyrus; also at Furnace I. and S. of Carraroe. **7**. Very frequent on dunes on the S. and W. coasts, northwards to Cleggan and Inishbofin; not seen recently on the N. coast, though there is an *Atlas* record for the Renvyle region.

First record: Foot (1864). Burren: 'a highly characteristic plant, occurring everywhere throughout the district'.

*S. forsteranum Sm.

. . . 4 . . 7 .   . .

Walls and stony places; very local.

4. On walls by the ruins of Menlough Castle. **7**. Abundant on the old railway-track at Recess. On a sodded wall N.W. of Ballynahinch L.

First record: Webb (1982).

**S. telephium* L. has been seen in a few places by ruined cottages or very close to gardens, but it cannot be regarded as effectively naturalized, especially as it has disappeared from the stations mentioned in the earlier literature.

A specimen collected from Inisheer in 1976 by DW as *S. anglicum* turned out to be in fact *S. dasyphyllum*. All attempts to re-find it have, however, been unsuccessful. For a fuller account of this tantalizing mystery, see Webb (1980*a*).

DROSERACEAE

Drosera   Sundew   *Druichtín móna*

Drosera rotundifolia, the commonest species, is easily distinguished by its leaves shaped like a squash-racket, with a circular blade. The other two species are most easily told apart by the fact that *D. anglica* occurs as solitary, isolated plants, with leaves up to 8 cm long and usually nearly erect, while *D. intermedia* forms mats of interconnected plants with nearly horizontal leaves seldom much over $2\frac{1}{2}$ cm long.

Drosera rotundifolia L.

1 . 3 4 5 6 7 8   . B

Bogs and wet heaths. Abundant in Connemara; local elsewhere.

1. Occasional. **3**. Bogs by Templebannagh L. and by the Ballyogan Loughs. Rinroe bog. **4**. On most of the bogs northwards from Moycullen. **5–8**. Abundant except in the extreme west of district 7; plentiful, however, on Inishbofin.

First record: Babington (1836). Bogs between Clifden and Roundstone.

D. anglica Huds.

1 . 3 4 5 6 7 8 . (*B*)

Bogs; rarely in fens. Very frequent in Connemara; rare elsewhere.

1. Occasional at the W. end of L. Goller. Moorland S. of Fisherstreet. *Slieve Elva* (Atlas). **3.** Locally frequent in Templebannagh bog. Cut-away bog at the Ballyogan Loughs. By the W. shore of the northern Ballyeighter L. A few plants in the fen S.W. of Mullaghmore. **4.** Occasional in bogs S. and E. of Carrowmoreknock. In calcareous fens by the N. and E. shores of Gortachalla L. **5–6.** Very frequent throughout. **7.** Very frequent in the east and south; rare in the west and north. *Formerly on Inishbofin, but apparently extinct from turf-cutting.* **8.** Frequent in the south; rare in the north.

First record: Wade (1802). Ballynahinch.

The station by Gortachalla L. in district 4 is remarkable; it is a highly calcareous fen which shows no sign of acidification. The *Drosera* is here accompanied by *Carex lepidocarpa, C. serotina, C. lasiocarpa, Schoenus nigricans, Phragmites australis* and *Euphrasia scottica*. Although *D. anglica* is recognized to be the most base-tolerant of the three species, it seldom ventures into such a calcareous habitat in the British Isles. DW has, however, seen it on pure calcareous marl in Gotland (S.E. Sweden).

D. anglica × D. rotundifolia (*D. × obovata* Mert. & Koch) has been tentatively recorded from a bog on the S. shore of Shanakeever L. (district 7); the record, however, requires confirmation. It is also recorded from the Maam Cross area in *Crit. Suppl.*

D. intermedia Hayne

1 . 3 4 5 6 7 8 . .

Wet bogs, bog-pools and lake-shores. Very frequent to abundant in Connemara; rare elsewhere.

1. Bog by Drumcullaun L. Summit of Slieve Elva. **3.** Frequent on Templebannagh bog. Cut-away bog at the Ballyogan Loughs. Rinroe bog. **4.** On several bogs in the north, from Ballycuirke L. to near Oughterard quay. **5–6.** Abundant throughout. **7.** Frequent, but not on Inishbofin. **8.** Very frequent.

First record: Balfour (1876). Ballynahinch.

HALORAGACEAE

Myriophyllum   Water milfoil

Although flowering plants are necessary for certain identification, the species of this genus can usually be determined by their leaves. *M. verticillatum* has mostly 5 leaves in a whorl; the others nearly always 4. *M. verticillatum* has rarely fewer than 13 segments on each side of the leaf; in *M. spicatum* they are usually 7–15 and in *M. alterniflorum* 4–8.

(*Myriophyllum verticillatum* L.)

. . . (*4*) . . [7] (*8*) . .

Lakes and ditches; very rare.

(*4*). *Ditch near Menlough* (Praeger & Carr, 1895). [7]. *In small lakes near Ballynahinch*

(Mackay, 1806). (*8*). *In the N.W. arm of L. Corrib, below Carnseefin* (Praeger & Carr, 1895).

A strongly calcicole species throughout its Irish distribution, and for this reason Mackay's record can be rejected as almost certainly erroneous. It is not clear whether Praeger & Carr's plant from below Carnseefin was actually rooted there or was merely washed up; if the latter, it may well have drifted across from the limestone shore outside our region, as the water in the north-western arm of the lake is slightly acid, and calcicole plants are scanty. The Menlough record is backed by a specimen in **DBN**, and the plant may well be there still.

M. spicatum L.

. **2 3 4 5** . **7** .   **A** .

Lakes, ponds, rivers and springs; rather rare.

2. In a spring feeding L. Aleenaun. Inishmore. **3**. Pond 3 km N.E. of Ruan. Coole L. **4–5**. In the R. Corrib and adjacent canals in Galway city. *Ross L.* (ITB). **7**. In the larger lake on Omey I. Doonloughan.

First record: Praeger (1900). Galway docks.

All stations are in calcareous waters – the lime in the case of the last two being derived from blown calcareous sand. Pearsall & Lind (1942) reported this species as present in many 'of the less calcareous lakes' in Connemara, such as L. Fee, Kylemore L. and L. Inagh, but neither we nor the recorders for the *Atlas* have been able to confirm this. An *Atlas* record for Maam Br. also requires confirmation.

M. alterniflorum DC.

1 . **3 4 5 6 7 8** . **B**

Lakes and rivers. Very frequent to abundant in N. Connemara; occasional to frequent elsewhere.

1. Stream at Derrymore Br. **3**. Frequent within a triangle defined by Castle L, Rinroe and Tubber; not seen elsewhere. **4**. R. Corrib, above Galway. L. Corrib, near Burnthouse. **5**. L. Bofin. L. Formoyle. Stream N.E. of Screeb. **6**. Effluent stream from Inver L. In several lakes northwards from Carraroe. **7**. Frequent, especially in the north and west. **8**. Abundant.

First record: Oliver (1851). Near Roundstone.

Rather commoner in acid than in calcareous waters, but occurs freely in both.

Gunnera

*Gunnera tinctoria (Molina) Mirbel (*G. chilensis*)   Giant rhubarb

. . . . . . **7 8** . .

Streamsides, grassy banks, roadside ditches and damp hollows. Abundantly naturalized in much of N. and W. Connemara, and spreading freely by seed.

7. Roadside bank S.S.W. of Canal Br. A few plants by the road on the N. shore of Clifden Bay, and in great profusion by the football-ground just outside the town. At several places around Moyard and Letterfrack, and at the E. end of Ballynakill L. Very luxuriant by a stream near the sea at Gowlaun. By L. Muck and L. Fee. **8**. In several places around Leenane, ascending the mountain streams W.S.W. of the village for some distance.

Fig. 13. *Gunnera tinctoria* ('giant rhubarb'), naturalized by a stream near Leenane.

First record: Praeger (1939), as *G. manicata*. 'Plentiful on rough hills on the south side of Killery Harbour below Leenane.'

The seeds seem to be spread partly by sheep and partly by water. Although it is steadily increasing, it is so far confined, in Connemara, to agriculturally useless land. In the S.E. part of Achill I. (Co. Mayo), however, it has become a serious weed in damp pastures.

It should perhaps be added that, despite the striking superficial resemblance, this plant is in no way related to the true rhubarb.

Earlier records of *G. manicata* André as naturalized in Ireland are to be discounted; it is rarely cultivated, and there are no reliable records of its naturalization.

HIPPURIDACEAE

Hippuris

Hippuris vulgaris L.   Mare's-tail

1 2 3 4 5 6 7 8   A .

Lakes, streams, marshes and turloughs. Frequent on the limestone; rather rare elsewhere, and absent from markedly oligotrophic waters.

1. Inchiquin L. R. Fergus above Killinaboy. **2.** Carran turlough. L. Aleenaun. L. Luick. Abundant in the large pond at Burren village. On all three Aran Is. **3.** Very frequent. **4.** Frequent. **5.** Marsh in the S.W. part of Galway city. Boggy fen N.W. of Rosscahill. **6.** Marsh by the 'coral strand' near Carraroe. **7.** Bunowen L., and in the smaller lake on Omey I. **8.** Abundant at Cong.

First record: Wright (1871). Aran Islands.

CALLITRICHACEAE

Callitriche   Water-starwort

A difficult genus; fruiting plants and considerable experience are necessary for distinguishing the species. For this reason the distributions given below are probably very incomplete.

Callitriche stagnalis Scop.

1 2 3 4 5 6 7 8   A B

Ponds, ditches and streams. Rare on the limestone; very frequent elsewhere.

1. Very frequent. **2.** Occasional on Aran. **3.** R. Fergus, N. of Dromore. **4.** Stream at Aughnanure Castle. Pool by L. Corrib, E. of Moycullen. **5.** In the river at Spiddal. In a stream 4 km N. of Gentian Hill. Abundant in a marsh near the E. end of L. Formoyle. **6–7.** Very frequent. **8.** Frequent in the E. half, occasional in the W.

First record: Census Catalogue (1972). Earlier records are ambiguous.

C. platycarpa Kütz. (*C. palustris, C. verna, C. polymorpha*)

1 (*2*) **3 . . . 7 8**   (*A*) **B**

Rivers, streams and other wet places; occasional.

1. R. Fergus at Killinaboy. *Near Lisdoonvarna* (Druce, 1931). (*2*). *Inishmore, 1866* (E. P. Wright, **TCD**). **3.** Tirneevin turlough. Stream by L. Cleggan. **7.** Inishbofin. **8.** Sandy flats at Killateeaun.

First record: Druce (1931). Near Lisdoonvarna.

C. obtusangula Le Gall

. 2 3 . . . .

Calcareous rivers and turloughs. Local and apparently rare.

2. L. Aleenaun, 1968 (F. H. Perring). **3.** Castledodge R., 1969 (R. Goodwillie). Very abundant in Coole L., 1970 (DW).

First record: Webb (1982).

C. brutia Pet. (*C. pedunculata, C. intermedia,* subsp. *pedunculata*)

. . 3 4 5 . 7 8   . B

Pools and streams; rather rare.

3. Ditch S. of L. Cullaun. **4.** Stream at Aughnanure Castle. **5.** Ditch 4 km S.E. of Costelloe. Pool S.E. of Lettercraffroe L. **7.** Stream 3 km N.W. of L. Inagh. Inishbofin (**TCD**). **8.** Ditch W. of Leenane.

First record: More (1876). Inishbofin. Recorded as *C. hamulata,* but it is fairly certain that this species was intended.

C. hamulata Koch (*C. intermedia,* subsp. *hamulata*)

. . . . . . 7 8   . .

Lakes and streams; very local.

7. L. Nacorrussaun, E. of Ballyconneely (**DBN**). **8.** Frequent around Maam, and thence eastwards to Clonbur and Killateeaun.

First record: Marshall & Shoolbred (1896). Maam; Clonbur.

C. hermaphroditica L. (*C. autumnalis*).

. . . . . . 7 . . .

Lakes; very rare.

7. Aughrusbeg L., 1968–9 (**DBN**). *Renvyle* (Sledge, 1942).

First record: Sledge (1942).

LYTHRACEAE

Lythrum

Lythrum salicaria L.   Purple loosestrife   *Earball caitín, Créachtach*

1 2 3 4 5 6 7 8   A B

Lake-shores, river-banks, marshes and wet grassland; very frequent to abundant.

The only areas where this species is comparatively scarce are the drier parts of the Burren and the mountainous centre of Connemara, from Recess to Leenane. In the coastal regions of Connemara it combines with *Filipendula ulmaria* and *Senecio aquaticus* to make a spectacular display in wet pastures in August.

First record: Oliver (1851). Letterfrack.

In one population S. of Ballyconneely, most of the plants have flowers of a clear pink colour, very different from the usual bright purple.

L. portula (L.) D. A. Webb (*Peplis portula*)

1 [2] 3 (*4*) . 6 7 8   . B

Wet places. Very rare on the limestone; occasional elsewhere.

1. Occasional to frequent. **3**. Locally abundant by Coole L. (*4*). *Bog N.E. of Ross L.* (P. cat.). **6**. Wet roadside N. of Carraroe. **7**. Occasional throughout. **8**. By a stream at the W. end of L. Nafooey. Pools in the sandy flats at Killateeaun. Near Gowlaun.

First record: Ogilby (1845). Near Roundstone.

The record for Aran in the *Atlas* is based on sheets in **TCD**, collected by Wright in 1866 and labelled with this name, but which are, in fact, *Callitriche platycarpa*.

It is characteristic of the acid/base anomalies of western Ireland that the largest population in our region of this normally calcifuge species should be on the limestone at Coole.

ONAGRACEAE

Epilobium   Willow-herb

The 8 species of this genus in the Flora region are not very hard to distinguish. *E. nerterioides* and *E. pedunculare* are distinct in their creeping habit; the former has leaves 6–8 mm wide, very indistinctly toothed, while in the latter they are about 12 mm wide and finely but sharply toothed. *E. hirsutum* and *E. parviflorum* are the only species with softly downy leaves; the former is a stout plant with flowers 20 mm across, the latter much smaller, with flowers only half that size. *E. angustifolium* is distinct in its handsome flowers 25 mm across, with somewhat unequal petals. *E. montanum* has leaves mostly 2–3 times as long as broad; in *E. obscurum* and *E. palustre* they are 4–8 times. These last two are easily confused, but in late summer they can be distinguished if the plant is pulled up; *E. palustre* has thread-like runners from the base; in *E. obscurum* there are short leafy shoots too short to be called runners.

Hybrids are common in the genus, but only two have been so far recorded for the region. Marshall & Shoolbred (1896) reported *E. obscurum × palustre* from a ditch S. of Oughterard, and DW collected *E. obscurum × parviflorum* from a wall E. of Clifden in 1969 (**TCD**).

*Epilobium angustifolium L. (*Chamaenerion angustifolium*)　　　Rose-bay

. . . . . . **7** (*8*) . .

Bushy places; very rare.

7. A few plants among brambles and bracken near Letterdife Ho., Roundstone, 1966. (*8*). *Near Leenane*, 1957 (Atlas).

First record: Atlas (1962).

In eastern Ireland this species, though nowhere as abundant as in most parts of Britain, is steadily increasing, but it seems doubtful that it will ever become very common in the west.

E. hirsutum L.

1 2 3 4 5 6 7 8　**A** .

Ditches, marshes, roadsides and damp waste ground. Widely but irregularly distributed; for the most part on limestone or near the sea.

1. Occasional, mainly in the west and north. **2**. Ballyvaughan harbour, and by a lane about 3 km N.E. of the village. By L. Luick. Roadside N.W. of Mullaghmore. Frequent on Aran. **3**. By L. Cleggan. Marshes and woods near Dromore. Marsh E. of Ballinderreen. **4**. Frequent around Galway, in various habitats. Disturbed ground by L. Corrib, E. of Moycullen. **5**. Waste ground 5 km W. of Spiddal. Near L. Naneevin. In the W. part of Galway city. **6**. 2 km S.E. of Carna. **7**. Common around Ballyconneely, and in several of the marshes on the Slyne Head peninsula. Roadside N.W. of Cleggan. **8**. Near the Hill of Doon. Frequent W. of Cong.

First record: Wright (1871). 'A few plants were met with in the damp ground to the west of Inishmore.'

E. parviflorum Schreb.

1 2 3 4 5 6 7 8　**A B**

Marshes, drains, seepage zones and damp roadsides. Frequent on the limestone; local elsewhere and mainly near the sea.

1. Occasional near the sea. Ennistymon, and thence on the road to Kilfenora. Marsh by L. Raha. N.E. of Kilnamona. **2–4**. Frequent. **5**. Occasional in the south and east. **6**. Occasional. **7**. Frequent near the coast. **8**. N. of Oughterard.

First record: Hart (1875). Aran Islands.

E. montanum L.

1 2 3 4 5 6 7 8　**A** (*B*)

Woods and other shady places; also on waste ground and as a garden weed. Frequent, but local.

1. Common around Ennistymon. Hedges N.E. of Kilnamona. Woods at Elmvale Ho.

2. Frequent. **3**. Frequent in the south; rare elsewhere. **4**. Very frequent from Galway north-westwards, but not seen around Galway Bay. **5**. Occasional near the coast. Moycullen woods. By the waterfall at Oughterard. **6**. Garden weed 5 km N. of Costelloe. Waste ground at Carna. **7**. Abundant on roadsides around Clifden and near Kingstown Bay. Frequent in the Ballynahinch woods. **8**. Very frequent except along the northern and western margins.

First record: Hart (1875). Aran Islands.

E. obscurum Schreb.

1 2 3 4 5 6 7 8   A B

Roadsides, walls, marshes and damp waste places. Very rare in the Burren; frequent to abundant elsewhere.

1. Frequent. **2**. Aran (rather rare); not seen elsewhere. **3**. Damp grassland near the Ballyogan Loughs. **4**. Occasional in the north. Marsh at Rinville. **5–8**. Very frequent to abundant.

First record: Hart (1875), as *E. tetragonum*. Aran Islands.

The scarcity of this species in the Burren is surprising, as its general distribution in Ireland does not suggest a calcifuge tendency; indeed, one of its most characteristic habitats is the mortared parapet of a bridge.

E. palustre L.

1 2 3 4 5 6 7 8   . B

Marshes, bogs, fens and ditches. Rare in the Burren; occasional to frequent elsewhere.

1. Frequent. **2**. By springs at Carran and on Moneen Mt. **3**. By the lake S.W. of Barefield. Marsh N.W. of Crusheen. Bogs at Templebannagh, Rinroe and Ballyogan. Fen 3 km W. of Ardrahan. **4**. Ditch W. of Carrowmoreknock. Fen at Coolagh. By Gortachalla L. **5**. Marshes and drains S.E. of Costelloe. Drain N.W. of Barna. **6**. Frequent, especially around Screeb. **7**. Occasional. **8**. Frequent around Maam; occasional elsewhere.

First record: More (1876). Inishbofin.

*E. nerterioides A. Cunn. (*E. nummulariifolium, E. brunnescens*)

. [2] . . 5 6 7 8   [A] .

Damp rocks, screes, roadsides, banks and quarries. Frequent in much of Connemara, and abundant in the north.

5. Frequent by the Oughterard–Costelloe road near its highest points. In a quarry near L. Bofin. **6**. Occasional in quarries and on roadsides around Glinsk and Cashel. **7**. Very abundant around Kylemore and Letterfrack, extending westwards to near Clifden, eastwards to Salruck, and southwards to roadsides W. of Toombeola. Up to 360 m (1200 ft) on the Twelve Pins. **8**. Very frequent almost throughout.

First record: Moon (1955). W. of Leenane.

There has been much confusion over the name of this and of the following species, both natives of New Zealand, originally cultivated in rock-gardens, where, however, they soon became intolerable weeds. This species has also been known as *E. pedunculare*, a name now applied to the following species, and it has recently been re-christened for the fourth time as *E. brunnescens*. We think it safer and more convenient to adhere for the time being to the nomenclature of *Flora Europaea*.

There is a specimen in **BM** collected in 1947 by J. G. Dony 'near Roughan House', but, as this lies at the junction of districts 1, 2 and 3, it cannot be assigned; nor has the plant been seen recently in the Burren. In 1981, H. Noltie reported a single plant growing on the road at the W. end of Inishmore. Whether it will establish itself on Aran or in the Burren seems doubtful; it is not often seen on dry calcareous habits.

First noticed as naturalized in Ireland in 1933, but we can trace no record of its occurrence in the region of this Flora before 1956, when it was noted in several places in the course of fieldwork for the *Atlas*.

*E. pedunculare A. Cunn. (*E. linnaeoides*)

. . . . . . 7 8 . .

Damp banks; very local.

7. In small quantity on roadside banks both E. and W. of Kylemore Abbey. **8**. Abundant on roadside banks for over 1 km W. of Leenane, and here and there between the village and the county boundary.

First record: Gordon (1954). Westwards from Leenane.

Circaea

Circaea lutetiana L. Enchanter's nightshade

1 2 3 4 5 6 7 8 A .

Woods, scrub, hedges and streamsides; frequent on the limestone, occasional elsewhere.

1. Abundant in woods at Ennistymon and at Elmvale Ho.; occasional elsewhere. **2**. Very frequent. **3**. Occasional to frequent throughout. **4**. Abundant north-westwards from Galway; not seen in the south. **5**. Woods near Barna and Spiddal. Moycullen woods. By the waterfall at Oughterard. **6**. Scrub at Flannery Br. Island on Athry L. Woods at Cashel Ho., perhaps introduced. **7**. Woods at Ballynahinch and Salruck. Wooded ravine W. of Letterfrack. Wooded defile N. of Dog's Bay. **8**. Leenane.

First record: Babington (1836). Leenane.

Mainly on the limestone; the stations overlying siliceous rocks are mostly in old-established woodland or on partly calcareous drift.

[Circaea intermedia Ehrh.]

Wade (1802) records *C. alpina* from the public road at Drimcong, between Galway and Oughterard, and also 'in various mountainous spots in Connemara'. Another record for the mountains above Ballynahinch is cited in CH1. It is now generally agreed that *C. alpina* does not occur in Ireland, and that plants so named are referable to *C. × intermedia*. Since these records have never been confirmed, however, and the plant is not known closer than Sligo (some 130 km distant), they may be confidently dismissed as errors.

Fuchsia

*Fuchsia magellanica Lam.

1 2 . . 5 . 7 8 A .

Planted for hedges very frequently in most of Connemara and occasionally elsewhere; seen also here and there surviving in the gardens of ruined cottages; rarely naturalized in scrub or by streams or on rocky ground.

1. In the wood at Moy Ho. **2.** Naturalized rather freely on Inishmaan and very sparingly on Inishmore. Ballyvaughan harbour. **5.** By a stream about 4 km S. of Oughterard. **7.** In scrub at Ardbear Br., and at the head of Streamstown Bay. **8.** Well naturalized at Ilion West, E. of L. Inagh. In scrub in several places around the N.W. arm of L. Corrib, and on one of the islands N.E. of Oughterard.

First record (as clearly naturalized): Webb (1980*a*). Aran Islands.

The spectacular sight of fuchsia-hedges well over 2 m high in full flower in August, which can be seen in many parts of Connemara, has led to the oft-repeated statement that 'Fuchsia grows wild in the west of Ireland'. The innocent are deluded into imagining it to be native, but even the more sophisticated (e.g. Perring & Walters in Atlas, p. xiii) imagine it to be very widely naturalized. Only a very little observation is needed to show that less than one plant in a thousand grows anywhere but where it has been planted. This species does not sucker widely, and in its commonest form it rarely, if ever, sets fruit, so that the occasional naturalized bushes must be derived mainly from detached twigs, thrown or blown around or washed down by a stream, which happen to take root.

Nearly all the fuchsia-hedges in W. Ireland are composed of var. *riccartonii*, a horticultural variety which apparently arose in a Scottish nursery before 1850. This has more globose buds than the typical variety, with wider sepals, and the calyx is of a richer crimson. It is almost invariably sterile. The typical variety, which has more elongated buds and calyx usually of a slightly duller and more pinkish red, sets seed freely, but is much less commonly planted. It has been noted near Ballynalackan Castle (district 1), on Inishmaan (district 2) and N.W. of Moyard (district 7). Only on Inishmaan, however, was there clear evidence of spread by seed.

UMBELLIFERAE

This is a difficult family in which to discriminate the genera, and a few hints may be of use. *Hydrocotyle* and *Sanicula* are distinct in the round outline of their leaves, and *Eryngium* in its very prickly leaves and blue flowers. *Pastinaca* and *Foeniculum* have fairly bright yellow flowers; in the latter the leaf-segments are very fine and hair-like. *Petroselinum*, *Smyrnium* and *Crithmum* have yellowish-green flowers; the first smells of parsley, the second is distinctive in its very shiny leaves and conspicuous black fruits, while the third has succulent, almost cylindrical leaf-segments. It is with the remaining, white-flowered genera that the difficulty begins. They may be divided into two groups, one growing in wet, the other in dry places; *Angelica* is somewhat intermediate, and it may be recognized by its nearly spherical, off-white umbels and large, elliptical regularly-toothed leaflets. In the wet group, all the species of *Apium* can be recognized by their smell of celery; *Sium*, *Berula* and *Oenanthe* have to be learnt from a Flora. In the dry-living group, four genera are hairy and five are hairless. Two of the hairy genera – *Daucus* and *Torilis* – have spiny or bristly fruits, and *Daucus* is recognized by its carroty smell. *Heracleum* and *Anthriscus* have smooth fruits; the former is much the coarser of the two, and its fruits are wide and flat (narrow and nearly cylindrical in *Anthriscus*). Among the hairless dry-living genera *Conium* is known by its purple-spotted stem, and *Aegopodium* by the fact that it spreads by rhizomes to form large, dense patches of foliage. This leaves *Pimpinella*, *Conopodium* and *Aethusa*, which cannot be separated without the use of somewhat technical characters.

Hydrocotyle

Hydrocotyle vulgaris L.   Marsh pennywort

1 2 3 4 5 6 7 8   A B

Ditches, lake-shores, marshes and wet meadows; abundant except in district 2 and the western part of district 8. In the former (where wet habitats are scarce), it has been recorded only from Aran and L. Luick. It seems to be unexpectedly scarce around Leenane and the Maumturk range.

First record: Hart (1875). Aran Islands.

Eryngium

Eryngium maritimum L.   Sea holly   *Cuillean trá*

1 2 (*3*) . . **6 7** .   **A B**

Sandy shores and mobile sand-dunes; local.

1. Dunes N. of Lehinch. **2**. Fanore; shore E. of Ballyvaughan. Aran. (*3*). *Near Kilcolgan* (ITB). **6**. N. end of Furnace I. S. of Kilkieran. S.E. of Carna. **7**. Very frequent in the west, from Gorteen Bay northwards to Aughrusbeg; especially abundant on the Slyne Head peninsula and N. of Aughrusbeg L. Also on Inishbofin and the strands E. of Gowlaun.

First record: Oliver (1851). Near Roundstone.

Sanicula

Sanicula europaea L.   Sanicle

1 2 3 4 5 6 7 8   A .

Woods, scrub, crevices in limestone pavement and other shady places. Frequent on the limestone; occasional elsewhere.

1. Hedges near Moy Ho. Scrub 1 km S. of L. Lickeen. **2**. Fairly frequent in crevices in the limestone pavement throughout; also in woods at Poulavallan and Clooncoose, in scrub S.W. of Ballyvaughan and in lanes on Aran. **3**. Frequent in the woods at Coole, and occasional in scrub further south. **4**. Frequent around Ross L., and occasional to the north and north-west of it. **5**. Occasional along the N.E. boundary, from Moycullen to Oughterard. **6**. Glendollagh wood, and in scrub at Flannery Br. **7**. Fairly frequent in suitable habitats, including some of the lake-islands. **8**. Frequent around the N.W. part of L. Corrib, and on some of its islands; not seen elsewhere.

First record: Foot (1864). Burren: 'everywhere and at all elevations'.

Conium

*Conium maculatum L.   Hemlock

1 2 3 4 . **6 7 8   A B**

Roadsides and waste ground, always near houses; occasional.

1. Liscannor. Beside O'Dea's Castle. **2**. Ballyvaughan harbour. New Quay. L. Muree. Inishmaan. **3**. Roadside by the sea 5 km E. of Aughinish. **4**. Occasional around Galway and to the N.W. of it. Carrowmoreknock. **6**. By the quay at the S.E. corner of

Gorumna I. S. of Lettermullan village. Near the end of the road on Mweenish I. **7.** Inishlackan. S.W. of Clifden. Near Aillebrack L. Bunowen quay. Inishbofin. **8.** Roadside on the N. shore of the 'Narrow Lake'.

First record: Hart (1875). Aran Islands.

Smyrnium

*Smyrnium olusatrum L.　　Alexanders

1 2 . . (*5*) . . . **A** .

Waste ground and roadsides; very local.

1. Occasional, especially around old castles. Abundant near Liscannor. **2.** Corcomroe Abbey. Near Fanore Br. Plentiful on Aran. *Burren village* (Atlas). (*5*). *Near Galway* (Praeger, 1934*c*). *Spiddal* (Atlas).

First record: Wright (1871). Aran Islands.

The great rarity of this species in Connemara is remarkable in view of its abundance in coastal districts over a large part of Ireland. It is to be accounted for partly, no doubt, by the scarcity of ancient settlements in Connemara, and partly by its tendency to thin out towards the north-west, both in Britain and Ireland.

Apium

Apium graveolens L.　　Wild celery

. (*2*) **3 4** . . **7** . (*A*) .

Saline marshes; rather rare.

(*2*). *Inishmore* (Wright, 1871; Hart, 1875); probably an escape from cultivation, and not seen since. **3.** S. side of Kilcolgan estuary. *Kinvara* (ITB). **4.** Plentiful on the N. side of the estuary at Oranmore. Sparingly on the sea wall E. of Galway station. **7.** Plentiful at Callow, and sparingly thence to Bunowen. Marsh near Doonloughan. Inishlackan.

First record: Wright (1871): near Kilronan. First confirmed record: ITB (1901). Oranmore and Kilcolgan.

A. nodiflorum (L.) Lag.

1 2 3 4 5 6 7 8　A B

Lake-shores, ditches, streams and marshes. Widespread and locally frequent; commonest on the limestone; elsewhere mainly near the sea or in places irrigated with base-rich water.

1. At the mouth of the stream S. of Hag's Head. **2.** In the Caher R. at Formoyle. Very frequent on Aran. **3–4.** Frequent. **5.** Occasional along the coast. **6.** Ditches near Costelloe. **7.** Frequent near the coast, and abundant around Ballyconneely. On an island in L. Inagh. **8.** Ditch N. of Maam Cross. Shore of L. Corrib at Glann, and in ditches between Glann and Oughterard.

First record: Mackay (1806), as *Sium repens*. 'On the banks of the river Fergus a little above the bridge at Ennis.'

Plants with the habit and general appearance of *A. repens* (Jacq.) Lag. have been seen W. of Kilmacduagh and between Corofin and Tulla, but they lack the long peduncles and numerous bracts of that species, and are best referred to one of the varieties of *A. nodiflorum*; the identity and nomenclature of these are too confused for us to hazard a name.

A. inundatum (L.) Rchb. fil.

1 2 3 4 . 6 7 8   A B

Lakes, marshes and slow streams. Generally frequent, but rare in S. Connemara.

1. Lickeen L. Ballycullinan L. **2**. L. Luick. L. Aleenaun. Aran. **3**. Frequent, mainly in the south, but also at Carrowgarrif, near Gort, and S. of L. Fingall. **4**. Frequent. **6**. E. end of Glendollagh L. Stream at Carna. In the Owengowla R., 3 km E. of Cashel. **7**. Frequent in the extreme west; occasional elsewhere. **8**. Occasional.

First record: Wright (1871). 'In some marshy ground at Bungowla, on the west of Inishmore.'

A. × *moorei* (Syme) Druce (*A. inundatum* × *nodiflorum*), a hybrid which is much commoner in Ireland than in Britain, is fairly frequent in the S. half of district 3, and has also been seen in districts 1 (W. of Corofin) and 4 (N. side of Kilcolgan estuary).

Petroselinum

*Petroselinum crispum (Miller) Hill   Parsley

. **2** [3] [4] . . . .

Walls of old castles; very rare.

2. In some abundance on Newtown Castle, 2 km S.W. of Ballyvaughan, 1958, 1980. [3]. *Dromore Castle* (Praeger, 1905); not there now. [4]. Walls of Aughnanure Castle; first noted here by Praeger (1906); 3 plants were still there in 1965 but none could be found in 1978.

First record: Praeger (1905). Dromore Castle.

The restoration and re-pointing of the castles at Dromore and Aughnanure destroyed all the plants growing on their walls. *Parietaria* has re-appeared at Aughnanure, but *Petroselinum* is unlikely to do so, partly on account of its biennial habit, but also because it grows mainly on old and crumbling mortar.

Sium

Sium latifolium L.

. . **3** . . . .

Rivers and ditches; very local.

3. Abundant in the R. Fergus for about 3 km northwards from Ennis. *Dromore L.*, 1905 (Praeger, **DBN**). *Ditches near Corofin* (Mackay, 1806).

First record: Mackay (1806). 'On the side of the river Fergus a little above the bridge at Ennis, also in ditches near Corrofin.'

A very local plant in Ireland. Although its range on the Fergus seems to be contracting, it is as abundant above Ennis as anywhere in Ireland.

Berula

Berula erecta (Huds.) Colville (*Sium erectum*)

. 2 3 4 . . 7 8 . .

Ditches, fens, marshes and streams; very local, but abundant in places.

2. Sparingly in the Caher R. above Formoyle. **3**. Frequent by streams near Cleggan L. By the R. Fergus above Ennis. Castledodge R., near Tubber. S. side of Kilcolgan estuary. **4**. N. side of Kilcolgan estuary. Fen on N.E. margin of Galway city. Near L. Corrib, E. of Moycullen. **7**. Abundant in drains, marshes and lakes on the Slyne Head peninsula. Omey I. N. shore of Cregduff L. E. shore of Ballyconneely Bay. **8**. Sparingly in a flush on the N. side of the Dooros peninsula, L. Corrib. Lake-shore near Ashford Castle.

First record: Praeger (1906). 'Dog's Bay, and frequent on the Bunowen peninsula, very dwarf.'

Aegopodium

*****Aegopodium podagraria** L.   Bishop's-weed

1 2 3 4 5 . 7 8   A .

Roadsides, waste ground, wood-margins and by ruins; occasional.

1. Lisdoonvarna. Liscannor. **2**. Ballyvaughan harbour. Inishmaan. **3**. Roadsides 5 km N.E. of Crusheen and 2 km W. of Gort. **4**. By a ruined cottage E. of Ross Ho. **5**. Roadsides N. and W. of Moycullen. Roadsides and woods at Spiddal. **7**. Letterfrack and Moyard. Ballyconneely. **8**. Leenane. Cong. Clonbur.

First record: ITB (1901). H16.

Probably overlooked in places, but certainly rarer than in most parts of Ireland, and almost confined to the immediate neighbourhood of towns and villages. It is found also, of course, as a weed in most old-established gardens.

Pimpinella

Pimpinella major (L.) Huds.

. 2 3 4 5 . . 8   A .

Roadsides, wood-margins and rocky pastures. Abundant in some places, mainly on the limestone, but very local.

2. Common on Aran, but not recorded from the mainland. **3**. Rocky ground S. of Ballyallia L. **4**. Frequent and locally abundant from near Galway northwards and north-westwards to L. Corrib and Oughterard. A few plants beside a wall W. of Carrowmore. **5**. In a hedge by Rosscahill, and along the old railway track nearby. **8**. Abundant along both roads leading from Cong to Clonbur, and sparingly on the S. shore of L. Mask. Here and there in small quantity between Oughterard and Glann.

First record: Wade (1802). 'Dangan, beyond Galway.'

P. saxifraga L.

. 2 3 4 5 . . . . .

Dry, calcareous grassland. Very frequent in the eastern part of the Burren; rare elsewhere.

2. Occasional in the north-east, as at Turlough village, by L. Luick, E. of New Quay and N. of Glencolumbkille Ho., but not seen in the west or on Aran. **3**. Very frequent in the north; rare in the south. **4**. By Kilcolgan estuary. *Ross L.* (P. cat.). **5**. Salthill and Gentian Hill.

First record: ITB (1901). H16.

The *Atlas* record for Aran is almost certainly an error for *P. major*.

Conopodium

Conopodium majus (Gouan) Loret   Pignut

1 2 3 4 5 6 7 8   A B

Woods, scrub, meadows, pastures, rocky ground and riversides. Frequent and locally abundant.

1. Frequent. **2**. Abundant, though rare on Aran. **3**. Frequent. **4**. Frequent around Ross L. and to the E. of it; not seen elsewhere. **5**. Frequent along the coast and the N.E. margin, but absent from most of the bog country. **6**. Very frequent around Lettermullan and by Flannery Br.; rare elsewhere. **7**. Frequent. **8**. Abundant.

First record: Nowers & Wells (1892). Inishmore.

Usually an indicator of the former existence of woodland or scrub, but capable of survival for a long time after clearance.

Chaerophyllum

(‡*Chaerophyllum temulentum* L.)   Wild chervil

. (2) . . . .

Roadsides; very rare, and perhaps extinct.

(2). *Finavarra*, 1895 (CH2); *same station, c.* 1936 (Praeger, 1939). Not seen recently.

First record: CH2 (1898). Finavarra.

Probably extinct, but we have not made a sufficiently exhaustive search at Finavarra to say so definitely. The station is in H9; Praeger (1939) by a slip attributes it to H15. He asserts (ITB) that the plant is undoubtedly native in parts of Ireland, but this does not seem to be true of the west (where it is very rare), and its steady decline, even in the east, suggests that it may not be true there either.

Scandix

[*Scandix pecten-veneris L.]   Shepherd's needle

Frequent as a weed and ruderal in districts 2, 3 and 4 in the latter part of the nineteenth century, but we can find no record for the Flora region later than 1899. It has long been extinct in the west of Ireland.

Anthriscus

Anthriscus sylvestris (L.) Hoffm.   Hedge-parsley, Queen Anne's lace

1 2 3 4 5 6 . (*8*)   **A** .

Hedges, roadsides, open woods and waste places. Occasional and locally frequent in the south and east; very rare in the west and north.

1. Very frequent at the foot of Clifden Hill; also between Kilfenora and Ennistymon. In smaller quantity in several other stations. **2.** Abundant on Aran; occasional elsewhere. **3.** Roadsides around Kinvara. **4.** Roadsides near the sea W. of Oranmore; also at Moycullen. By a ruin on the N. shore of Ross L. **5.** Frequent all along the coast; also near Rosscahill. **6.** By walls at the S.E. corner of Gorumna I. (*8*). *N.W. of Oughterard* (P. cat.).

First record: Wright (1871). Aran Islands.

The *Atlas* records for the Roundstone, Ballyconneely and Clifden areas all require confirmation.

* *A. caucalis* Bieb. was recorded by Hart (1875) for Inishmore (as *A. vulgaris*), but has not been seen since. It is precariously established on the east coast and in the midlands, but has never been more than a casual in the west.

Foeniculum

*Foeniculum vulgare Miller   Fennel

. **2 3** . (*5*) . . .   . .

Roadsides and waste ground; rare.

2. Waste ground near Finavarra, 1970 (DW). **3.** Roadside S. of Ruan, 1969 (DW). (*5*). *Near Galway city* (CH1).

First record: Sim (1859). Near Galway.

Perhaps only casual in the region, but further observations are required.

Crithmum

Crithmum maritimum L.   Samphire

. **2** . . **5 6 7** .   **A B**

Maritime rocks and boulder-beaches. Locally frequent on the more exposed coasts; rather rare elsewhere.

2. Frequent on Aran; abundant at Poulsallagh on limestone cliffs up to 300 m from the sea; occasional elsewhere. **5.** Sparingly at Spiddal harbour. **6.** Sparingly on a stony shore at the S.E. corner of Gorumna I. **7.** Frequent on the Slyne Head peninsula, and thence northwards to Aughrusbeg L. and Inishbofin.

First record: Graham (1840). Near Clifden, sparingly.

Oenanthe

Oenanthe fistulosa L.

(*1*) . .**4** . . . .   . .

Fens, marshy river-banks and lake-shores; very local.

(*1*). *Ennistymon*, 1899 (ITB). **4.** Marshy banks of the R. Corrib above Galway. Abundant in Coolagh fen. On the shore of L. Corrib, $1\frac{1}{2}$ km W. of Carrowmoreknock.

First record: ITB (1901). Ennistymon.

O. lachenalii Gmel.

1 . 3 4 5 6 7 .   . .

Salt-marshes and saline meadows; local.

1. N. of Lehinch. **3.** S. side of Kilcolgan estuary. **4.** Frequent from Kilcolgan to Galway. **5.** Gentian Hill. **6.** E. of Carna, and on the W. side of Mweenish I. **7.** Frequent on the Slyne Head peninsula and on the E. shore of Ballyconneely Bay. On the mainland opposite Omey I.

First record: Balfour (1876). W. of Galway.

O. crocata L.

1 [2] 3 4 5 6 7 8   [A] B

Riversides, lake-shore, ditches and streams. Frequent in the extreme south; widespread but rather rare elsewhere.

1. Frequent and locally abundant. **[2].** A single, small plant without flowers was seen by Goodwillie in 1974 on a boulder-beach on Inishmaan, but the identification, though probable, is not certain, and confirmation is desirable. **3.** Frequent by the R. Fergus from Corofin southwards; also by the river at Kiltartan. **4.** A few plants in the R. Corrib at Galway, opposite the Cathedral. In the stream at Aughnanure Castle. **5.** In the river at Spiddal near its mouth, and one plant in a stream 1 km to the east. **6.** Abundant among rocks near the shore at the S.E. corner of Gorumna I. **7.** One large plant S. of Callow. By the lake at Renvyle Hotel. Streamstown Bay. W. of Letterfrack. Inishbofin. *Roundstone* (ITB). **8.** By the river W. of Cornamona. In a small stream on the W. side of the Maumturk Mts., E. of Lissoughter. *Clonbur and Leenane* (ITB).

First record: ITB (1901). Roundstone, 1896.

The distribution of this species in our region is most peculiar. It is rarer in Connemara than in any other extensive tract of siliceous hill-country in Ireland, and its abundance in calcareous water in a few stations in districts 3 and 4 contrasts strangely with its undoubted calcifuge tendency in most of Britain and Ireland.

O. aquatica (L.) Poir.

. . 3 4 . . . .

Streams, river-sides and lake-margins; local.

3. Abundant in and beside the R. Fergus above Ennis, and frequent thence northwards to Tubber and Rinroe. In a stream 3 km W. of Gort. **4.** In a stream 3 km N.N.E. of Galway. *Aughnanure Castle* (Praeger, 1906).

First record: Praeger (1906), as *O. phellandrium*. Aughnanure Castle.

Aethusa

*Aethusa cynapium L.   Fool's parsley

. 2 3 4 5 . 7 (8)   A (B)

Cultivated fields, gardens, roadsides and waste places. Rather rare.

2. Aran; rare, and perhaps only casual. **3.** Field 1 km E.N.E. of Corofin. **4.** Field at Ardnasillagh. Waste ground at Galway docks. **5.** Roadside 2½ km S.E. of Moycullen. **7.** Inishlackan. *Inishbofin* (Praeger, 1911). **(8).** *Near Maam* (Atlas).

First record: Hart (1875). Inishmore.

Probably inconstant in many of its stations, and apparently rarer than formerly. We have omitted some old records which do not alter appreciably its distribution pattern.

Ligusticum scoticum L., the lovage, has twice been reported from our region. Balfour (1853) thought he saw it on Aran, but admitted that he was not certain of the identification. Sledge (1942) reported it, without comment or further details, from 'Slyne Head'. As no specimen can be traced and the nearest known station is 250 km distant, in N. Donegal, this record cannot be accepted without confirmation.

Angelica

Angelica sylvestris L.   Wild angelica

1 2 3 4 5 6 7 8   A B

Damp meadows, marshes, riversides and ditches; more rarely in open woods or on roadsides, cliff-ledges, maritime rocks or limestone pavement. Abundant on acid soils; rather local on the limestone.

1. Abundant. **2.** Occasional in scrub, by turloughs or on limestone pavement; abundant in the marsh at Formoyle. **3.** Occasional in the south and east. **4.** Occasional in the north; not seen S. of Ballycuirke L. **5–6.** Very frequent. **7–8.** Abundant.

First record: Hart (1875). Aran Islands.

Tutin (1980) mentions a dwarf, glabrous form growing around turloughs, but we have not seen any very distinctive variant there. There is, however, a rather striking variant in the crevices of the limestone pavement on the cliff-top on Aranmore, near Dun Aengus; here it grows 15–25 cm high with dark green, succulent, very shining leaves.

Pastinaca

*Pastinaca sativa L.   Wild parsnip

. 2 . 4 5 . 7 .   A B

Cultivated and waste ground; rarely in other semi-open communities. Rather rare.

2. Inishmore and Inishccr. **4.** In an oatfield near Aughnanure Castle. On an esker N.W. of Gortachalla L. N.E. of Moycullen. **5.** Cultivated field 5 km S.S.W. of Costelloe Br. *About 6 km N.W. of Galway* (Atlas). **7.** Tidal marsh on the E. side of Ballyconneely Bay. Inishbofin.

First record: Hart (1875). Near Kilmurvy, Inishmore.

Probably no more than casual in some of the above stations, but near Ballyconneely, in the area S.E. of Oughterard, and on Inishmore it seems fairly well established.

Heracleum

Heracleum sphondylium L.   Hogweed

1 2 3 4 5 6 7 8   A B

Meadows, pastures, roadsides and sand-dunes. Very frequent on the limestone and shale; occasional to frequent in Connemara.

1–4. Very frequent. **5.** Occasional near the coast. **6.** Furnace I. S.E. corner of Gorumna

I. Dunes at Moyrus. **7.** Frequent, mainly in the west. **8.** Frequent around Clonbur and Cornamona; rather rare elsewhere.

First record: Hart (1875). Aran Islands.

*H. mantegazzianum Somm. & Lev.　　Giant hogweed

. . . . **5** . . .　. .

Rocky ground and roadsides; rare.

5. Well naturalized on rocky ground about 1 km W. of Costelloe Lodge, and sparingly on roadsides near the lodge, 1968, 1977 (DW).

First record: Webb (1982).

Daucus

Daucus carota L.　Wild carrot

1 2 3 4 5 6 7 8　A B

Pastures, meadows, roadsides and sand-dunes. Very frequent on the limestone and near the coast; rather rare elsewhere.

1. Frequent by the coast; not seen inland. **2.** Frequent. **3.** Abundant. **4.** Very frequent. **5.** Occasional by the coast. On the old railway-track near Oughterard. **6.** Frequent by the coast in the south and east. **7.** Frequent by the coast. **8.** Frequent around Maam, and thence eastwards to Cong, but rare over much of the district.

First record: Wright (1871). Aran Islands.

　The distribution of this species in Connemara is rather capricious. It is frequent in most of the coastal strip, but rare or absent around Screeb, Kilkieran, Carna and Cashel. Inland it is confined to calcareous soils except for a number of stations near L. Mask and L. Corrib where no basic influence is apparent.

Torilis

Torilis nodosa (L.) Gaertn.

. **2 3 4** . . . .　**A** .

Roadsides, waste ground and stony places; rare, and only on the limestone.

2. Rocks at Black Head. Roadsides on Aran (commonest on Inishmaan). **3.** Roadsides at Aughinish. *Kinvara* (ITB). **4.** Just above the shore-line on the N. side of Tawin I. Waste ground near the shore, E. of L. Atalia.

First record: Wright (1867). Aran Islands.

T. japonica (Houtt.) DC.

1 2 3 4 5 6 7 8　A .

Roadsides, hedges and wood-margins; rarely in meadows. Frequent and locally abundant on the limestone; occasional to rather rare elsewhere.

All stations recorded off the limestone are coastal, except for two in district 1 (by Clifden Ho., and near L. Raha). Both of these, however, are fairly close to the boundary with the limestone.

First record: Wright (1871). Aran Islands.

ARALIACEAE

Hedera

Hedera helix L.   Ivy   *Eidhneán*

1 2 3 4 5 6 7 8   A B

Rocks, cliffs, hedges and woods; abundant almost throughout.

First record: Wright (1871). Aran Islands.

CORNACEAE

Cornus

Cornus sanguinea L. (*Thelycrania sanguinea*)   Dogwood

. **2 3** . . . . . **A** .

Hedges and rocky places on the limestone; rare.

2. Two bushes on the E. side of Mullaghmore, 1969. Sparingly on Turlough hill, 1963. Aran; formerly frequent, but now rare. *Near Ballynalackan* (Corry, 1880). **3**. Hedge half-way between Ballinderreen and Kinvara, 1971. *Rocks at L. Fingall*, 1899 (**DBN**).

First record: Wright (1871). 'Very common amid the crevices of the rocks on Inishmore.'

Praeger (ITB) describes the species as 'frequent' in the Burren, but gives no localities. It is not mentioned by Foot. In view of its undoubted decline on Aran, it may well have diminished on the mainland also, but the explanation is not obvious.

*C. sericea L. (*C. stolonifera*)

There is a large thicket of this shrub on the W. bank of the R. Corrib, below the Franciscan monastery on the N.W. margin of Galway city (district 5). It is steadily spreading among native vegetation by suckering, and may by now be regarded as naturalized.

CAPRIFOLIACEAE

Sambucus

*Sambucus ebulus L.   Danesweed

. **2 3 4** . . **7** . **A** .

Roadsides and pastures; occasional, but often in some abundance where it occurs.

2. Roadsides near Ballyvaughan. 2 km N.W. of Bealaclugga. N. of the field-station at Carran. Inishmore. **3**. 1 km S. of Corofin, and 3 km E.N.E. of it. Near the W. end of Dromore L. **4**. Abundant at Ballynacloghy, W. of Clarinbridge. **7**. Abundant in bracken at the eastern foot of Bunowen Hill.

First record: Foot (1864). 'Grows on some crags near Ballyvaughan.'

An alien with remarkable powers of persistence, once established.

‡**S. nigra** L.　　Elder　　*Trom*

1 2 3 4 5 6 7 8　A .

Hedges, woods and waste ground; locally frequent, but rare over wide areas.

It is very difficult to distinguish between planted trees and those that are naturalized, and in many of the lists which were sent in to us the status was not made clear. On the whole it would seem that the elder is naturalized fairly extensively in districts 1–3, more sparingly in 4, and only rarely in Connemara. Whether it is truly native in any part of the Flora region is very doubtful, as it is very seldom seen far from houses.

In view of the failure of all early authors to distinguish between planted and self-sown trees we cannot cite a first record.

Viburnum

Viburnum opulus L.　　Guelder-rose

. 2 3 4 5 6 7 8　A .

Woods, hedges, scrub and limestone pavement. Occasional and locally frequent.

2. Fairly frequent throughout. **3.** Occasional in the south and west; not seen elsewhere. **4.** Frequent in the north; not seen east or south of Galway. **5.** Occasional along the N.E. margin; also near L. Bofin and on an island in Knocka L., N.N.E. of Spiddal. **6.** Woods at Glendollagh. **7.** Woods at Ballynahinch, Kylemore, and E. of Clifden. Islands in L. Inagh and Athry L. **8.** Woods by L. Corrib N. of Oughterard, and on the Hill of Doon; frequent also on the islands of L. Corrib. Woods by L. Mask, N. of Clonbur.

First record: More (1860). Canova I.

Lonicera

Lonicera periclymenum L.　　Woodbine, Honeysuckle　　*Taith-fheithleann, Feithleóg*

1 2 3 4 5 6 7 8　A B

Hedges, walls, woods and rocky ground. Very frequent throughout and locally abundant.

Especially characteristic of the limestone pavement of the Burren and Aran. The only region in which it is rather scarce is the neighbourhood of Galway city and the drift-covered country at the head of Galway Bay.

First record: Andrews (1845). Inishmore.

Symphoricarpos

*****Symphoricarpos albus** (L.) Blake (*S. rivularis*)　　Snowberry

Occasionally planted for hedges, and tending to spread by suckering, but whether this leads to naturalization depends mainly on whether the ground adjacent to the hedge is kept mown or grazed or not. The only places where it has been seen indubitably naturalized are by the river below Ennistymon (district 1), in the woods at Coole (district 3), by the ruins of Menlough Castle (district 4) and by the river at Maam (district 8).

Leycesteria

*Leycesteria formosa Wallich

1 . . . . . . 8 . .

Hedges, woods and scrub; rarely on walls. Rather rare, but increasing.

1. At the base of Clifden Hill. **8.** On walls in Oughterard. Several self-sown plants in the woods at Gortdrishagh, and in a hedge near the entrance-gate. One large plant among gorse at Glann. Plentifully naturalized in woods and scrub near Ebor Hall, between Cornamona and Cong.

First record: Webb (1982).

This species is steadily establishing itself in several parts of Ireland, mainly in the south and west.

RUBIACEAE

Rubia

Rubia peregrina L.   Wild madder

. **2 3 4** . . (7) .   **A** .

Walls, hedges and crevices in limestone pavement. Frequent in most of the Burren and on Aran; occasional on limestones N. of Galway.

2–3. Very frequent almost throughout. **4.** Hedges 2–3 km N. of Galway. On the hill N.E. of Menlough. Occasional in the country N. of Ross L. *By Ballycuirke L.*, 1906 (**DBN**). (7). *Salruck* (Hart, 1883).

First record: Mackay (1806). 'On the south isles of Arran, among limestone rocks.'

Even if the stations for Co. Mayo and for Salruck (where it has not been seen recently) are discounted, this Mediterranean–Atlantic species reaches its northern limit on the limestones N. of Ross L. Its distribution is limited fairly accurately by the January isotherm for 5 °C.

Galium   Bedstraw

Apart from *G. verum*, distinguished by its yellow flowers, the bedstraws may in the first instance be separated into those which never have more than 6 leaves in each whorl (and usually only 4–5), and those in which at least some of the whorls have 6 leaves and others up to 8. The first group contains *G. boreale*, with stiffish, erect stems and fruit covered with hooked bristles, and *G. palustre* with weak, straggling stems and smooth fruit. In the second group we can distinguish two with hooked bristles on the fruit: *G. aparine*, with minute, off-white flowers and rough, straggling stems, and *G. odoratum*, with conspicuous, pure white flowers and smooth, erect stems. This leaves only two common species (with smooth fruits and at least 6 leaves in a whorl): *G. saxatile* and *G. sterneri*. They are very similar, but *G. sterneri* has a long, slender bristle at the tip of each leaf; in *G. saxatile* it is shorter and stouter.

Galium boreale L.

. **2 3 4** . . . **8** A .

Frequent on the shores of limestone lakes and around turloughs; more rarely on limestone pavement, rocky grassland or mountain cliffs.

2. On the S. side of the saddle between Carnsefin and Gleninagh Mt. Turloughs at Carran and Turlough village. Widespread on Aran, but not common. **3**. Frequent. **4**. Frequent by L. Corrib. **8**. Occasional by the shores of L. Corrib and L. Mask, and frequent on the islands of L. Corrib, on the upper part of the stony shore. Cliffs in the N.E. corrie of Bunnacummeen. Sparingly at the Maumeen gap.

First record: (Wade, 1802). 'Bilberry Island and other islands on Lough Corrib, abundant, and about the shore of L. Corib.'

G. verum L.   Ladies' bedstraw

1 2 3 4 5 6 7 8   A B

Banks, roadsides, rocky ground, pastures and stabilized dunes. Abundant on limestone and calcareous sands; on acid rocks frequent near the sea, otherwise only rarely on roadsides, where it may be introduced.

1. Frequent by the coast north and south of Lehinch, and on limestone on the N.E. margin of the district and N.E. of Kilnamona. **2–4**. Abundant. **5**. Occasional by the coast. Roadside near L. Bofin. **6**. Grassland at S.E. corner of Gorumna I. N. of Carraroe. **7**. Frequent near the coast, from Roundstone to Omey I., and locally abundant on stabilized dunes and on the Slyne Head peninsula; rarer in the north. Roadside near Ballynahinch. **8**. Grassland at Ferry Br., and along the S. shore of the 'Narrow Lake'. Rather sparingly on the limestones near Cong.

First record: Foot (1864). 'Everywhere, and spreading over the coal measures.'

*G. mollugo L.

. (*2*) 3 (*4*) . . . .

Grassland and banks; a rare casual, precariously established in one or two places.

(*2*). *Near Kilmurvy, Inishmore*, 1894 (**DBN**, O'Kelly). *Ballyvaughan*, 1891 (**DBN**, Levinge; 1938 (**BEL**, Kerr). **3**. Near the N. shore of L. Atedaun, 1966 (MS and E. Booth), 1980 (D. Lambert). (*4*). *Abundant in a meadow N.E. of Galway* (Phillips, 1924).

First record: CH2 (1898). Kilmurvy.

G. saxatile L.

1 . . **4 5 6 7 8** . B

Heaths, banks, bog-margins and grassland. Abundant on acid soils; unknown on the limestone.

The records from the mainly calcareous district 4 are from the margins of the bogs which overlie the limestones N. of Moycullen.

First record: More (1860). From an island called 'Canova' in L. Corrib.

G. sterneri Ehrend. (*G. pumilum*)

. **2 3** (*4*) . . [7] .   A .

Limestone pavement and dry grassland. Common in the Burren and Aran; very rare elsewhere.

2–3. Very frequent throughout. (*4*). *Near Gortachalla L.* (Praeger, 1906).

First record: Mackay (1806), as *G. pusillum.* 'Abundant among the lime-stone rocks at Magherinraheen, near Corrofin.'

The *Atlas* record for 93/06 is probably an error; that for 84/72 (district 7) is based on a specimen in **CGE** labelled 'Clifden', collected by J. Luddington in 1911. We suspect an error in labelling.

G. palustre L.

1 2 3 4 5 6 7 8   A B

Marshes, lake-shores, ditches and other wet places; common throughout.

First record: Hart (1875). Aran Islands.

[G. uliginosum L.]

Record in 1959 from a turlough W. of Mullaghmore (district 3) during the fieldwork for the preparation of the *Atlas*, but as the species has often been misinterpreted and the locality has frequently been visited since, we are inclined to think that the record should not be accepted without confirmation.

G. aparine L.   Cleavers, goose-grass   *Garbhlus*

1 2 3 4 5 6 7 8   A B

Hedges, scrub, cultivated ground, roadsides and sand-dunes; frequent to abundant in all districts.

First record: Hart (1875). Aran Islands.

Rare only in the mountains and in areas of extensive blanket-bog. On Aran, where there are no hedges, it is frequent, growing through stone walls.

G. odoratum (L.) Scop. (*Asperula odorata*)   Woodruff

. 2 3 4 . . 7 8   . .

Woods and scrub. Local, and commoner on limestone than on siliceous rocks.

2. S. and S.W. of Ballyvaughan. Glensleade Castle. W. of Mullaghmore. E. end of the Glen of Clab. **3**. In two places W. of Crusheen. Garryland wood. **4**. Rather frequent in the N.E. corner of the district; not seen elsewhere. **7**. Occasional in woods W. of Ballynahinch L. Derreen wood, N.W. of Clifden. **8**. Woods 4 km N.N.W. of Oughterard. On the Hill of Doon. Copse 4 km W. of Oughterard. On an island in L. Shindilla, and several islands in L. Corrib. Near Cong.

First record: Carter (1846). Near Kiltartan.

Asperula

Asperula cynanchica L.   Squinancy-wort

1 2 3 4 5 6 7 .   A .

Limestone pavement, calcareous grassland and sand-dunes; frequent and locally abundant.

1. Frequent on the dunes N. of Lehinch. **2–4**. Very frequent. **5**. Occasional on Gentian Hill. *Near Inveran* (Atlas). **6**. Grassland behind the 'coral strand' S.W. of Carraroe. **7**. Dog's Bay. Abundant on stabilized dunes on the Slyne Head peninsula, and thence in several places northwards to Aughrusbeg L.

First record: Wade (1802). 'On the sandy banks along the sea shore at Aughris.'

Tutin & Chater (1974) have referred some specimens from Dog's Bay to *A. occidentalis* Rouy. This is supposed to differ from *A. cynanchica* in possessing orange stolons, shorter leaves, more constantly sessile flowers and slightly shorter corolla-tube. It is said to be confined to maritime sands. In the Burren, the populations of the dunes are continuous with those of the limestone pavement, and a sampling along a transect failed to reveal any consistent difference in these characters, except for the orange stolons, and we believe that these may be a phenotypic response to sandy soil. It is conceivable that *all* Irish plants might be referred to *A. occidentalis* by Ehrendorfer (who first made the distinction in *Flora Europaea*), but for the time being we think it best to interpret *A. cynanchica* in a comprehensive sense.

Sherardia

†**Sherardia arvensis** L.   Field madder

. 2 3 4 5 6 7 . A B

Roadsides, walls, rocks, sand-dunes and dry grassland. Widespread, but nowhere very common.

2. Dunes at Fanore, and sandy ground at Rine Point. Roadside at Ballyvaughan harbour. Walls near Carran. Roadside near Turlough village. Frequent on Aran. **3**. Fairly frequent, mainly in the north and east. **4**. Occasional. **5**. Dunes W. of Spiddal. Thin turf on maritime rocks S.E. of L. Nagravin. **6**. Occasional by the coast, mostly in the E. half. **7**. Occasional and locally frequent by the coast, mainly on sandy ground.

First record: Hart (1875). Aran Islands.

Several botanists regard this species as an alien in N.W. Europe, but its wide distribution in semi-natural habitats in remote parts of W. Ireland makes this hard to believe.

VALERIANACEAE

Valeriana

Valeriana officinalis L.   Wild valerian

1 2 3 4 5 (*6*) 7 8   A .

Marshes, damp woods, lake-shores and other damp places, but also not infrequently in hedge-banks, below walls, in crevices of limestone pavement and other dry habitats. Widespread, and locally very frequent.

1–2. Very frequent. **3**. Frequent in the extreme south; not seen elsewhere. **4**. Occasional near Galway, and northwards to near Moycullen. **5**. Near Rosscahill and Moycullen. (*6*). *Near Lettermullan* (Atlas). **7**. Occasional in the north; rare in the south. **8**. Very frequent and locally abundant.

First record: O'Mahony (1860). 'On the high mountain tops of Burren.'

***V. pyrenaica** L.

. . . . . . 7 (*8*) . .

7. Well naturalized at the base of walls in two places W. of Clifden. (*8*). *Near the coastguard station at Leenane*, 1935 (**DBN**, Mrs Gough).

First record: Webb (1982).

A robust, handsome plant, with large, nearly round leaves and dense inflorescences of pink flowers. Cultivated at Clifden Castle *c*. 1835 (Graham, 1840).

Centranthus

*Centranthus ruber (L.) DC.   Red valerian

1 2 3 4 5 6 7 . . .

Walls, waste ground and limestone pavement. Well naturalized in several places.

1. On the bridge over the R. Fergus, W. of Corofin. **2**. Abundant on limestone pavement E. of Burren village. **3**. Abundant on the sea-wall at Kinvara. A few plants by the cross-roads 2½ km N. of Tirneevin. On the railway-bridge near Tubber. **4–5**. By the Cathedral, and elsewhere, in Galway city. **6**. Walls by the sea at Cashel. **7**. In several places W. and S.W. of Clifden.

First record: Foot (1864). 'Everywhere to be met with in abundance, extending on to the coal-measures.'

Valerianella

Valerianella locusta (L.) Latt. (*V. olitoria*)   Lamb's lettuce

1 2 3 . . . **7 8   A** .

Walls, roadsides, stabilized dunes and semi-open grassland. Occasional in the south, and locally frequent by the coast in W. Connemara; very rare elsewhere.

1. Occasional on sand-dunes and roadsides near Lehinch. S.E. of Roadford. **2**. Stabilized dunes at Fanore. Roadside near Black Head. Frequent on Aran. **3**. Walls S.E. of Ruan. *Garryland* (Praeger, 1934*c*). **7**. Very frequent by the coast from Dog's Bay to the Slyne Head peninsula. Abundant on walls, stabilized dunes and heathy ground near the beaches E. and W. of Gowlaun. **8**. Roadside W. of Cong.

First record: Hart (1875). Aran Islands.

Doubtfully native in many of its stations.

DIPSACACEAE

Dipsacus

†Dipsacus fullonum L. (*D. sylvestris*)   Teasel

. . . **4** . . **7** . . .

Waste ground and roadsides; very rare.

4. On waste ground at the back of a shingle beach half-way between Galway and Oranmore, 1966, 1972 (DW). Seen in the same station by Praeger in 1899. **7**. Roadside E. of Claddaghduff, 1972 (DW).

First record: ITB (1901). Between Galway and Oranmore.

Although fairly frequent in the E. and S. of Ireland, this species is absent from the north-west except as a rare casual, and the stations here listed are the most northerly in western Ireland. It is difficult to be certain of its status in our region, but its persistence in the same spot in district 4 inclines us to believe that it is native.

Knautia

Knautia arvensis (L.) Coulter   Scabious

. **2 3 4 5** . . .   (*A*) .

Banks, walls, roadsides and meadows. Occasional to frequent on the limestone; unknown elsewhere.

2. Frequent on banks near the dunes E. of Ballyvaughan. 3 km N.of Glencolumbkille Ho. *Inishmore* (Praeger, 1895*b*; Atlas); searched for in vain, 1972–6. **3**. Frequent in the north; not seen S. of Kilmacduagh. **4**. Frequent around Moycullen and northwards. *Rinville* (P. cat.). **5**. A few plants on a wall on the W. side of the main road N.W. of Moycullen.

First record: Corry (1880). In a wheat-field between Ballyvaughan and Black Head.

Succisa

Succisa pratensis Moench   Devil's bit   *Urach bhallach*

1 2 3 4 5 6 7 8   A B

In a very wide variety of habitats, ranging from bogs and marshes to limestone pavement; abundant throughout.

First record: Hart (1875). Aran Islands.

One of the most ecologically tolerant species in the flora; there are very few communities in which it cannot be found, at least occasionally.

COMPOSITAE

Eupatorium

Eupatorium cannabinum L.   Hemp-agrimony

1 2 3 4 5 . **7 8   A** .

Lake-shores, fens, marshes and crevices in limestone pavement. Frequent on Aran and in the western part of the Burren, also locally by L. Corrib and L. Mask; rare elsewhere.

1. By the R. Fergus just below Inchiquin L. **2**. Frequent in crevices in the pavement, especially in the western part and on Aran; more rarely in damp grassland. **3**. Pavement near Ballyogan Loughs. *L. Fingall* (P. cat.). **4**. Stream by Aughnanure Castle. Marsh by L. Corrib, E. of Moycullen. Near L. Corrib at the Menlough quarries. **5**. Edge of a fen by L. Naneevin. **7**. Waste ground at Clifden. *Roundstone* (P. cat.). **8**. Frequent by the shore of L. Mask; also around the N.W. part of L. Corrib and on several of its islands.

First record: Wade (1802). 'Bilberry-island, L. Corib.'

On the limestone pavement this species is seen mostly as vigorous but very dwarf plants. 'Droll little specimens . . . , a couple of inches high, are constantly arresting the attention' (Hart, 1875).

Solidago

Solidago virgaurea L.   Golden rod

1 2 3 4 5 6 7 8   A B

Limestone pavement, rocky ground, lake-shores, walls and heathy grassland. Frequent, but somewhat local.

1. Wall by the river at Ennistymon. **2.** Very frequent. **3.** Frequent, mainly in the west and north. **4.** Frequent N. and E. of Ross L.; rare elsewhere. **5.** Occasional. **6–8.** Very frequent to abundant.

First record: Foot (1864). Burren: 'everywhere to be met with in abundance'.

At Poulsallagh, on limestone pavement near the sea, there is an interesting variant which is dwarf and very early-flowering. Both characters are preserved in cultivation. Dwarf forms are also found on mountains, but their performance in cultivation has not been tested.

Bellis

Bellis perennis L.   Daisy   *Nóinín*

1 2 3 4 5 6 7 8   A B

Pastures, roadsides, stabilized sand-dunes and other grassy places. Abundant throughout, especially in overgrazed pastures on clay soil, or on trodden areas of grassland. On the sand-dunes and sandy pastures at Killeany (Inishmore) the daisy becomes dominant on rabbit-tracks, which are visible as white streaks at a distance of over a kilometre.

First record: Wright (1871). Aran Islands.

Aster

Aster tripolium L.

1 2 3 4 5 6 7 8   A B

Salt-marshes; less commonly on maritime rocks or cliff-ledges. Frequent, but in some districts rather local.

1. By the estuary N. of Lehinch. **2.** Frequent all along the coast, and abundant on some of the more sheltered parts. **3.** Aughinish and the head of Aughinish Bay. Kinvara Bay. **4.** Frequent; abundant round L. Atalia. **5.** Gentian Hill. Spiddal. E. side of Cashla Bay. **6.** Abundant. **7.** Locally frequent on the W. coast and on Inishbofin; rare in the north. **8.** Leenane.

First record: Babington (1836). On the shore at Leenane.

On Inishmore the ligules on many of the plants are white, instead of the usual lilac.

*A. novi-belgii L.   Michaelmas daisy

. . . . **8** . .

This species (or possibly a garden hybrid in which it is predominant) is established in fair quantity on the banks of the Oughterard river near its mouth, and on the adjacent shore of L. Corrib, 1972–80 (DW). It does not flower freely, and most of the plants are rather dwarfed.

First record: Webb (1982).

Erigeron

Erigeron acer L.

. (*2*) . **4** . . . .

Dry calcareous soils; very rare.

(*2*). *Near Ballyvaughan*, 1895 (CH2). **4**. N. of Ross L., 1976 (D. Lambert). *Near Galway* (Atlas).

First record: CH2 (1898). 'Sparingly near Ballyvaughan; *N.C.*'

There is also a record in CH2 for 'Golden Bay, L. Corrib'. We have failed to locate this, and suspect that it is outside our region. It is possible that the station 'near Castle Taylor' cited in CH1 is in our district 3.

Filago

Filago vulgaris Lam. (*F. germanica*)

. (*2*) . . (*5*) **6** (*7*) (*8*)   (*A*) .

Sandy or stony ground; formerly occasional, now very rare.

(*2*). *Inisheer* (Hart, 1875). *Ballyvaughan* (Foot, 1864). (*5*). *By the railway W. of Oughterard* (Praeger, 1900). **6**. By the E. shore of Loughaunwillin, near Carraroe, 1964 (DW and I. K. Ferguson). (*7*). *Near Dog's Bay*, 1905 (**DBN**). (*8*). *W. of Clonbur* (Marshall & Shoolbred, 1896).

First record: Foot (1864). 'Roadsides about Ballyvaughan, by no means common.'

In FWI Praeger describes this species as frequent, though local, in W. Galway; this is certainly not true today.

Logfia

Logfia minima (Sm.) Dum. (*Filago minima*)

. . . . (*5*) . . **8**   . .

Sandy ground and rocks; occasional in the extreme north-east.

(*5*). *By the railway W. of Oughterard* (Praeger, 1900). **8**. Occasional around the S.W. corner of L. Mask, mostly near the Owenbrin R. A few plants S.W. of Cornamona.

First record: Marshall & Shoolbred (1896). 'Plentiful, but stunted, on the sandy shore of L. Mask at Maamtrasna.'

Omalotheca

Omalotheca sylvatica (L.) Schultz-Bip. & Schultz (*Gnaphalium sylvaticum*)

[1] . . . **5** . (*7*) (*8*)   . .

Pastures; very rare.

[1]. Given by ICP (p. 226) for 'a clayey field between Lisdoonvarna and Poulsallagh', but the associated species suggest strongly that this was a *lapsus calami*, and that *Gnaphalium uliginosum* was intended. **5**. On abandoned farmland at Bunnagippaun, W. of Rosscahill, 1966 (DW). (*7*). *Glen Inagh*, 1898 (ITB). *Benlettery* (Wade, 1802). (*8*). *In several places around Clonbur and Cong* (Marshall & Shoolbred, 1896). *Glann* (P. cat.).

First record: Wade (1802). 'On Lettery mountain and other places about Ballynahinch.'

The *Atlas* records 80 stations for this species in Ireland, but in only 19 of them has it been seen since 1930. This diminution may be in part illusory, because of the plant's tendency to occur here and there in very small populations or as single plants, but there is little doubt that it is much less common than it was.

Gnaphalium

Gnaphalium uliginosum L. (*Filaginella uliginosa*)   Marsh cudweed

1 (2) 3 . 5 6 7 8   (A) B

Lake-shores, roadsides, cart-tracks, field-entrances, forest rides, turloughs, cultivated fields, and other damp, disturbed habitats. Fairly frequent on acid soils; very local on the limestone.

1. Occasional. (2). *Inishmore* (Hart, 1875). Not seen since; probably a casual introduced with turf. 3. Damp ground by L. Cleggan. Frequent by the Kiltartan R. and at Coole L. Tirneevin turlough. 5. Tonabrocky. Bog-margin E.S.E. of Costelloe. Roadside 5 km W. of Rosscahill. 6. Abundant. 7–8. Occasional.

First record: Hart (1875). Inishmore. First confirmed record: More (1876). Inishbofin.

We have departed from the name given to this species in *Flora Europaea* because it seems to be generally agreed that *G. uliginosum* must be regarded as the type-species of the genus *Gnaphalium*, and cannot, therefore, be transferred to another genus.

Antennaria

Antennaria dioica (L.) Gaertn.   Mountain everlasting, Cat's foot

. 2 3 4 5 6 7 8   A B

Limestone pavement and other rocks; stony or gravelly pastures, heaths, screes and sand-dunes. Abundant in the Burren and on the L. Corrib limestones, and frequent on Aran; widespread but local elsewhere.

2–3. Abundant. 4. Abundant in the north. Rinville. N. shore of Dunkellin estuary. 5. Frequent. 6. Heaths W. of Glinsk, N.E. of L. Nafurnace, and S.W. of Carraroe. At the head of the inlet 2 km S.E. of Cashel. 7. Frequent in the S. half and in the Gowlaun area. Screes on the S.W. side of Benchoona. Near Dawros. Inishbofin. 8. Screes on Lissoughter. Occasional on the Maumturk Mts. Pasture on the highest point of the Dooros peninsula. Limestone pavement near Cong.

First record: Wade (1802). 'Ballynahinch, and on many mountains, Cunnemara.'

Although it is by no means rare on siliceous rocks it is a most unpredictable plant when off the limestone. In one area it will be plentiful on heathy ground; from another, in what seems to be an identical habitat, it will be absent. Although it grows freely on limestone in several parts of Britain it is reckoned a calcifuge in most parts of the Continent. Braun-Blanquet, therefore, when he saw it growing on calcareous gravels on his first visit to Ireland in 1949, was so scandalized that he decided it must be another species, and described the limestone plant as *A. hibernica* (Braun-Blanquet, 1952). His supposed differential characters are, however, neither clear-cut nor constant, and plants from acid and calcareous soils are indistinguishable when cultivated side by side.

Inula

*__Inula helenium__ L.   Elecampane

__1 2 3 4 . (6) 7 .   . B__

Field-margins and roadsides; occasional.

__1__. Roadside at Lower Ballycotteen, E. of the Cliffs of Moher. Near O'Brien's Br. __2__. 1 km N. of Bealaclugga. __3__. 2 km N.W. of Boston. Ballyclery, N.E. of Kinvara. 2 km S. of Ruan. __4__. In several places between Clarinbridge and Tawin I. Between Aughnanure Castle and the Galway–Oughterard road. *(6)*. *On an island 3 km S. of Cashel* (Babington, 1836). *St Macdara's I.*, 1896 (ITB). __7__. On the Dog's Bay promontory. 1 km N. of Bunowen Castle. By a ruined cottage near Dawros Br. Inishbofin. Inishnee (Ogilby, 1845).

First record: Babington (1836). 'Cruig Neit, a rocky island near the head of Bertraghboy Bay.'

It is often stated that this plant is found mainly associated with old monastic sites, but in the Flora region this is not the case.

We have omitted a number of old stations cited in CH1 but not recently confirmed; they do not provide a significant extension of range.

Pulicaria

__Pulicaria dysenterica__ (L.) Bernh.   Yellow fleabane

__1 2 3 4 5 6 7 .   A B__

Marshes, wet grassland and fen-margins; on Aran on limestone pavement. Widespread, but nowhere very common.

__1__. Occasional S. of Lehinch. By the E. side of Inchiquin L. __2__. Roadside 1½ km E. of L. Aleenaun. Widespread on Aran. __3–4__. Occasional. __5__. Roadside 3 km N.N.E. of Spiddal. __6__. Occasional near the coast. __7__. Frequent on the Slyne Head peninsula. Marshy ground on Inishlackan. Inishbofin.

First record: Hart (1875). Aran Islands.

Bidens   Bur-marigold

__Bidens cernua__ L.

__. 2 3 4 . 6 7 .   . .__

Marshes and pond-margins. Rare, and mainly on the limestone.

__2__. By a pond 2 km E. of Kilfenora. __3__. Marsh near L. Cleggan. Edge of Templebannagh bog. __4__. By a pond near Parkmore. Bog-pool by L. Corrib, E. of Moycullen. Marsh W. of Ballycuirke L. __6__. Marsh 2 km S.E. of Carna. *Recess* (ITB). __7__. By the shore of L. Nalawny, N.E. of Cleggan. By Ballynakill L.

First record: ITB (1901). Recess.

__B. triparrita__ L.

__. . 3 (4) . 6 7 8   . B__

Damp, open habitats; rare.

__3__. Occasional by Kiltartan R. *(4)*. *Ballyloughan* (P. cat.). __6__. By a lake 2 km N.E. of

Carraroe. **7**. Inishbofin (rare). *Near Clifden* (Graham, 1840). *Glen Inagh* (ITB). **8**. Sparingly by L. Corrib 3 km N.W. of Oughterard. A single plant on the sandy flats at Killateeaun.

First record: Babington (1836). L. Corrib, near Oughterard.

Achillea

Achillea millefolium L.   Yarrow

1 2 3 4 5 6 7 8   A B

Roadsides, pastures and rough grazing; occasionally on walls, banks, sand-dunes or sea-cliffs. Common everywhere.

First record: Wright (1871). Aran Islands.

A. ptarmica L.   Sneezewort

1 (*2*) **3 4 5 6 7 8**   (*A*) **B**

Marshes, lake-shores, turloughs and damp meadows; rarely on sand-dunes. Frequent, but rather local; commonest on limestone or near the coast.

1. Frequent, mainly in the southern half. (*2*). *Inishmaan and Inisheer* (Hart, 1875). **3**. Abundant around Coole L., and occasional at other turloughs, mainly in the N. half. **4**. Occasional by L. Corrib. **5**. Shore of L. Bofin, sparingly. **6**. Occasional near Carraroe. S.E. corner of Gorumna I. **7**. Frequent in the west; not seen in the east. **8**. Frequent by L. Corrib and occasional elsewhere.

First record: Hart (1875). Aran Islands. First confirmed record: More (1876). Inishbofin.

Chamaemelum

Chamaemelum nobile (L.) All. (*Anthemis nobilis*)   Chamomile

. . . . (*5*) (*6*) (*7*) **8**   . .

(*5*). *Inveran* (ITB). (*6*). *Near Cashel*, 1918 (**DBN**). (*7*). *Near Renvyle* (Wade, 1802). *Kylemore L.*, 1875 (Barrington, **DBN**). *Rather common about Clifden and Roundstone* (Druce, 1911). **8**. *North shore of the 'Narrow Lake'* (Marshall & Shoolbred, 1896). In fair quantity on grassy flats by L. Mask below Killateeaun, 1970–80 (DW).

First record: Wade (1802). 'Abundantly at Glogen, within a few miles of Renvi, Cunnamara.' We have failed to locate 'Glogen'.

It is hard to see why this easily recognized and fairly conspicuous plant cannot be confirmed in any of its old stations; these also include 'Oughterard, 1899' (ITB), which might be in districts 4, 5 or 8, and an *Atlas* record for the Cashel area, the basis for which is unknown to us.

Matricaria

Matricaria maritima L. (*Tripleurospermum maritimum*)

1 2 3 4 5 6 7 .   A B

Sea-shores, cliff-ledges and waste ground near the sea. Frequent and locally abundant, especially in the west.

1–3. Frequent, but rather local. **4–5**. Frequent. **6–7**. Abundant.

First record: Hart (1875). Aran Islands.

*M. perforata Mérat (*Tripleurospermum maritimum*, subsp. *inodorum*)
. 2 . . (*5*) . 7 .   A .

Cultivated and waste ground; frequency uncertain, but probably occasional to rather rare.

2. Inisheer, 1976. Waste ground near Burren village, 1969. (*5*). *N.W. of Inveran*, 1892 (Levinge). **7.** S. side of Slyne Head peninsula, 1975.

First record: Webb (1980*a*).

The plant, which resembles the preceding in many characters, but is erect, of annual habit, and with small but distinct differences in the fruit, has only fairly recently been recognized as a separate species. We have, therefore, no field-records; the four stations cited above are from herbarium specimens, the first two in **TCD**, the other two in **DBN**. It is possible, but by no means certain, that early records of *Matricaria maritima*, var. *salina*, given in FWI for Fanore, Salthill and Roundstone, may be referable to this species; unfortunately we can find no voucher specimens.

*Anthemis cotula L., somewhat similar in general appearance, was formerly occasional as a ruderal, but has long been extinct in the region.

Chamomilla

*Chamomilla suaveolens (Purch) Rydb. (*Matricaria matricarioides*)
Pineapple-weed

1 2 3 4 5 6 7 8   A B

Farmyards, roadsides, and waste ground; occasionally on sand-dunes or as a weed of cultivated ground. Frequent to abundant.

Varies considerably in its frequency in different parts of the region, but not absent from any considerable area. It is commonest in districts 3, 4 and 6, scarcest in district 2.

First record: ITB (1901). Claddagh, 1900.

For some years it was believed that this species, although it soon established itself in abundance all over Ireland soon after its introduction around 1893, was replaced in the Galway region by a related species, *Matricaria occidentalis* Greene. Whether this is really a distinct species seems doubtful; at all events plants from the Galway region today are no different from those from other regions.

Chrysanthemum

‡**Chrysanthemum segetum** L.   Corn marigold   *Liathán*
(*1*) (*2*) **3** . **5 6 7** .   (*A*) **B**

Cultivated fields and waste ground; persisting in small quantity in S. Connemara; extinct or casual elsewhere.

(*1*). *Waste ground near L. Goller*, 1955 (DW). (*2*). *Inishmore* (Hart, 1875). **3.** A single plant between Corofin and L. Atedaun, 1980. **5.** Derroogh, 1968. By the school N. of Bovraughaun Hill, 1968. *Near Tonabrocky*, 1955 (DW). **6.** Frequent near Lettermullan Br., 1967. **7.** Inishbofin: a single plant on decaying thatch, 1967.

First record: Wright (1871). 'Only in two or three patches of cultivated ground on Inishmore.'

Tanacetum

*Tanacetum vulgare L. (*Chrysanthemum vulgare*)   Tansy

. 2 3 . . . 7 . . . (*A*) (*B*)

Roadsides; rare.

2. In several places around Burren village, Finavarra and Bealaclugga. **3.** By the lane leading from Corofin to L. Atedaun. **7.** On a by-road 1 km N. of Bunowen Castle.

First record: More (1872). 'Middle Island of Aran (H.C.H.).'

In several other places the plant is established in the immediate neighbourhood of houses, but can hardly be regarded as effectively naturalized. It was recorded for Aran and Inishbofin by earlier authors, but is not to be found there now.

*T. parthenium (L.) Schultz-Bip. (*Chrysanthemum parthenium*), the feverfew, is seen in several districts in hedges or on waste ground close to cottages, but nowhere can it be regarded as effectively naturalized.

Leucanthemum

Leucanthemum vulgare Lam. (*Chrysanthemum leucanthemum*).   Dog-daisy, Ox-eye daisy   *Nóinín mór*

1 2 3 4 5 6 7 8 A B

Meadows, pastures, roadsides, walls and limestone pavement. Common everywhere except in the mountainous areas of Connemara.

First record: Wright (1871). Aran Islands.

The infra-specific taxonomy of this variable species is still in a state of flux, and we cannot say whether more than one distinct variant occurs in our region.

Artemisia

Artemisia vulgaris L.   Mugwort

. 2 . 4 5 6 7 8   A (*B*)

Waste ground and roadsides; rather rare.

2. Occasional in and around the villages on Inishmaan and Inisheer. *Near the mouth of the Caher R., and near Glencolumbkille Ho.* (Atlas). **4.** Roadside N.E. of Moycullen. Bog N.E. of Ross L. **5.** Roadside W. of Salthill, and in the western suburbs of Galway. **6.** Roadside 4 km S.S.W. of Carraroe. E. side of Mweenish I. **7.** Tilled ground at Doonloughan. *Inishbofin* (More, 1876; Praeger, 1911). **8.** Waste ground at Cong. One plant on roadside E. of Bohaun.

First record: Hart (1875). Aran Islands.

Probably a casual in many of its stations, but apparently established in small quantity on Aran, in the western part of Galway and at Cong.

*A. absinthium L.   Wormwood

. (*2*) (*3*) (*4*) . **6** (*7*) .   (*A*) (*B*)

Waste places and roadsides; formerly occasional, now almost extinct.

(*2*). *Inishmore* (Wright, 1871). (*3*). *Kilcolgan*, 1899 (ITB). (*4*). *Kilbeg ferry*, 1899 (ITB). **6.** Near the S. end of Mweenish I., 1980 (Robinson). (*7*). *Inishbofin* (Wade, 1802).

First record: Wade (1802). 'Abundant on Buffin-island.'

A. maritima L.   Sea wormwood

. **2 3 4 5** . . . . .

Sea-shores; local.

2. One large patch on Rine Point. **3.** E. end of Aughinish. Near the head of Corranroo Bay. W. side of Kinvara Bay. **4.** N. side of Kilcolgan estuary. Oranmore. Abundant from Parkmore and Carrowmore westwards. **5.** Barna, and very locally westwards to near Spiddal.

First record: Corry (1880). Near Gleninagh (= Rine Point).

A very local plant in Ireland; on the west coast it is confined to Galway Bay and the Shannon estuary.

The *Atlas* record for 93/17 (N. of Poulsallagh) seems to be an error.

Tussilago

Tussilago farfara L.   Colt's-foot

1 2 3 4 5 6 7 8   A B

Roadsides, stony sea-shores, cliffs, waste ground and field-margins. Widespread, but seldom in great abundance.

First record: Hart (1875). Inishmore.

Less common than in most parts of Ireland, and relatively rare in most of central Connemara and the southern part of the Burren. Its scarcity is doubtless in part the result of a scarcity of heavy boulder-clays, which provide it with its most congenial habitat, but in several districts it gives the impression of being a fairly recent introduction.

Petasites

*Petasites fragrans (Vill.) Presl.   Winter heliotrope

1 2 3 4 5 . (*7*) **8** . .

Woods and shady banks. Occasional, and locally frequent.

1. Abundant 1–2 km W. of Corofin. Between Ennistymon and Lehinch, and between Liscannor and the Cliffs of Moher. **2.** 3 km from Ballyvaughan on the Lisdoonvarna road. Near houses in New Quay and Burren village. By an old graveyard 7 km W. of Carran. **3.** Occasional in the woods at Coole. **4.** By the ruins of Menlough Castle. **5.** Frequent by the coast road, westwards from Galway; also around Moycullen and Oughterard. (*7*). *Roundstone* (P. cat.). **8.** N. of Oughterard. 3 km W. of Cong.

First record: Praeger (1906). Oughterard.

†P. hybridus (L.) Gaert.   Butterbur   *Gallán mór*

1 2 3 (*4*) . . . **8** . .

Roadsides and damp waste places; very local.

1. Occasional to frequent. **2.** Roadside N. of Carran. **3.** 1½ km S.S.W. of Gort. In several places around Tubber. (*4*). *Galway city* (Atlas). **8.** Near the county boundary W. of Cong.

First record: ITB (1901). H16.

Plate 1. The Twelve Pins from the south; in the foreground low-level blanket-bog cut for fuel.

Plate 2. (*a*) *Daboecia cantabrica*. (*b*) *Saxifraga spathularis*. (*c*) *Erica mackaiana*; *E. tetralix*, with withered flowers, in the right foreground.

Senecio

Senecio vulgaris L.   Groundsel   *Gronnlus*
1 2 3 4 5 6 7 8 A B

Sand-dunes, shingle, cultivated ground, waste places and limestone pavement.

Very frequent throughout most of the region, but less ubiquitous than in most other parts of Ireland, and relatively scarce in districts 1 and 8.

First record: Wright (1871). Aran Islands.

The variant with short ligulate florets was collected in 1924 (Stelfox, **DBN**) on the railway S. of Lehinch, but has presumably disappeared, along with the railway. Its taxonomic status is uncertain.

S. sylvaticus L.

. . . (4) 5 6 (7) .   . (B)

Heathy ground, among rocks, and on boulder-clay cliffs. Rare, and recently seen only in S. Connemara.

(4). *Kilbeg ferry* (P. cat.). **5**. Boulder-clay cliffs 3 km W. of Spiddal. In two places to the S. and E. of Costelloe. *L. Bofin* (P. cat.). **6**. Among granite boulders S. of Carraroe. 1 km S. of Carna. (7). *Roundstone*, 1913 (**BEL**). *Inishbofin* (Praeger, 1911).

First record: ITB (1901). H16.

S. jacobaea L.   Ragwort   *Buachalán buí*
1 2 3 4 5 6 7 8 A B

Pastures, roadsides, sand-dunes and limestone pavement. Very frequent to abundant on the limestone and in much of the coastal belt of Connemara; rather local elsewhere.

1. Abundant near the coast; fairly frequent inland. **2–4**. Very frequent and locally abundant. **5**. Occasional by the coast; very rare elsewhere. **6**. Occasional. **7**. Abundant in the extreme west, thinning out eastwards. **8**. Frequent.

First record: Graham (1840). 'At Roundstone...in considerable quantity...without ray.'

On Aran and on most of the Connemara dunes the variant without ligulate florets (var. *flosculosus* DC.) predominates, but every grade of intermediate between it and the typical form is usually to be found near by.

S. aquaticus Hill
1 2 3 4 5 6 7 8 A B

Marshes, lake-shores and damp grassland; occasionally on roadsides or in relatively dry grassland. Abundant on acid soils; fairly frequent in the Burren, but rare around the head of Galway Bay.

First record: Wright (1871). West end of Inishmore.

The distributions of this species and the preceding are to some extent complementary, *S. aquaticus* being common in districts 1, 5, 6 and 8, and largely replacing *S. jacobaea* as a weed of pastures. In districts 2, 3 and 4, *S. jacobaea* is the commoner; in district 7, they are both very frequent. Fairly often, and especially in districts 3 and 4, they grow close together, and the hybrid (*S.* × *ostenfeldii* Druce) is frequent, usually accompanied by back-crosses, so as to form a hybrid swarm.

*S. fluviatilis Wallr. (*S. sarracenicus*)

Established on a roadside between Clonbur and Ferry Br. (district 8) where it has persisted for some 40 years, but shows little tendency to spread. It is a native of C. and E. Europe.

*S. mikanioides Walp.

. 2 . . . . 7 . A .

Hedges and rocky ground; rare.

2. On broken rocky ground, used as a rubbish dump, W. of Kilronan, Inishmore, 1976 (MS). **7.** By the track leading from Roundstone to the saddle of Errisbeg, at some distance from the village, 1980 (MS).

First record: Webb (1982).

   This is a climbing plant from South Africa, with somewhat ivy-like leaves; it must be regarded as only rather precariously naturalized. Praeger (1939) records a report, almost certainly false, of its presence on an island in L. Anaserd. It is only rarely seen in gardens in the region.

Arctium

Arctium nemorosum Lej.   Burdock

1 2 3 4 5 6 7 8   A B

Roadsides, farmyards, sand-dunes, pastures, cultivated fields and other open habitats. Evenly but thinly distributed over the whole region, and often seen as single, isolated plants.

First record: Hart (1875). Aran Islands.

   Only a few specimens have been examined critically, but it seems probable that all burdocks from the region are referable to this species rather than to *A. minus*, which is in Ireland, as far as is known, mainly a plant of the south-east.

A note on thistles

Only six species of thistle are at all common in the region, and they are not difficult to tell apart. One (*Carlina vulgaris*) has straw-coloured flowers; in two they are pale lilac; in the other three they are fairly dark purple. Of the lilac-flowered two, *Carduus tenuiflorus* is distinguished by being annual, and in having green, spiny wings all down the stem; the other (*Cirsium arvense*) has a creeping rhizome and a stem without wings. Among those with deep purple flowers *Cirsium palustre* has flower-heads less than 25 mm across, usually crowded together in groups of 3 or 4; the other two have flower-heads well over 25 mm across, solitary or in pairs. *Cirsium vulgare* has very spiny leaves – spiny on the upper surface as well as on the margin; *C. dissectum* has leaves with only weak spines on the margins and none on the surface.

Carduus

Carduus tenuiflorus Curtis

. 2 3 4 . . . . A .

Gravel-beaches, seaside pastures and waste places; very local, and mainly around Galway Bay.

2. A few plants by the roadside W. of New Quay. Locally frequent on Aran. **3.** At the back of the storm-beach on Aughinish. **4.** Frequent on the stony shores of L. Atalia, and on waste ground near by. Oranmore, near the castle. Frequent in pastures just behind the beach on the N. side of Tawin I.

First record: Wright (1871). Aran Islands.

The *Atlas* records for Fanore and the Killinaboy region require confirmation.

C. acanthoides L. (*C. crispus*) was collected near Corofin by Levinge in 1892 (CH2), and a small colony was seen by DW on Inishmore in 1972 in a newly laid-out garden, but the species is only casual in the Flora region, as in most of Ireland. The *Atlas* record for 93/48 is an error for 93/47, which lies outside our region.

C. nutans L. was formerly well established around Galway Bay and on Aran, but it has not been seen in the district since 1931 (Gentian Hill, **DBN**), and is probably extinct.

Cirsium   *Feochadán*

Cirsium vulgare (Savi) Ten.   Spear thistle

1 2 3 4 5 6 7 8   A B

Pastures, roadsides, waste places and sand-dunes. Common throughout the region.

First record: Hart (1875). Aran Islands.

C. palustre (L.) Scop.   Marsh thistle

1 2 3 4 5 6 7 8   . B

Marshes and damp grassland; common except in parts of the extreme west.

This generally ubiquitous species is unexpectedly rare west of a line drawn from Toombeola to the mouth of Killary harbour. It occurs on Inishbofin, but in very small numbers, and has not been seen on Aran.

First record: More (1876). Inishbofin.

C. dissectum (L.) Hill   Bog thistle

1 2 3 4 5 6 7 8   . B

In a wide variety of habitats; very frequent almost throughout.

1. Very frequent. **2.** Frequent in the west (though absent from Aran), rather rare in the east. **3–4.** Frequent. **5–8.** Very frequent and locally abundant.

First record: Balfour (1853).

This species may be found in marshes, fens and bogs, around turloughs, and in meadows and rough grazing on peaty soils. Only where the soil is very dry and shallow, or where conditions are inimical to peat-formation, is it absent. Its curious distribution in the British Isles, however (mainly western in Ireland, mainly south-eastern in Britain), shows that it has subtle ecological preferences which we do not yet understand.

C. dissectum × *palustre* (*C.* × *forsteri* (Sm.) Loud.) has been seen in abundance in a boggy meadow N.W. of Moyard (district 7), and doubtless occurs elsewhere.

C. arvense (L.) Scop.   Creeping thistle

1 2 3 4 5 6 7 8   A B

Pastures and roadsides; common.

It avoids very base-poor soils, but even in blanket-bog areas it is quite common on grassy roadsides, where it derives some base from the road-metal. It is rather rare as a weed of arable land.

First record: Wright (1871). Aran Islands.

*Silybum marianum (L.) Gaertn., the milk thistle, has been recorded from Galway, Inishmore and Burren village, but the most recent record dates from 1895, and the plant has never been more than a casual in the region – as, indeed, is true for most of Ireland.

Carlina

Carlina vulgaris L.   Carline thistle

. **2 3 4 5** (*6*) **7 8   A** .

Limestone pavement, dry, calcareous grassland and sand-dunes. Abundant on the limestone; very rare elsewhere.

2–3. Abundant throughout. **4.** Frequent in the north; rather rare in the south. **5.** Gentian Hill. (*6*). *Lettermullan* (Atlas). **7.** Widespread, though in small quantity, on sandy or rocky ground on the Slyne Head peninsula. *Roundstone* (P. cat.). **8.** On limestone pavement 2½ km N.W. of Cong.

First record: Andrews (1845). Inishmore.

Saussurea

Saussurea alpina (L.) DC.

. . . . . . **7** .   . .

Mountain cliff-ledges; very rare.

7. Sparingly on the N. side of Muckanaght, 1883–1980. On the cliffs on the W. side of Bengower, 1981 (Roden).

First record: Hart (1883). Benlettery.

Hart's record for Benlettery, as is the case with many alpines ascribed to that mountain, almost certainly refers to the Bengower station.

Centaurea

Centaurea nigra L.   Knapweed, Blackheads   *Mullach dubh*

1 2 3 4 5 6 7 8   A B

Meadows, pastures and roadsides; very frequent to abundant in all districts.

First record: Wright (1871). Aran Islands.

Somewhat commoner on the limestone than off it, but absent only from the most mountainous regions of Connemara and the areas of uninterrupted blanket-bog.

C. scabiosa L.

. 2 3 4 5 6 7 8 A B

Meadows, rough grazing, roadsides, walls and rocky ground. Occasional to frequent on the limestone; elsewhere only on calcareous sands or limestone walls.
2. At the back of the dunes E. of Ballyvaughan. Meadows near the E. end of L. Aleenaun. Meadows 1–2 km N. of Glencolumbkille Ho. Rough grazing 2½ km S.W. of Ballyvaughan. Locally frequent on Aran. **3–4.** Frequent. **5.** Roadside at Gentian Hill. On a limestone wall on the S.W. side of the Galway–Moycullen road. **6.** Grassland behind the 'coral strand' near Carraroe. Pasture by the sea on Mweenish I. 3 km S.E. of Carna. **7.** Abundant in a meadow behind the 'coral strand' near Ballyconneely. By the graveyard at Gorteen Bay. Inishbofin (rare). **8.** Limestone rocks by the road W. of Cong.

First record: Graham (1840). On limestone near Galway.

Cichorium

[*Cichorium intybus L.]   Chicory

Some half-dozen records for this species exist for districts 2, 3 and 4, all dating from the period 1895–1900, but already in 1909 Praeger (FWI) dismissed it as 'little more than casual', and it has not been seen since then. Several recent sightings have been reported to us, but they all turn out, on investigation, to refer either to *Lactuca tatarica* or to *Cicerbita macrophylla* (qq.v.).

Note on 'yellow composites'

Members of the next eleven genera (apart from *Lactuca* and *Cicerbita*) have yellow, more or less dandelion-like flowers, and constitute a notoriously troublesome group for amateur botanists. We therefore provide a simplified key for their identification, which avoids overtechnical terms and very minute or inconspicuous characters. A few rare species are omitted.

1*a*	Flower-heads solitary on completely leafless stems	
2*a*	Flower-stems hollow and hairless	*Taraxacum* spp.
2*b*	Flower-stems solid, more or less hairy	
3*a*	Leaves white on lower surface; flowers lemon-yellow	*Hieracium pilosella*
3*b*	Leaves green on lower surface; flowers golden-yellow	
4*a*	Flower-heads at least 25 mm across; flower-stalks very hairy	*Leontodon hispidus*
4*b*	Flower-heads less than 25 mm across; flower-stalks only sparsely hairy	
		Leontodon taraxacoides
1*b*	Flowering stems bearing leaves or scales, and usually 2 or more heads	
5*a*	Leaves narrow, neither toothed nor lobed	*Tragopogon pratensis*
5*b*	Leaves lobed or toothed	
6*a*	Stems bearing only small scale-leaves	
7*a*	Leaf-lobes pointing somewhat towards base of leaf; plant coarsely hairy; usually in fairly dry situations	*Hypochoeris radicata*
7*b*	Leaf-lobes pointing straight outwards; plant with few hairs; usually in damp grassland	*Leontodon autumnalis*
6*b*	Stems bearing at least one well-developed foliage-leaf	
8*a*	Leaves distinctly hairy	
9*a*	Leaves deeply lobed	*Crepis vesicaria*
9*b*	Leaves entire or toothed, but not lobed	*Hieracium* spp.
8*b*	Leaves without hairs, or with very few	

10*a*   Each flower-head usually with only 5 florets   *Mycelis muralis*
10*b*   Each flower-head with at least 10 florets
  11*a*   Leaves prickly   *Sonchus asper*
  11*b*   Leaves not prickly
    12*a*   Flower-heads about 40 mm across   *Sonchus arvensis*
    12*b*   Flower-heads not more than 25 mm across
      13*a*   Flower-heads 8 mm across; fruit naked, without hairs   *Lapsana communis*
      13*b*   Flower-heads 12–25 mm across; fruits crowned with a plume of hairs, as in
                 the dandelion
        14*a*   Leaves toothed but not lobed   *Crepis paludosa*
        14*b*   Leaves lobed
          15*a*   Leaves bluish-green, mostly 2–3 times as long as broad; stem sappy and
                     brittle   *Sonchus oleraceus*
          15*b*   Leaves pure green, mostly 4–6 times as long as broad; stem fairly tough
                     and wiry   *Crepis capillaris*

Lapsana

Lapsana communis L.   Nipplewort   *Duilleóg mhaith*

1 2 3 4 5 6 7 8   A B

Hedges, wood-margins, roadsides and disturbed ground. Frequent on the limestone; occasional elsewhere.

1. Abundant E.N.E. of Kilnamona. Near Elmvale Ho. Derrymore Br. **2.** Frequent. **3.** Very frequent. **4.** Frequent. **5.** Occasional by the coast. **6.** Occasional around Carraroe, and at the S.E. corner of Gorumna I. **7.** Potato-field at Doonloughan. Hedges at Letterfrack. Cultivated ground near Gowlaun school. Garden weed on Inishbofin. **8.** Occasional.

First record: Hart (1875). Aran Islands.

Hypochoeris

Hypochoeris radicata L.   Cat's ear

1 2 3 4 5 6 7 8   A B

Grassland of all types, sand-dunes, rocky bluffs and limestone pavement. Abundant throughout.

First record: Hart (1875). Aran Islands.

Leontodon

Leontodon taraxacoides (Vill.) Mérat (*L. leysseri*)

1 2 3 4 5 6 7 8   A B

Dry grassland and sand-dunes. Locally abundant on limestone and near the sea; rather rare elsewhere.

1. Near O'Brien's Br. **2.** Frequent on stabilized dunes, both on Aran and the mainland; elsewhere only in a few stations in pastures on the E. margin. **3–4.** Abundant. **5–6.** Frequent, mainly near the coast; **7.** Abundant near the coast; rare elsewhere. **8.** Occasional by L. Corrib and L. Mask.

First record: CH1 (1866). 'Garryland, Galway (*A.G.M*).'

L. hispidus L.

(*1*) **2 3 4** . . . .   **A** .

Dry grassland and rocky ground on limestone; rather rare.

(*1*). *Sand-dunes at Lehinch*, 1901 (Praeger, 1903*a*). **2**. Roadside 1 km W. of New Quay. Occasional on Aran. **3**. Frequent on roadsides 3 km N.N.E. of Coole. Pasture at Carrowgarriff. **4**. Frequent in pastures 1½ km N. of Ross L. Abundant on rocky ground above Menlough.

First record: CH1 (1866). 'Near the town of Galway but rather rare; *Prof. Melville.*'

Strongly calcicole, but for some reason nearly confined to the central plain and relatively scarce on the western limestones.

L. autumnalis L.

1 2 3 4 5 6 7 8   A B

Grassland, especially the damper types, and lake-shores. Abundant in Connemara; occasional to frequent elsewhere.

First record: Wright (1871). Aran Islands.

Commonest in district 6; scarcest in district 1.

Tragopogon

‡**Tragopogon pratensis** L.   Goat's beard

. . (*3*) **4** . . . .

Waste ground; very rare.

(*3*). *Garryland*, before 1865 (CH1). **4**. By the railway 1–2 km E. of Galway station, 1965, 1972 (DW). Roadside 3½ km N.W. of Galway, 1980 (D. Lambert).

First record: CH1 (1866). 'At Garryland, near Gort, Galway; *A.G.M....* near the town of Galway; *Prof. Melville.*'

Sonchus   Sow-thistle   *Bainne muice*

‡**Sonchus oleraceus** L.

1 2 3 4 5 6 7 8   A B

Roadsides, farmyards, limestone pavement and open habitats of all kinds, especially near houses. Frequent on the limestone; occasional elsewhere.

1. Occasional, mainly near the coast or on limestone. **2–4**. Frequent. **5**. Occasional near the coast; not seen inland. **6**. Occasional. **7**. Fairly frequent. **8**. Rare: seen only at Leenane and Clonbur.

First record: Hart (1875). Aran Islands.

Much more obviously dependent on human activity and disturbance than is *S. asper*, and perhaps more strongly nitrophile. It seldom has the appearance of a native.

S. asper (L.) Hill

1 2 3 4 5 6 7 8   A B

Cultivated ground, limestone pavement, sand-dunes, walls, roadsides and other semi-open communities. Abundant on the limestone and frequent elsewhere.

First record: Hart (1875). Aran Islands.

S. arvensis L.

1 2 3 4 5 6 7 .   A .

Stony sea-shores, roadsides and cultivated ground. Very frequent over most of the limestone; occasional elsewhere.

1. N.E. of Kilnamona. In several places on the coast. **2.** Frequent. **3–4.** Very frequent. **5.** Occasional on the coast; also in a sand-pit W. of Rosscahill. **6.** Frequent in the south and east; rare elsewhere. **7.** Very frequent on the Slyne Head peninsula. On the stony shore of L. Athola. Inishlackan. Not seen in the north or east.

First record; Wright (1871). Aran Islands.

Cicerbita

*Cicerbita macrophylla (Willd.) Wallr.

1 2 . . 5 . . .   . .

Hedges, waste ground and at the base of walls; very local.

1. *c.* 5 km W. of Corofin, on the road to Ennistymon, 1969. **2.** Frequent in the village of Ballyvaughan, extending westwards to the harbour and eastwards to a by-road 2½ km E. of the village, 1967–80. Roadside bank just N. of Bealaclugga, 1965–6. **5.** Waste ground W. of the river in Galway city, 1965.

First record: Webb (1982).

This Russian plant appears to be steadily gaining ground, although we have never seen it in gardens in the region. It has several times been reported to us as 'chicory', but apart from its much coarser growth and very much larger leaves, its flowers are dull violet instead of clear blue. Our plants are probably referable to subsp. *uralensis* (Rouy) Sell.

Lactuca

*Lactuca tatarica (L.) Meyer

. . . 4 . . . .   . .

Stony sea-shores; rare.

4. Frequent over a short stretch of the shore E. of Galway, about ½ km E. of the mouth of L. Atalia, 1969–80.

First record: Praeger (1933).

Established at least since 1923 (Mrs Evans, **BM**). Collected several times by C. Pearson, 1929–33 (**DBN, TCD**), who noted that there was a wreck of a sailing-ship nearby, and believed that the plant had been derived from its ballast. It still persists in fair quantity.

It is a weedy-looking plant 20–30 cm high, with dull lilac-blue flowers, spreading by underground stolons. It is native to E. Europe, Siberia and C. Asia, and is widely naturalized in N. Europe.

Mycelis

*Mycelis muralis (L.) Dum. (*Lactuca muralis*)   Wall lettuce

. 2 3 4 . . . . A .

Limestone pavement. Very frequent in the Burren; very rare elsewhere.

2–3. Very frequent throughout, except in the eastern margin of district 3; recorded for Aran in small quantity for the first time in 1981. **4.** Sparingly on pavement near. L. Corrib 7½ km E.S.E. of Oughterard, 1967 (DW).

First record: Praeger (1939). Between Ballyvaughan and Black Head.

This plant is so abundant and apparently so much at home on the barer parts of the Burren that most people find it very difficult to believe that it is not native. But the fact remains that there is no mention of it in the literature before 1939, and no herbarium specimens of earlier date have come to light. It seems almost impossible that a native and conspicuous plant could have been so long overlooked in such a well-worked region, and the hesitation expressed by Webb (1962) must now be abandoned in favour of a clear declaration of alien status. Elsewhere in Ireland it is a local and relatively rare weed of walls and gardens; a hundred years ago it was known from only 3 vice-counties, but is now known from 19, indicating its steady spread. As far as we know it has never been seen in Ireland in what is in Europe its characteristic habitat as a native – in the ground-flora of base-rich woods.

Unfortunately there is no record of its arrival or spread in the Burren; it probably appeared around 1930, and was certainly widespread by 1947.

Taraxacum   Dandelion

Thanks largely to the devoted work of A. J. Richards, our knowledge of this genus in the Flora region is more complete than in most other parts of Ireland. Much collecting is still needed, especially in the mountains and on roadsides, before generalizations can be made with any safety; but it seems probable that our knowledge of the dandelion-flora of sand-dunes, fens and limestone pavements is reasonably satisfactory. All the records below have been determined or confirmed by Richards; the specimens on which they are based are largely in **DBN** or **TCD**, though some are in Belfast, Lund, Oxford or London. For further details about recent finds, see Lambert & Hackney (1976, 1979) and Lambert (1980).

Sect. Erythrosperma

T. brachyglossum (Dahlst.) Dahlst. **2.** Frequent on dunes and pavement. **3.** On pavement S.W. of Mullaghmore. **4.** Pavement N.E. of Oughterard. **6.** By the sea E. of Carna. **8.** Gravel-pit N.W. of Oughterard.

T. silesiacum Hagl. *(7). Roundstone, 1931 (**OXF**).*

T. gotlandicum (Dahlst.) Dahlst. **2.** Dunes at Fanore; not uncommon. Elsewhere known only from E. Scotland, Norway and the Baltic region. It was collected by the two authors independently at Fanore in 1956 and 1972.

T. laetiforme Dahlst. **2.** Dunes at Fanore.

T. fulvum Raunk. **2.** Limestone talus in the Glen of Clab. **7.** Dunes N.E. of Gowlaun.

T. fulviforme Dahlst. **5.** Roadside in Oughterard. **7.** Mannin Bay and Slyne Head peninsula. **8.** Roadside N.W. of Oughterard.

T. oxoniense Dahlst. **2.** Poulsallagh, Fanore and Black Head. Inishmore. **7.** Dunes W. of Cleggan.

T. glauciniforme Dahlst. *(2). Rocks at Kilronan, Inishmore, 1933 (Haglund, 1935).*

T. degelii Hagl. **2.** Fanore and Poulsallagh. *Kilmurvy, Inishmore, 1933 (Haglund, 1935).* **6.** Mweenish I. *(7). Roundstone, 1928 (**OXF**).*

T. arenastrum Markl. **2.** Fanore dunes.

T. pseudolacistophyllum van Soest. **7.** Dunes near Aillebrack L.

Palustria

T. palustre (Lyons) Symons. **2–3.** In considerable abundance around most of the turloughs, from Carran and Turlough village to Ballinderreen, Gort and Mullaghmore; also in the fen at the N. end of L. Bunny. *Marshy grassland near the Kilcolgan estuary*, 1899 (**DBN**). **4.** *By a lake near Galway*, 1946 (**TCD**). S.W. shore of L. Corrib.

Known from a few stations in the midlands of Ireland, but rare and diminishing in the rest of its range (England and N.W. Europe). Before its abundance in the Burren lowlands was discovered, Richards (1972) wrote that 'the world population may well number less than 1,000 plants'.

T. webbii Richards. **2.** At the edge of the *Potentilla fruticosa* hollow S.W. of Ballyvaughan. **3.** By L. Bunny. Fen E. of Ballinderreen.

Elsewhere known so far only from L. Carra (Co. Mayo).

Some plants in **TCD** collected near Galway by Pearson in 1943–6 are strongly suggestive of *T. anglicum* Dahlst., but the specimens are not good enough for confident identification.

Spectabilia

T. unguilobum Dahlst. **2.** Fanore. **5.** Railway cutting S.W. of Oughterard. *Galway city*, 1933 (Haglund, 1935). **7.** Near Clifden.

T. fulvicarpum Dahlst. **2.** Roadside W. of Ballyvaughan. **4.** Roadside E. of Rosscahill.

T. landmarkii Dahlst. **7.** Sandy bank at Dog's Bay.

T. faeroense (Dahlst.) Dahlst. **2.** Damp grassland at 180 m, E. of Black Head. **8.** By a stream N.E. of Leenane.

T. maculosum Richards. **2.** At the edge of the *Potentilla fruticosa* hollow S.W. of Ballyvaughan.

T. laetifrons Dahlst. **2.** Caher R. valley above Fanore. (**4**). *By L. Atalia*, 1933 (Haglund, 1935). **7.** Dunes at Gorteen Bay.

T. nordstedtii Dahlst. **2.** Poulsallagh. Pavement at 90 m, above Black Head. Caherconnell. **3.** Shore of L. Bunny, at N. end. Fen S.W. of Kilmacduagh. **7.** Near the 'coral strand' at Ballyconneely. Grass above the beach at Renvyle. **8.** Roadsides N.W. of Oughterard, and on the shore of L. Corrib near by. Maamtrasna and Killateeaun.

T. adamii Claire. **2.** Poulsallagh. Fanore. Hillside 4 km S.S.W. of Ballyvaughan. **3.** Fen S. of Kilmacduagh. **7.** Sandy pasture on Slyne Head peninsula.

T. britannicum Dahlst. (*1*). *Roadside in Lisdoonvarna*, 1933 (Haglund, 1935).

T. hibernicum Hagl. **2.** Fanore dunes. *Roadside at Kilronan* (Haglund, 1935). **7.** Meadow N.E. of Gowlaun.

Taraxacum

T. subcyanolepis Chr. **2.** Stable dunes E. of Ballyvaughan. *Kilronan* (Haglund, 1935). **3.** Roadsides W. of Gort and near Kinvara. (*4*). *Roadside near Merlin Park* (Haglund, 1935). **5.** Disturbed ground by the river, W. of Oughterard.

T. sellandii Dahlst. **7.** Roadside N.W. of Moyard.

T. pannulatiforme Dahlst. **2.** Caherconnell.

T. pannucium Dahlst. **5.** Garden weed in Oughterard. (*7*). *Roundstone* (Haglund, 1935).

T. croceiflorum Dahlst. **5.** Garden S. of Oughterard.

T. insigne Raunk. **7.** Seaside pasture near Cleggan.

T. cordatum Palmgr. (*2*). *Kilronan* (Haglund, 1935). **3.** *Juniperus*-heath W. of Gort.

T. longisquameum Lindb. **1.** Roadside N. of Lisdoonvarna.

T. dahlstedtii Lindb. **1.** Roadside near Lisdoonvarna.

T. subhamatum Chr. (*T. marklundii*). **2.** Roadside in the Caher R. valley. **7.** Roadside bank in Renvyle village.

T. fusciflorum Øllg. (**5**). Garden weed in Oughterard.

T. brachylepis Markl. (**5**). Garden weed in Oughterard.

Richards informs us that a plant recorded from near University College, Galway, as *T. crispifolium* Dahlst. is really an undescribed species in the section *Taraxacum*.

Crepis

*Crepis vesicaria L. (*C. taraxacifolia*)

. **2 3 4** . . **7** (*8*) . .

Roadsides, waste ground, walls and rocks. Occasional and locally frequent on parts of the limestone; very rare elsewhere.

2. Roadsides on both sides of Ballyvaughan. Deelin Beg. Near Glencolumbkille Ho. Frequent on Aran. **3**. Waste ground at Tullaghnafrankagh L. Roadside near L. Briskeen. **4**. Frequent at Menlough quarries. Wall by the railway E. of Galway. Roadside E. of Ross L. Pasture near Burnthouse. On rocks at Gortachalla L. **7**. Roadside bank opposite the lane leading down to Dog's Bay, 1980. (*8*). *Bridge at Ashford Castle* (ITB).

First record: CH2 (1898). 'Roadside near Ballyvaughan, 1896; *P. B. O'Kelly.*'

Until recently a mainly eastern species in Ireland, but it has evidently been spreading fairly rapidly in the west over the past 30 years or so, and is perhaps commoner in the Burren by now than these records would indicate.

As elsewhere in the British Isles all plants belong to subsp. *haenseleri* (DC.) Sell.

**C. biennis* L., another alien sometimes confused with the above, is not reliably reported from the region, although it has been seen just outside it, near Gort and Tulla. The *Atlas* record for the Black Head area appears to be an error.

C. capillaris (L.) Wallr.

1 2 3 4 5 6 7 8 A B

Walls, banks, roadsides, meadows and waste places. Common over most of the region, though somewhat unevenly distributed.

1. Frequent, especially near the coast. **2**. Occasional. **3–4**. Abundant. **5**. Frequent. **6**. Abundant. **7**. Frequent in the west, rare in the east. **8**. Abundant in the east, rare in the west.

First record: Hart (1875). Aran Islands.

The most characteristic habitat of this species in our region is on sodded walls.

C. paludosa (L.) Moench

. . . . **5 6 7 8** . .

Damp woods, streamsides, mountain cliffs and shady grassland. Frequent in Connemara; not recorded elsewhere.

5. In several places along the N.E. margin, from Moycullen to Oughterard. **6**. Frequent in the Glendollagh woods. Islands in Athry L. **7**. Frequent in the north; rarer in the south. **8**. Very frequent, especially in the east.

First record: Babington (1836). On mountains behind the inn at Maam.

Hieracium

This genus is rather sparsely represented in Ireland in comparison with Britain, and, partly for this reason, has attracted less detailed study. Our records are, therefore, probably somewhat incomplete. All have been determined by Sell, or by Sell & West, except for a few specimens of widespread species of which they had already determined several examples.

All species (other than *H. pilosella*) grow on rocks and cliffs; habitat notes are, therefore, omitted.

The order of species follows that of *Crit. Suppl.*

H. ampliatum (Linton) Ley. **2**. Wood-margin on Mullaghmore, 1973 (**TCD**). **8**. Gully W. of L. Nadirkmore, 1968 (**CGE**).

H. langwellense Hanb. **7**. Above the waterfall in the ravine on the S.W. side of L. Fee, 1969 (GH). **8**. Cliffs above Coolin L., 1965 (GH). N. side of Benbeg, 1969 (GH).

H. anglicum Fr. **2**. Widespread and locally frequent. Rare on Aran. **3**. S.W. of Mullaghmore. *Garryland*, 1900 (**DBN**). **7**. Bengower. Muckanaght. Benchoona. Walls near Clifden. **8**. N. of the Maumeen gap. *Lissoughter* (Praeger & Carr, 1895).

H. iricum Fr. **2**. Occasional. (*3*). *Garryland, and S. of Kinvara* (ITB). **7**. Benchoona and Garraun. (*8*). '*Mount Gable*' (Marshall & Shoolbred, 1896). *Benwee* (Colgan, 1900).

The records in *Crit. Suppl.* for 94/12 (N.W. corner of L. Corrib) for both *H. iricum* and *H. anglicum* appear to be errors arising from a confusion between the Carnseefin in this square with the Carnsefin of district 2.

H. basalticola Pugsl. **2**. Formoyle, 1965 (**CGE**). (*4*). *Shore of L. Corrib near Burnthouse*, 1899 (**DBN**).

(*H. hypochoeroides* Gibson). (*2*). *At the foot of the Slieve Carran cliffs* (Corry, 1880). *Black Head*, 1950 (**CGE**).

(*H. schmidtii* Tausch). (*4*). *Menlough*, 1899 (ITB).

H. argenteum Fr. **7**. Muckanaght, 1966. **8**. Cliffs S.E. of L. Nadirkmore, 1968 (**CGE**). *Maumeen*, 1882 (**DBN**).

H. sanguineum (Ley) Linton. **2**. Fairly frequent, mainly in N. half. (*3*). *L. Bunny*, 1933 (Pugsley, 1948).

*H. maculatum Sm. **2**. At the base of the Slieve Carran cliffs, 1966 (E. Booth). Rather a surprising locality for a species generally agreed to be an alien, but another alien (*Gaudinia fragilis*) was found nearby.

(*H. vulgatum* Fr.) (*4*). *Walls in Galway city, near the river*, 1944 (**TCD**).

H. sparsifolium Lind. **8**. Island in L. Corrib S.W. of Inchagoill (**CGE**).

(*H. stewartii* (Hanb.) Roffey). (*7*). *Twelve Pins* (**BM**).

H. umbellatum L. **6**. Islands in Athry L., 1961 (DW). **7**. Round I. in Derryclare L., 1961 (DW). *L. Inagh and E. slopes of Twelve Pins* (ITB). (*8*). *Maam* (ITB).

In the first two stations, and perhaps some of the others, the plant is referable to subsp. *bichlorophyllum* (Druce & Zahn) Sell & West.

H. pilosella L. (*Pilosella officinarum*)

1 2 3 4 5 6 7 8   A B

Banks, roadsides, rocky ground, sand-dunes and rough grazing. Very frequent to abundant throughout, except in the E. part of district 7 and the W. part of district 8, where it is rather scarce.

First record: Hart (1875). Aran Islands.

Of the 8 subspecies recognized in *Flora Europaea*, 5 have so far been recorded for the region: *concinnata*, *trichosoma*, *tricholepia*, *melanops* and *euromota*. The records are, however, insufficient to disclose any clear geographical and ecological pattern in their distribution.

CAMPANULACEAE

Lobelia

Lobelia dortmanna L. Water lobelia

. . . . **5 6 7 8** . **B**

Lakes; frequent and locally abundant in Connemara; not recorded elsewhere.

5. Very frequent in the north; rather rare near the coast. **6.** Very frequent. **7.** Abundant. **8.** Very frequent.

First record: Wade (1802). 'There are very few fresh water loughs in Cunnamara that this singular and elegant plant does not adorn.'

Strictly calcifuge, but this applies to the substratum, not the water. In the N.W. part of L. Corrib it is frequent at Glann and near the tip of the Dooros peninsula; the rocks here are siliceous, but the water is calcareous, with a pH of about 7.6. The same phenomenon has been observed at the S. end of L. Conn in Co. Mayo.

When not in flower this species can be distinguished from other submerged rosette-plants by its *blunt* leaves traversed by two large air-canals.

Jasione

Jasione montana L. Sheep's-bit

1 . . . **5 6 7 8** . **B**

Banks, sea-cliffs, heathy ground, walls and rocks. Absent from the limestone; frequent and locally abundant elsewhere.

1. Occasional near the coast; not seen far inland. **5.** Frequent. **6–8.** Very frequent to abundant.

First record: More (1876). Inishbofin.

Var. *latifolia* Pugsl., distinguished chiefly by its large involucral bracts, has been reported from the Cliffs of Moher by J. Parnell, and may, perhaps, occur elsewhere.

Campanula

Campanula rotundifolia L. Harebell *Méaracán gorm*

1 2 3 4 5 . **7 8 A B**

Limestone pavement, rocky ground, lake-shores, sand-dunes and mountain cliffs and screes. Abundant in the Burren and on sand-dunes in W. Connemara; frequent around L. Corrib; occasional elsewhere.

1. Roadside by the dunes N. of Lehinch. **2–3.** Abundant. **4.** Frequent. **5.** Hedge-banks at Gentian Hill. Sand-dunes W. of Spiddal. **7.** Abundant on most of the dunes from Gorteen Bay northwards to Aughrusbeg L. and Inishbofin; also N.E. of Gowlaun. In a meadow behind the 'coral strand' near Ballyconneely. Cliffs and screes on Muckanaght, Benchoona and Altnagaighera. **8.** Cliffs at the Maumeen gap and above L. Nadirkmore. Abundant on screes on Lissoughter. Frequent on the rocky shore of L. Corrib, on the Dooros peninsula, from Glann to Oughterard, and on several of the islands.

First record: Wade (1802). 'Maam Turk, Cunnemara.'

ERICACEAE

Vaccinium

Vaccinium vitis-idaea L.

. . . . **8** . .

Mountain cliffs; very rare.

8. One plant on the N. face of the Benbeg–Bencorragh ridge, N.E. of the summit of Benbeg, at *c*. 500 m (1600 ft) (Roden, 1978). A few plants at *c*. 600 m (2000 ft) on the E. side of Benwee, 1981 (Roden).

First record: Roden (1978).

Mainly in the mountains of the north and east; these are the most southerly stations known in the western half of Ireland.

V. myrtillus L.   Bilberry, Frochan   *Fraochán, Fraochóg*

1 2 . . **5 6 7 8** . **B**

Banks, bog-margins, heathy ground and rocky ridges; occasionally in woods. Generally frequent to abundant on acid soils; very rare on limestone.

1. Fairly frequent inland; not seen near the coast. **2**. Frequent in scrub at Poulavallan. In small quantity near the summit of Gleninagh Mt., in *Dryas*-heath. **5**. Frequent. **6**. Occasional. **7**. Frequent, mainly in the east; abundant in woodland on most of the lake-islands. **8**. Abundant.

First record: Wade (1802). 'Bilberry Island, L. Corib.'

Rather intolerant of exposure, so that, although it is found throughout Connemara, it is not very common on the western and south-western margin.

V. oxycoccos L. (*Oxycoccus palustris*)   Cranberry

1 . . . .

Wet bogs and lake-margins; occasional in a restricted area of the Clare shales.

1. Abundant around the small lakes 2½ km N.E. of the Cliffs of Moher, 1971 (DW). Occasional at the W. end of L. Goller (W. A. Watts). In a small bog 3 km N.W. of Lisdoonvarna, 1969 (R. M. Wadsworth).

First record: ITB (1901). Lisdoonvarna.

The last station cited above is probably identical to that recorded in ITB. The cranberry is mainly a plant of the midland bogs and the Wicklow Mts., but is locally abundant in Co. Clare.

Arctostaphylos

Arctostaphylos uva-ursi (L.) Sprengel   Bearberry

. **2 3** (*4*) **5** (*6*) **7** . . .

Rocky ground and thin, peaty soil. Locally abundant, but apparently diminishing.

2. Locally abundant on and near the summits of Carnsefin, Gleninagh Mt. and Cappanawalla, especially the last, and more sparingly on their lower slopes. *Turlough Hill*, 1892 (**DBN**). **3**. Rocky pastures 3 km W. of Ardrahan. *L. Fingall* (P. cat.). (*4*).

Near L. Corrib, E. or S.E. of Moycullen (Praeger, 1934*d*). **5**. On a boulder 2 km S.E. of Screeb. On a boulder by the road 3 km N. of Spiddal. (*6*). *Rocks near Cashel*, 1928 (**BM**). **7**. On the islands of two lakes W.N.W. of Toombeola (Webb & Glanville, 1962). Fairly widespread on the southern and eastern peaks of the Twelve Pins. *Diamond Hill*, 1892 (**BM**). *Roundstone* (P. cat.). *Inishbofin* (Wade, 1802).

First record: Wade (1802). 'Common throughout the western isles bordering on Cunnamara, as Bufin, &c., and on all the mountains about Ballynahinch, Streamstown, Aughrus, Renvi, &c.'

The nineteenth-century records suggest a far greater abundance of this species than has been observed recently, especially on the low ground of Connemara. Even on the mountains it does not seem to justify today the indications of abundance indicated by Wade, or even by Hart. But it is not easy to suggest an explanation for its decline. The *Atlas* record for 94/12 (north-western arm of L. Corrib) is an error.

Calluna

Calluna vulgaris (L.) Hull   Common heather, Ling   *Fraoch*

1 2 3 4 5 6 7 8   A B

Moors, bogs and heaths; also on limestone pavement. Abundant on acid soils, unless very waterlogged; locally frequent on the limestone.

First record: Babington (1836). Bogs between Clifden and Roundstone.

The frequency with which this normally calcifuge species occurs on shallow peat overlying limestone in the Burren, and to a lesser extent on Aran, is one of the more remarkable features of the region. It is only from areas covered mainly by calcareous boulder-clay that it is absent.

Erica   Heather   *Fraoch*

This genus is of special interest in the region of the Flora as three of the species represented are very rare and two of them are absent from Britain. *E. erigena* is distinct in flowering in spring, and in having blackish anthers protruding from the mouth of the pale mauve corolla; *E. cinerea* in its bright reddish-purple flowers and needle-like leaves in tufts, not (as in the remaining three) in whorls of 3 or 4. These three need a closer look for their discrimination, but *E. ciliaris* is distinct in its late flowering season (end of August to October), in its longer corolla (almost 12 mm), which is slightly asymmetrical, in its flowers strung out over an inflorescence 2–5 cm long instead of being bunched together in an umbel, and in having its leaves usually in whorls of 3 (not 4). The distinction between *E. tetralix* and *E. mackaiana* needs a good lens for absolute certainty, and the situation is confused by the frequency of the hybrid between them, but in general *E. mackaiana* can be recognized by its brighter pink and later (July and early August) flowers, and by the fact that the leaves immediately below the flowers stand out straight from the stem, whereas those of *E. tetralix* point upwards.

Erica tetralix L.

1 . 3 4 5 6 7 8   . B

Bogs, acid marshes and wet, heathy grassland. Abundant in suitable habitats, but necessarily absent from most of the Burren.

1. Frequent. **3**. Bogs at Rinroe, Templebannagh and Ballyogan. In acid fen by a small lake S.W. of Barefield. **4**. Frequent in the bogs N. of Moycullen. By Ballycuirke L. In a transition from fen to bog by Ballindooly L. **5–8**. Abundant throughout.

First record: Babington (1836). Bogs between Clifden and Roundstone.

Foot's statement cited above must be interpreted as referring to Slieve Elva and the moorlands around Lisdoonvarna, for, although many calcifuge species occur in peaty pockets on the limestone, *E. tetralix* has never been recorded from them.

E. mackaiana Bab.   Mackay's heath

. . . . . **6 7** . . .

Bogs (mainly on the drier parts) and lake-margins; very local (Plate 2).

6. In rather small quantity on a bog 1½ km S.E. of Carna. **7**. In considerable abundance over an area of about 3 sq. km around L. Nalawney and the adjacent hillock, known in the literature as 'Craiggamore'; from here extending discontinuously but fairly frequently north-westwards to a point 3½ km E.S.E. of Clifden, and southwards to near Letterdife and to the region of L. Nalawney on the N. flank of Errisbeg.

First record: Hooker (1835), as *E. mackaii*.

The area delimited by Webb (1954) represents with fair accuracy the continuous distribution of the species in the Roundstone area, but a number of outlying stations have been added since, starting with T. A. Barry's discovery of the species more than 5 km distant to the north-west. For a summary of our present knowledge of the distribution, see Nelson (1981a).

In addition to the two areas described above and one in Co. Donegal, this species is known only from N. Spain, where its distribution coincides very nearly with that of the province of Oviedo (Asturias). It does not occur in France or Britain, and serves, therefore, as an exemplar of the most extreme form of Hiberno-Cantabrian distribution. By a strange coincidence it was discovered in Spain by Durieu and Gay and in Ireland by William McCalla almost simultaneously in the summer of 1835.

The Irish populations are sterile, for reasons not yet understood, although the plant sets seed well in Spain. It may be that the plant is self-sterile, and that, as it spreads readily by layering, populations over a considerable area consist of a single clone.

Despite its sterility, it hybridizes readily with *E. tetralix*, and, as the latter always accompanies it, hybrids are common. They are completely (or almost completely) sterile and no back-crossing appears to take place. The hybrid was until recently known as *E. × praegeri* Ostenfeld, but a recent unfortunate change in the International Code of Nomenclature means that it must now be called *E. × stuartii* Linton. This name was originally given to a peculiar teratological mutant of the hybrid, with deformed corolla and exserted anthers, which has been perpetuated vegetatively in cultivation from the single specimen collected about 1908. It was for many years misinterpreted as a hybrid between *E. mackaiana* and *E. erigena*.

Double-flowered variants of *E. mackaiana* have occasionally been found and are in cultivation.

E. ciliaris L.   Dorset heath

. . . . . . **7** . . .

Only one small colony is known, approximately 6 km S.E. of Clifden.

First record: Newman (1846). Reported as found by Mr Bergin not far from Clifden.

Although it is quite possible that further colonies of this plant may await discovery in the vast tract of largely unexplored blanket-bog between Clifden and Roundstone,

Fig. 14. *Erica tetralix* and *E. mackaiana*. The two erect stems on the left are *E. tetralix*, the remainder *E. mackaiana*. The widely spaced and more erect leaves of the former contrast clearly with the more crowded and outstanding leaves of *E. mackaiana*.

the fact that the only known population covers no more than a few square metres makes it undesirable to disclose its position exactly. For an account of its discovery, loss and re-discovery in 1965 see Webb (1966). There is some doubt as to whether the colony seen by Balfour is the same as that which is known today; it was, however, collected in its present site by T. F. Bergin in 1846 and identified by William McCalla of Roundstone (Mackay, 1859).

Although on the Continent and in England it is mainly a plant of fairly dry heaths, it grows here in a rather damp hollow on the edge of blanket-bog, surrounded by *Juncus effusus* and *Molinia caerulea*. These sometimes threaten to swamp it, and we have occasionally done a little discreet 'gardening' in its interest. It managed, however, to survive a bad fire in 1966. It seems never to set seed. The Irish population is distinct in never having glands on the tips of the stout marginal hairs on the leaf; this condition occurs occasionally in English plants, but gland-tipped hairs are much commoner.

A single plant of the hybrid with *E. tetralix* (*E. × watsonii* Benth.) was found by MS in 1971, immediately beside the colony of *E. ciliaris*.

E. cinerea L.

1 2 3 4 5 6 7 8   A B

Heaths, banks, and other peaty and fairly dry habitats; less often on wet moorland, drier parts of blanket-bog or limestone pavement.

1. Very frequent. **2**. Occasional in peaty pockets on limestone pavement. **3**. Bogs at Rinroe and Ballyogan, and on pavement in several places between Gort and Slieve Carran. **4**. Frequent on and around the bogs in the north. **5–8**. Abundant.

First record: Babington (1836). Bogs between Clifden and Roundstone.

This species is, generally speaking, much more intolerant of calcareous conditions than is *Calluna vulgaris*; its presence, therefore, on shallow soil overlying limestone pavement is one of the more striking anomalies of the Burren flora.

E. erigena Ross (*E. mediterranea, E. hibernica*)   Mediterranean heath

. . . . . . 7 .   . .

Lake-shores, streamsides and wet bog; very local.

7. In fair quantity around L. Nalawney, spreading up the feeding streams for some distance, and rather sparsely down the effluent stream and over the saddle to L. Bollard, where there are a few plants on the S. shore; there is also an outlying colony on wet, level bog about 1 km S.W. of L. Nalawney. By drains and among rocky outcrops in bog on the promontory 2 km N.E. of Gowlaun, extending southwards nearly to the road. *Mountains about Kylemore; on each side of Killary Bay; and about Salruck pass; J. MacKinnon* (CH2).

First record: Mackay (1830). 'On a declivity, by a stream, in boggy ground, at the foot of Urrisbeg mountain...on its western side.'

All likely habitats in the mountains north of Kylemore have been searched several times in vain, as has also the 'pass of Salruck'. There may, however, be stations not recently recorded on the S. side of Killary harbour, E. of Salruck.

For the justification of the confusing changes of name which this species has suffered in recent years see Ross (1967) and Dandy (1969). The disappearance of the familiar epithet *mediterranea* has, however, one advantage, in that it removes a misunderstanding, for the distribution of the plant is Atlantic, not Mediterranean. There are only two places in S. Spain where it is found (and in both cases in small quantity) anywhere near the Mediterranean Sea. Its main headquarters is in Co. Mayo; outside Ireland it is found in fair quantity in Portugal, and sparsely and widely scattered in various parts of Spain. One station was known in S.W. France (N.N.W. of Bordeaux), but the plant is probably extinct there now. It does not occur in Britain or in N.W. France.

The normal flowering season is from late March to the end of April, but a few flowers are often to be seen open in late autumn or winter. Unlike the other two rare heathers of Connemara, it is fully fertile and reproduces well by seed.

Daboecia

Daboecia cantabrica (Huds.) Koch   St Dabeoc's heath   *Fraoch gallda, Fraoch na h-aon choise*

. . . **4 5 6 7 8** . .

Heaths, rough grazing, grassy banks and gorse-thickets; more rarely on cliffs or rock-outcrops. Frequent to abundant in suitable habitats throughout Connemara; not known elsewhere (Plate 2).

4. Locally abundant by a stream and by a lane beside it above the N. end of Ballycuirke L. About 1 km S. of Oughterard quay. **5–8.** Very frequent to abundant, but absent from Inishbofin.

First record: Petiver (1703).

Outside the region of the Flora this species is known from S. Mayo, but from nowhere else in Ireland or Britain. On the Continent it ranges from C. Portugal through N. Spain and W. France as far as the Loire. In Ireland its area is limited on the south and west by the sea, and on the east by the limestone, on to which it never ventures. The basis for its fairly sharp northern boundary is not so clear.

Like many heathers it produces occasional albino mutants with white flowers and pale green foliage; these are cultivated, as is also a variant (cv. 'praegerae'), with reddish pink instead of purple flowers. The latter was discovered in Connemara by Mrs Praeger.

St Dabeoc worked mostly in Co. Donegal, and why his name should be associated with a Connemara plant, or why Linnaeus transposed its two middle vowels are questions which we cannot answer.

Rhododendron

*Rhododendron ponticum L.

1 . 3 . **5 6 7 8** . .

Woods, hedges, rocky mountainsides and bogs. Very rare on the limestone; elsewhere widespread and locally abundant.

1. A few bushes by Drumcullaun L. Frequent in the woods on Clifden Hill. **3.** A few bushes by a lake S.E. of Tubber; overlying limestone but on deep, partly leached soil. **5.** Occasional on bogs near the N.E. margin; also near L. Formoyle. **6.** Occasional in the woods at Glendollagh; also in the plantation S.E. of the road opposite the gate of Ballynahinch Castle. **7.** Extremely abundant around Kylemore, extending eastwards to L. Fee and westwards to beyond Letterfrack; also, more sparingly from Letterfrack to Renvyle. Ballynahinch woods. Ballinaboy. **8.** Occasional throughout, including a few of the islands in L. Corrib, and locally abundant, as near Cong and in woods near Teernakill Br.

First record: Atlas (1962).

It is only in the Kylemore area that this species is spreading with the aggressiveness which it shows at Killarney; here, however, it has spread far up the hillside and obliterated much of the original vegetation in the woods. It has also taken complete possession of a small rocky island in L. Fee. Although seedlings and young plants are very common on wet bogs they do not thrive and seldom reach flowering size; for successful growth reasonably good drainage is needed.

Although there are no early records it seems probable that it is only in the last hundred years that the plant has become widely naturalized. Unfortunately it was not, until about 1950, considered a 'respectable' alien, and Floras before that date, which chronicled the wallflower and the cornflower with loving care, made no mention of the rhododendron.

It is of interest to note that its remains have been found in an inter-glacial deposit near Gort, only just outside the Flora region. As a native plant today, however, its nearest stations are in S. Portugal.

Pernettya

*Pernettya mucronata (L. fil.) Spreng.

. . . . **5** . **7** . . .

Moorlands and woods; rare.

5. A few plants established on moorland near Formoyle L. **7**. Well naturalized in woods at the W. end of Ballynahinch L.

First record: Webb (1982).

PYROLACEAE

Pyrola   Wintergreen

Pyrola media Sw.

. **2** . . . .

Peaty pockets on limestone pavement on the hills; rare.

2. Occasional on the summit plateaus of Gleninagh, Carnsefin and Cappanawalla Mts., descending the S.W. slope of Gleninagh to about 215 m (700 ft). *On the summit plateaus of Slieve Carran, Gortaclare, the hill S.E. of Ballyvaughan and the northern (limestone) shoulder of Slieve Elva* (Foot, 1864).

First record: Foot (1864). 'It grows in tolerable abundance in many places...always ...in heathery spots, and generally along with *Empetrum nigrum* and *Rubus saxatilis.*'

Foot's stations in the south and east of the district have not been exhaustively searched, and it is probable that the plant survives in at least some of them. The Burren hills represent its southern limit in Ireland, and in Britain it is almost confined to Scotland.

Monotropa

Monotropa hypopitys L.   Yellow bird's-nest

. . **3 4** . . . .

Woods and scrub on basic soils; very rare.

3. Garryland wood, under beech trees, 1979 (J. Cross). **4**. On a limestone island (Ilaunmahon) in the S.W. part of L. Corrib, E. of Gortachalla L., 1974 (Roden, **DBN**).

First record: Roden (1979). Ilaunmahon.

A very rare plant in Ireland; it may be noted, however, that it is known from two stations which lie only just outside the limits of this Flora: Castle Taylor (N. of Ardrahan), and Ballybeg L. (S. of Ennis; mis-spelt 'Ballybay' in Praeger, 1934*d*).

The habitat on the L. Corrib island is unusual, as the plant is normally associated with either pine or beech; here, it was found in hazel-scrub. There is, however, a record from a similar community at Dooney Rock, by L. Gill, Co. Sligo (ITB).

EMPETRACEAE

Empetrum

Empetrum nigrum L.   Crowberry

1 2 . . [5]. **7 8** . **B**

Rocky ground or moorland on mountains; more rarely on lowland heaths near the sea. Frequent in parts of the Burren and Connemara, but local.

1. Heathy ground near the Cliffs of Moher. **2**. Abundant on the high ground of the northern Burren hills, from Carnsefin eastwards to Moneen; much more sparingly on low ground above Poulsallagh and Black Head. **7**. Bengower and Benlettery. *Muckanaght* (ITB). Benchoona. Occasional on heathy ground near sea-level from near Cleggan to Salruck. Inishshark. *Inishbofin* (Praeger, 1911). **8**. Frequent on Leckavrea, and N. of the Maumeen gap on the Maumturk range, mostly above 535 m (1750 ft). Occasional on the Maamtrasna plateau.

First record: Wade (1802). 'Mountainous heaths, Cunnamara on Buffin Island.'

Reported from Gentian Hill (district 5) in 1885 by Johnston (**DBN**), but the record requires confirmation, because although Johnston's plants are rightly named many of them are inaccurately labelled as regards locality.

PLUMBAGINACEAE

Limonium   Sea lavender

Limonium humile Miller

. **2 3 4 5 6 7** . . .

Salt-marshes and muddy or stony sea-shores. Occasional to frequent around Galway Bay; rare elsewhere.

2. Rine Point. Frequent at Finavarra and New Quay. **3**. S.E. side of Aughinish, and at the head of Corranroo Bay. **4**. Frequent from Kilcaimin southwards. **5**. Gentian

Hill. **6.** At the end of the Carraroe peninsula. By the bridge at the N. end of Gorumna I. **7.** By the salt lake at Ardbear, and by L. Athola. Not seen N. of Clifden.

First record: CH1 (1866), as *Statice bahusiensis*. 'Reaches as far north as Clifden, Connemara, on the west...side of Ireland.'

L. transwallianum (Pugsl.) Pugsl.

. 2 . . . . . . A .

Maritime limestone rocks; very local.

2. In fair quantity at Poulsallagh, and more sparingly here and there northwards towards Black Head. Locally frequent on cliff-tops on the S.W. side of Inishmore.

First record: More (1860), as *Statice occidentalis*. Black Head.

This plant, originally described from Pembrokeshire, S. Wales, is one of a number of apomictic segregates from *L. binervosum*; authorities are not agreed on the best taxonomic treatment for them, as it would appear that almost every local population of *L. binervosum* is slightly different from all others. The plants from the mainland and Aran appear to be identical, but they differ slightly from those from Wales, and recent work suggests that they should, perhaps, be referred to *L. paradoxum* Pugsl.

It seems unlikely that *L. binervosum* in the narrower sense grows in the region of this Flora. Moore & More (CH1) say of the records of *Statice occidentalis* from Black Head and Aran that 'the form found in the last two localities differs remarkably in appearance from the ordinary state of the species..., being only about half the size, with a less branched panicle'. This makes it clear that they are talking of *L. transwallianum*. They also say that it is not hardy in Dublin; this is contrary to our experience, as it has not only survived several winters at the Trinity College Botanic Garden, but has seeded itself freely.

The *Atlas* record for 93/28 (coast near Gleninagh Castle) needs confirmation. It is attributed to *L. binervosum* in an inclusive sense, and it might be based on a misidentification of *L. humile*; more probably it represents an extension of *L. transwallianum* a little further to the east than is given by recent observations.

For a full discussion, see Baker (1954).

Armeria

Armeria maritima (Miller) Willd.   Thrift, Sea-pink

1 2 3 4 5 6 7 8 A B

Salt-marshes, maritime rocks, grassy cliff-tops and other places by the sea; also on rocky mountain summits. Very frequent and locally abundant by the sea; rare on the mountains.

Especially abundant in district 6 and the western part of district 7, but there is no substantial stretch of coast from which it is absent. On mountain-tops it appears to be confined to the Twelve Pins, where it has been recorded from Muckanaght, Benbaun, Benlettery, Bengower and Benbreen.

First record: Babington (1836). On the shore at Leenane.

Some plants from the Burren, Aran and Connemara are rather densely grey-pubescent, and have sometimes been referred to *A. pubigera* (DC.) Boiss. This is, however, a very obscure taxon, originally described from a cultivated plant of unknown origin, and supposedly distinguished by a shrubby habit and short, broad, triangular, stiff leaves. It seems more likely that the hairy Irish plants (which have particularly narrow leaves) come under the head of one of the numerous varieties of *A. maritima* to which the epithet *pubescens* has been affixed by various authors. In view of the great polymorphism of the species it is impossible to delimit this satisfactorily, even as a variety.

PRIMULACEAE

Primula

Primula vulgaris Huds.   Primrose

1 2 3 4 5 6 7 8   A B

Woods, scrub, hedges, pastures, banks, sea-cliffs and sand-dunes. Very frequent to abundant throughout; absent only from areas of continuous blanket-bog.

First record: Wright (1871). Aran Islands.

In western Ireland the primrose is much less dependent on shade than in most of Britain, and is often to be seen in abundance in exposed habitats.

P. veris L.   Cowslip

1 2 3 4 5 . 7 .   A .

Pastures, meadows and roadsides; frequent on the limestone, rare elsewhere.

1. By a stream S. of Hag's Head. Frequent on roadsides 3–5 km E. of Inagh. Roadsides between Ennistymon and Lehinch, and occasional on the dunes at Lehinch. **2–3.** Frequent throughout, though doubtfully native in Aran. **4.** Frequent around Galway; occasional elsewhere. **5.** Dunes near Spiddal. Pastures W. of Salthill. Pastures half-way between Galway and Moycullen. On very damp and apparently acid grassland near Tonabrocky, N.W. of Galway. **7.** Behind the 'coral strand' at Ballyconneely.

First record: CII1 (1866). 'Professor Melville finds the Cowslip to the west of the town of Galway.'

Hybrids between the cowslip and primrose (*P. × tommasinii* Gren. & Godr.) are found almost throughout the range of the former in the region. They are especially frequent in the Burren – perhaps more so than anywhere else in the British Isles – and have also been recorded from Roadford and W. of Inagh (district 1), between Galway and Oranmore (district 4) and near Gentian Hill (district 5).

As in several other parts of Ireland, occasional plants of the cowslip have orange-brown instead of golden-yellow flowers.

*P. japonica Gray

. . . . . . 7 .   . .

7. Naturalized in a meadow behind the Youth Hostel N.N.W. of Salruck, 1975 (M. S. and D. McClintock).

First record: Webb (1982).

Lysimachia

Lysimachia vulgaris L.   Yellow loosestrife

1 2 3 4 5 . . 8   . .

Lake-shores, fens, river-banks and wet woods; local.

1. By L. Raha. Swamp by the E. side of Inchiquin L. **2**. Fen near Turlough village. **3**. By the R. Fergus at Dromore and above Ennis. Around several of the turloughs, as at Coole, Gort, and S. of Mullaghmore. Castle L. **4**. Frequent. **5**. In acid fen by a lakelet N.W. of Rosscahill. **8**. Fairly frequent along the shore of L. Corrib from Cong to Oughterard, and on several islands; also on the S. shore of L. Mask, and in marshes near these lakes.

First record: Wade (1802). 'Bilberry Island, Lough Carib.'

L. punctata L., a somewhat similar species from E. Europe, has been planted on roadsides between Moyard and Letterfrack and shows some tendency to spread; it may soon become naturalized.

L. nemorum L.   Yellow pimpernel

1 2 3 4 5 6 7 8   A B

Woods, scrub, ditches, banks and damp grassland. Occasional on the limestone; frequent and locally abundant elsewhere.

1. Frequent. **2**. Occasional, mainly in the eastern half; rare on Aran. **3**. Woods at Dromore and Garryland. **4**. Woods and scrub by Ross L. and north of it. **5–6**. Occasional. **7**. Frequent. **8**. Abundant.

First record: Babington (1836). Mountains above Maam.

‡**L. nummularia** L.   Creeping Jenny

. . . . . . [7] **8** . .

[7]. *Streamstown* (Wade, 1802). **8**. Wet grassland and stony ground by the mouth of the Oughterard R., and along the lake-shore to Oughterard quay; also in wet woods at Currarevagh.

First record: Webb (1982). Wade's record has usually been dismissed as an error.

It is very difficult to decide whether the plant is native here or not. In many of its Irish stations it is certainly an escape from gardens, but it seems fairly certain that it is native in many places in the north, in habitats similar to those it occupies at Oughterard.

Glaux

Glaux maritima L.   Sea milkwort

1 2 3 4 5 6 7 8   A B

Salt-marshes; less often on wet, rocky or shingly places near the sea. Abundant in most of Connemara, more local elsewhere.

1. Frequent around Lehinch. **2**. Frequent in the north-east; not seen W. of Rine Point. **3**. Frequent. **4**. Carrowmore. Estuary of the Dunkellin R. L. Atalia. **5**. Frequent. **6–7**. Abundant. **8**. Leenane.

First record: Babington (1836). On the shore at Leenane.

Anagallis

Anagallis arvensis L.   Scarlet pimpernel

1 2 3 4 5 6 7 (*8*) **A B**

Sand-dunes, cultivated fields, roadsides, and in a very wide variety of dry, more or less open habitats. Frequent over much of the region, but rare over considerable areas.

1. Roadsides S. of Lehinch. *Near Ballynalackan Castle* (Atlas). **2**. Rine Point. Near Bealaclugga, and in a farmyard 3 km to the S. Tilled fields near Corcomroe Abbey. Abundant on Aran. **3–4**. Frequent. **5**. Sandy shore 4 km W. of Spiddal. Old railway track at Oughterard, 1965. Roadsides near Gentian Hill. **6**. Frequent. **7**. Frequent in the extreme west, and occasional on the coast elsewhere. *(8)*. *Clonbur and Cornamona* (Atlas).

First record: Wright (1871). Aran Islands.

Nearly always in calcareous or submaritime habitats.

A. tenella (L.) L.   Bog pimpernel

1 2 3 4 5 6 7 8   A B

Bogs, lake-shores, acid fens and marshes, and wet peaty soil in many situations.

1. Frequent near the coast, rather rare inland. **2**. Abundant on damp grass near Glensleade Castle. In the marsh S. of Feenagh. In flushes above Black Head and on Moneen Mt. Occasional on Aran, sometimes in relatively dry situations. **3**. Occasional throughout. **4**. Frequent. **5–8**. Abundant.

First record: Ogilby (1845). 'To be found running at the bottom of every ditch-bank [near Maam Cross]'.

Although some flourishing populations can be seen in very calcareous habitats, it is enormously more abundant in acid conditions.

A. minima (L.) Krause (*Centunculus minimus*)   Chaffweed

1 . . . 5 6 7 8   . B

Bare, peaty soil, usually near the sea, usually mixed with *Radiola linoides*. Rather rare, but doubtless often overlooked.

1. Frequent on banks in a lane between Kilnamona and Inagh. Cliffs of Moher. **5**. By the sea S.W. of Costelloe Br. **6**. Near the 'coral strand' S.W. of Carraroe. **7**. Near Derryclare L. Locally frequent on both sides of Ballynakill harbour. Inishbofin. *Clifden* (ITB). **8**. On peaty sand near the road at Killateeaun. *Plentiful on the shore of L. Corrib, N. of Oughterard* (Praeger, 1906).

First record: Wade (1802). 'In the watery, sandy heaths near Ballynahinch.'

Samolus

Samolus valerandi L.   Brookweed

1 2 3 4 5 6 7 8   A B

Marshes, lake-shores, margins of turloughs, ditches, edges of salt-marshes and other damp places. Very frequent in district 4 and the northern part of 3; also in district 6 and the western part of 7; occasional elsewhere.

First record: Wright (1871). Aran Islands.

In Ireland this species occurs mainly near the sea or near one of the large limestone lakes. Most of the stations in the Flora region conform to this pattern, but there are some half-dozen inland stations which are not close to L. Corrib or L. Mask.

OLEACEAE

Fraxinus

Fraxinus excelsior L. Ash *Fuinnseóg*

1 2 3 4 5 6 7 8 A .

Woods, hedges and rocky ground. Abundant on the limestone; occasional to frequent elsewhere.

1. Occasional, mainly in the east. **2–4**. Very frequent to abundant, except in the most exposed situations. **5**. Frequent. **6**. Roadside W. of Maam Cross. Woods at Glendollagh. Between Cashel and Toombeola. **7**. Occasional. **8**. Frequent to abundant in the east, rarer in the west.

First record: Foot (1864). 'Very abundant in the dwarf form.'

Ligustrum

*Ligustrum vulgare L. Privet

1 . 3 . 5 . . 8 .

Sea-cliffs, wood-margins and scrub; rare.

1. On low cliffs 5 km W. of Liscannor. **3**. Incipient scrub on rough grazing at Carrowgariff. *Garryland* (ITB). **5**. Self-sown in woods at Spiddal. **8**. Wood-margin at the Hill of Doon. Surviving by a ruined cottage at Maam. Frequent on rocky outcrops by the road W. of Cong.

First record: ITB (1901). H16.

The above are the only stations observed where the privet can be regarded as naturalized. It is occasionally seen as a planted hedge, mainly in districts 1 and 5, but not nearly as frequently as in most of Ireland.

GENTIANACEAE

Blackstonia

Blackstonia perfoliata (L.) Hudson Yellow-wort

. 2 3 4 5 . 7 8 . .

Dry grassland, sand-dunes, limestone pavement and stony places. Widely distributed on the limestone, though seldom very common; very rare elsewhere.

2. In wettish grassland S.S.E. of Feenagh. Sand-dunes at Rine Point. Abundant on the dunes at Fanore. Grassland S.W. of Ballyvaughan. S.S.E. of the Slieve Carran cliffs. Occasional on Aran. **3–4**. Frequent. **5**. On the old railway 1½ km N. of Rosscahill, and near the former station at Oughterard. Gentian Hill. **7**. Occasional on the Slyne Head peninsula. In a meadow by the 'coral strand', Ballyconneely. On the promontory beyond Dog's Bay. **8**. On an island in L. Corrib N.N.W. of Oughterard.

First record: Mackay (1836). 'South Isles of Arran; *Messrs. Ball* and *Thomson.*'

Not seen on the limestone W. of Cong, where it might have been expected. The *Atlas* record for 93/16 (Ennistymon region) is an error.

Centaurium

Centaurium erythraea Rafn (*C. minus*)   Centaury   *Dréimire Mhuire*

1 2 3 4 5 6 7 8   A B

Pastures and other grassland; frequent on limestone and near the sea; rather rare elsewhere.

1. Dunes at Lehinch. Cliffs of Moher, and by the quarries at Knocknalarabana. Near Moy Ho. **2–7.** Frequent. **8.** Occasional near L. Corrib and L. Mask; not seen elsewhere.

First record: More (1860). 'On an island called "Canova" in L. Corrib.'

Gentiana

Gentiana verna L.   Gentian

. 2 3 4 5 . . .   A .

Limestone pavement, pastures and stabilized dunes. Very frequent in the Burren; widespread but less frequent on Aran; occasional thence on the limestone northwards to near Oughterard.

2–3. Very frequent, but thinning out towards the eastern margin. **4.** Knolls above Menlough. Around Ross L. Eskers S. of Carrowmoreknock. By L. Corrib, E.S.E. of Oughterard. **5.** Frequent on Gentian Hill. *Knocknagreena* (P. cat.).

First record: How (1650). 'On the mountains between Gort and Galloway abundant, Mr. Heaton.'

This species is the most celebrated member of the Burren flora, and is rightly regarded as its chief glory. It flowers from mid-April to early June, with a maximum usually around 10–15 May. When out of flower it is very easily overlooked. Most plants have flowers of an intense azure-blue, but in a few they are much paler.

Praeger (1907) records it as growing 'on peaty banks on the overlying shales, amid a calcifuge flora, 700–800 ft. elevation, in several spots between Ballyvaughan and Lisdoonvarna'. There is only one short stretch of the road between these places that is above 700 ft, and we have searched the banks here in vain. It seems probable that it was introduced with limestone road-metal and did not persist.

From the Flora region, the gentian extends northwards to near L. Carra (Co. Mayo) and eastwards to Athenry; it is not known elsewhere in Ireland, and in Britain is confined to a small area of the N. Pennines. Although it is such a conspicuous component of the alpine pastures in all the principal mountains of Europe it is not an exclusively 'alpine' plant, as it is found at relatively low levels in C. Germany and (more significantly perhaps) on the karst of N.W. Yugoslavia. Its occurrence at sea level in the Burren is, therefore, although remarkable, not quite so paradoxical as many people think. Still less can it be regarded as an arctic–alpine, for, apart from one small area in the extreme north-east of European Russia, it is unknown in the Arctic, and also in Scandinavia.

Gentianella

Gentianella campestris (L.) Börner

(*1*) 2 . 4 5 6 7 .   A B

Heathy pastures. Fairly frequent near the sea in Connemara; very rare elsewhere.

(*1*). *Slieve Elva* (Atlas). **2**. Mullaghmore. *Near Black Head* (Atlas). Occasional on Aran. **4**. W. of Ballycuirke L. N. of Ross L. **5**. W. of Spiddal. *Gentian Hill*, 1899 (**DBN**). **6**. Frequent in S.E. part. **7**. Locally frequent near the sea, and on Inishbofin.

First record: Mackay (1806). 'On a limestone soil near Oughterard.'

G. amarella (L.) Börner

. **2 3 4** (*5*) . **7** .   . .

Stabilized dunes and dry, calcareous pastures, mainly around Galway Bay; occasional.

2. Dunes at Fanore, and on both sides of Ballyvaughan. By Muckinish Bay. **3**. Frequent by the E. shore of Ballyallia L. N.N.W. and E.N.E. of Kinvara. *N.E. of Corofin* (Atlas). **4**. N. shore of the Dunkellin R. estuary. N.E. of Ross L. Shore of L. Corrib near Burnthouse. *Near Menlough*, 1899 (**DBN**). (*5*). *Gentian Hill*, 1900 (ITB). **7**. By Aillebrack L.

First record: Mackay (1806). 'Very plentiful on a limestone soil between Gort and Corofin.'

A remarkable range of colour variants of this species may be seen on the dunes E. of Ballyvaughan, where, in addition to the normal dull violet-blue, numerous plants may be seen with clear, pale blue or with flesh-pink flowers. The pink form has also been noted on the shore of L. Corrib near Burnthouse. We can find no reference to these variants in the literature, although yellow- and white-flowered variants are recorded from Central Europe.

Irish plants tend to have larger corollas and more numerous internodes on the stem than do English ones, and for this reason have been distinguished as subsp. *hibernica* by Pritchard (1959). There is, however, a large overlap in both characters, and we are reluctant to recognize a subspecies which is defined purely on statistical biometric characters.

MENYANTHACEAE

Menyanthes

Menyanthes trifoliata L.   Bogbean   *Bearnán lachan*

1 2 3 4 5 6 7 8   A B

Marshes, bogs, fens and other very wet places. Abundant almost throughout, and without any clear preference for acid or calcareous substrata.

Widespread in all districts except 2, where suitable habitats are rare. It is, however, enormously abundant in the large pond at Burren village, occupying some hundreds of square metres to the exclusion of other species. There are also 2 colonies on Inishmore.

First record: Hart (1875). Inishmore: L. Atalia.

APOCYNACEAE

Vinca Periwinkle

*Vinca major L.

. 2 . . . . . . **A** .

This cottage-garden favourite is often seen in walls or hedges immediately beside or opposite the garden gate, but can hardly be regarded as naturalized there. Levinge (**DBN**) reported it as established on old walls W. of Ballyvaughan in 1892, but it does not seem to have persisted. In 1976, however, DW noted a large clump at the base of a small inland cliff near Gortnagapple on Inishmore, at some considerable distance from houses; it had probably been recorded from the same place by the *Atlas* recorder in 1955.

It does not set seed in Ireland, and its prospects, therefore, as a permanent member of the naturalized flora are somewhat speculative.

BUDDLEJACEAE

*Buddleja

*Buddleja davidii Franch.

. **2** . **4** . **6** . **8** . .

Waste ground, river-banks and lake-shores. Rather rare, but increasing.

2. Ballyvaughan harbour, and elsewhere in the village, 1978–9. **4**. Frequent on waste ground E. of Galway railway station; also by the river-bank near University College, 1969. **6**. Shore of a small lake N.E. of Carraroe, 1964. **8**. Hedges at Glann, 1967. By the river at Maam, 1968. Recess, 1979. Roadside E. of Maam Cross, 1971. Waste ground at Oughterard, 1980.

First record: Webb (1982).

Increasing rapidly in S. and E. Ireland; rather more slowly in the west and north.

**B. globosa* Hope, with spherical heads of orange flowers, has been seen in hedges or thickets at some distance from houses in districts 2 (S. of Ballyvaughan) and 8 (Glann), but the trees might have been planted there, and there is no firm evidence of its seeding itself among native vegetation.

BORAGINACEAE

Symphytum Comfrey *Lus na gcnámh mbriste*

‡Symphytum officinale L.

. **2 3 4** . . . . **A** .

In various habitats, usually subject to disturbance; rather rare.

2. Roadsides on Inishmore. Near a farm in the Caher R. valley. Roadside W. of Leamaneh Castle. **3**. Meadow *c.* 3 km N. of Ennis. By a farmyard S.W. of L. Bunny. **4**. Lake-shore near the ruins of Menlough Castle.

First record: Hart (1875). Inishmore.

Although doubtless indigenous in some parts of Ireland, in none of the stations cited above, except possibly that near Ennis, does this plant have the appearance of a native.

*S. × uplandicum Nyman (*S. peregrinum*)

1 2 3 . 5 . 7 . A .

Roadsides; rare.

1. E. of Lisdoonvarna. **2**. Near Kilronan. **3**. *c*. 3 km S. of L. Bunny. **5**. On the W. side of the Galway–Moycullen road, 8 km from Galway. **7**. Near Aughrusbeg L. Omey I. *Near Clifden* (Bangerter, 1954).

First record: Bangerter (1954).

A hybrid between the preceding species and *S. asperum* Lep., from S.W. Asia. It was originally introduced as a fodder-plant, but has never been widely cultivated in Ireland.

Pentaglottis

*Pentaglottis sempervirens (L.) Bailey (*Anchusa sempervirens*)

. 2 . 4 . . . .

2. Well naturalized along a lane 2½ km E. of Ballyvaughan; also on a roadside W. of New Quay. **4**. Waste ground in Galway city, near the river.

First record: Webb (1982).

Myosotis   Forget-me-not

The five species of forget-me-not found in the Flora region need a care in their identification. *M. discolor* is distinct in its very small flowers, which open pale yellow and then turn blue. *M. arvensis* differs from the remainder in having hooked hairs on the calyx, and in growing in dry places. Of those which grow in wet places *M. secunda* is usually easy to tell by the rough, outstanding hairs which clothe the lower part of the stem. *M. scorpioides* usually differs from *M. laxa* in its larger flowers, shorter pedicels and less deeply divided calyx, but there is some overlap in all these characters. The only certain distinction is in the length of the style (3–4 mm in *M. scorpioides*; 1–1½ mm in *M. laxa*).

Myosotis laxa Lehm. (*M. caespitosa*)

1 2 3 4 5 6 7 8   A B

Wet places; occasional to frequent.

Rather evenly distributed through the region; commonest in districts 4 and 7; scarcest in 2 and 5. It avoids the most acid soils, but is otherwise tolerant.

First record: Babington (1836). Maam.

M. scorpioides L. (*M. palustris*)

1 2 3 4 . . 7 8 [A] .

Marshy lake-shores, riversides and turloughs. Frequent on the limestone; rare elsewhere.

1. Occasional in the north, as by the river at Killinaboy, by the E. side of Inchiquin

L., and at Drumcullaun L. **2**. Abundant in the Carran turlough at Castletown, and frequent at L. Aleenaun. **3**. Very frequent. **4**. Occasional. **7**. By the lake at Bunowen. **8**. Frequent around Cong, and occasional by the shore of L. Corrib N.W. of Oughterard.

First record: ITB (1901). H16. Hart (1875) recorded it for Aran, but almost certainly in error for *M. laxa*.

M. secunda Murray (*M. repens*)

1 . . 4 5 6 7 8 . B

Ditches, marshes and streamsides. Frequent, though local, in acid habitats; very rare elsewhere.

1. Frequent. **4**. Ditches near Ballindooly L. **5**. Occasional. **6**. Frequent. **7**. Frequent in the extreme west; not seen elsewhere. **8**. Frequent around Maam; rare elsewhere.

First record: Corry (1880). Boggy ground near Lisdoonvarna.

There is no obvious explanation for the apparent discontinuities in the distribution of this species in Connemara. There has been so much confusion between it and the two preceding species that some of the *Atlas* records may well be erroneous.

M. arvensis (L.) Hill

(*1*) 2 3 4 5 6 7 8 A B

Roadsides, sand-dunes, cultivated ground, thin grassland, and a wide variety of other semi-open habitats. Widespread, but nowhere very common.

(*1*). *Lehinch region* (Atlas). **2**. Fairly frequent. **3**. Occasional in the N. half; not seen in the south. **4**. Frequent. **5**. Frequent by the coast. **6**. N. end of Furnace I. **7**. Occasional by the coast, from the Slyne Head peninsula to Inishbofin. **8**. By L. Mask, N. of Ferry Br.

First record: Hart (1875). Aran Islands.

M. discolor Pers. (*M. versicolor*)

. (*2*) 3 4 5 6 7 . (*A*) B

Cultivated fields, roadsides and other dry, open habitats; also in wet meadows, where it can grow up to 65 cm high. Widespread, but rather rare.

(*2*). *Kilmurvy* (Hart, 1875); perhaps only casual. **3**. By the ruined castle S.E. of Ruan. Grassland N.N.E. of Tulla. Roadside S.E. of Tubber. **4**. Roadside by the N. shore of Ross L. **5**. By the coast E. of Inveran. In a potato field S.E. of Costelloe. **6**. Frequent on the E. shore of Kilkieran Bay, as around Kilbrickan and south of it. **7**. Frequent in wet meadows on Inishbofin. *S.E. of Ballyconneely*, 1957 (DW).

First record: Hart (1875). Inishmore: about Kilmurvy.

Lithospermum

Lithospermum officinale L.

. 2 3 4 . 6 [7] . A .

Roadsides, waste places and river-gravels. Rather rare, and nearly always on limestone.

2. Sparingly in two places on Inishmore. By the Caher R. at Fanore Br. In a hedge near Burren village. *Roadsides near Ballyvaughan* (Foot, 1864). **3**. Roadside 3 km S.

of Kinvara. **4**. N.E. of Menlough. By a ruined cottage at Srue. *Aughnanure Castle*, 1906 (**DBN**); collected here also by McNab in 1832, but not seen recently. **6**. In a lane by an old churchyard near the coast 5 km N.W. of Carna.

First record: Foot (1864). Near Ballyvaughan.

A record by Isaac Carroll for Bunowen (district 7) is cited in CH1, but as the plant was said to grow on limestone rocks and there is no limestone here, it is best discounted.

CONVOLVULACEAE

Calystegia   Bindweed

Calystegia sepium (L.) R. Br.

This species occurs as two different subspecies, which are best treated separately. Subsp. *sepium* has white flowers, and is common throughout Ireland as a weed; subsp. *roseata* has pink flowers and slightly inflated bracteoles; it seems to occur only near the W. coast. The latter is certainly native in the region; the former only doubtfully so.

‡Subsp. **sepium**

1 2 3 4 5 6 7 8   A .

Widely distributed, and occasional to frequent in most districts, but chiefly in those that are extensively settled and farmed. It is frequent on Aran, though obviously introduced; it has not been seen on Inishbofin.

Subsp. **roseata** Brummitt

. 2 3 4 5 6 7 8   . B

2. Fairly frequent by the N. coast, from Ballyvaughan to Abbey Hill. **3**. 2 km S.W. of L. Bunny. *Near Kinvara*, 1958 (**TCD**). **4**. In Galway city, and *c*. 6 km from the city on the Oughterard road. **5**. Around Costelloe. In a marsh N. of Kilbrickan. **6**. Lettermullan. **7**. Occasional along the coast, from Dog's Bay to Renvyle. In marshes on Inishbofin. **8**. Near Leenane, and in several places around Clonbur.

Although it is doubtless spread by human agency, this plant seems certainly native on Inishbofin and N. of Kilbrickan.

First record (for the species): Wade (1802). '...with flesh-coloured flowers. Buffin Island.'

***C. pulchra** Brummitt & Heywood (*C. dahurica*)

. 2 . 4 . . . . . .

Hedges; very rare.

2. Near Turlough village. **4**. At Aughnanure Castle.

First record: Webb (1982). Earlier records probably refer to *C. sepium* subsp. *roseata*.

We follow for this genus the taxonomy of *Flora Europaea*, but it seems to us that this plant differs from *C. sepium*, subsp. *roseata* less than the latter does from subsp. *sepium*, and its claim to specific status is therefore dubious.

Plate 3. (*a*) *Spiranthes romanzoffiana*. (*b*) *Neotinea maculata*. (*c*) *Orobanche alba*. (*d*) *Potentilla fruticosa*.

(a)

(b)

(c)

Plate 4. (a) The Cliffs of Moher from the south. The tower stands at about 165 m (540 ft) above sea level. (b) *Rhodiola rosea* (roseroot) at the Cliffs of Moher. (c) *Adiantum capillus-veneris* (maiden-hair fern) at Poulsallagh.

*C. silvatica (Kit.) Griseb. (*C. sylvestris*)

. 2 . . . . 7 . . .

Hedges; very local, but abundant in two places.

2. S. of Ballyvaughan, on the Corofin and Lisdoonvarna roads. **7.** Westwards from Letterfrack to Ballynakill harbour.

First record: Atlas (1962).

C. soldanella (L.) R. Br.

. 2 . . . . 7 . A .

Sandy shores and unstable dunes; rather rare.

2. Occasional on the dunes at Fanore, and on all the principal dunes on Aran. *Fisherstreet*, 1954 (DW). **7.** On the Slyne Head peninsula, W. of Aillebrack L. and elsewhere. Sandy shores S. of Claddaghduff, and in some abundance on dunes N. of Aughrusbeg L.

First record: Andrews (1845). Inishmore.

It is very curious that this plant which, at least in mid-summer, is conspicuous and unmistakable, should have escaped detection in Connemara for so long. The credit for its discovery there belongs to Sir Alan Hodgkin, who observed it near Claddaghduff in 1963 and showed it to DW in 1964.

Evans & Turner (1977) publish a photograph of *Eryngium maritimum*, allegedly taken by R. J. Welch at Dog's Bay, in which *Calystegia soldanella* is prominent in the foreground. We find it very difficult to believe that the latter can have been overlooked in a place so thoroughly searched in the past, not only by Praeger, but by almost every botanist who has visited Connemara, and where, despite its recent discovery further north, the plant has not been recorded in recent times either. We think it more probable that there has been some confusion over the *provenance* of the photograph; it may well have been taken at Lady's Island Lake, Co. Wexford, where both species grow together in abundance, and which Welch certainly visited to photograph *Otanthus maritimus*.

It is a rare species on the west coast; the next station northwards is on the Mullet in Co. Mayo, and southwards in N. Kerry. The *Atlas* gives a rather misleading impression of its abundance in the Burren, for the dots for 93/06 and 93/07 both refer to the limited station at Fisherstreet (which falls on a grid-line), and those for 93/17 and 93/18 to the dunes at Fanore.

Convolvulus

Convolvulus arvensis L.

. 2 3 4 5 . 7 8 A B

Roadsides, waste places, sand-dunes, limestone pavement, and at the base of walls. Frequent in the E. part of the Burren; occasional elsewhere.

2. In a hedge at New Quay. By a track near the dunes E. of Ballyvaughan. Occasional on Aran. **3.** Very frequent. **4.** Esker-gravels by the N. shore of Gortachalla L. By the N. shore of the Kilcolgan estuary. **5.** Roadside near Gentian Hill. *Near Inveran* (Atlas). **7.** At the base of wall near Clifden and on Omey I. Locally abundant on sand-dunes S. of Kingstown Bay, and on the Slyne Head peninsula near Truska L. Inishbofin. **8.** Near Cong.

Fig. 15. *Cuscuta epithymum* (dodder) on the dunes at Fanore.

First record: More (1872). 'In the great Island of Aran; *H.C.H.*'

All stations are either on limestone or near the sea.

Cuscuta

‡**Cuscuta epithymum** (L.) L.   Dodder

. 2 . . . . . . A .

Stabilized sand-dunes; locally abundant.

2. Very abundant on the dunes at Fanore. Frequent on dunes at Portmurvy and Killeany on Inishmore, and N. of the landing-stage on Inishmaan.

First record: Nowers & Wells (1892). Kilmurvy and Killeany.

Parasitic mainly on *Thymus praecox* and *Lotus corniculatus*, but also to some extent on *Galium verum*, *Asperula cynanchica* and a few other species. The stems may be either yellow or pink; both seem about equally common at Fanore.

A rare plant in Ireland, and nowhere certainly native. Although the habitats in our region suggest native status, the fact that it was not recorded on Aran by Hart (who certainly visited the Killeany dunes), nor from the mainland until nearly 20 years after its discovery on Aran (Druce, 1909) strongly suggests introduction, first to Aran and thence later to Fanore. Druce gives as his station 'a pasture about a mile south of Black Head'. As the Fanore dunes lie some 2 miles (*c*. 3 km) further south it seems possible that it existed for a short period in the pasture before establishing itself permanently on the dunes.

SOLANACEAE

Solanum

Solanum dulcamara L.   Woody nightshade   *Dréimire gorm*

. **2 3 4** . . **7 8**   A .

Shingle-beaches and stony lake-shores; less commonly in ditches or thickets or on river-banks.

2. Abundant on boulder-beaches on Aran, and occasional on roadsides. In a wall near Turlough village. *Fanore* (Barrington, 1877). **3**. Very abundant on a rocky bluff by the S.W. corner of Coole L., and occasional elsewhere nearby; also at Caherglassaun turlough. In a ditch N.E. of Dromore. **4**. Frequent E. of Galway, and in the city itself. On an island in L. Corrib, *c*. 7 km E. of Oughterard. **7**. On the shingle-beach at Renvyle. **8**. Frequent on the stony shores of the N.W. part of L. Corrib, from Cong to within 3 km of Oughterard, and also on many of the islands; especially abundant at the tip of the Dooros peninsula. In a gorse-thicket at Maam Cross.

First record: Wade (1802). 'On the beach at Renvi, Connemara.'

In some places in and around Galway all plants are white-flowered; in others white- and violet-flowered plants grow side by side. Some of the white-flowered plants are very hairy. Most of the maritime specimens approximate to var. *maritimum* Bab.

Hyoscyamus

*Hyoscyamus niger L.   Henbane

. . . **4** (*5*) . . .   . .

Stony sea-shores; very rare.

4. In fair quantity on the shore of Galway Bay E. of L. Atalia, 1969–80, but varying in abundance from year to year. *Oranmore and Rinville* (ITB). (*5*). *Shore W. of the river-mouth at Galway* (ITB).

First record: ITB (1901). 'Shore at Galway, '99–P.'

Casual or intermittent in its appearance in many of its Irish stations; it seems, however, to have established a precarious foothold for itself E. of Galway.

SCROPHULARIACEAE

Verbascum

Verbascum thapsus L.   Mullein

1 2 3 4 . . . **8**   A .

Roadsides, gravel-pits and other dry, open habitats. Locally frequent in the E. and S.E. parts of the Burren; rare elsewhere.

1. About half-way between Ennis and Corofin, on the W. side of the road. E.N.E. of Kilnamona. **2**. Frequent on Inishmore. Waste ground on the S. side of Ballyvaughan. **3**. Frequent. **4**. By Aughnanure Castle. Gravel-pit E. of Rosscahill. Porridgetown. **8**.

Near the shore of L. Corrib 3 km N.W. of Oughterard, and also 2½ km S.W. of Clonbur.

First record: Foot (1864). 'Very abundant in the low ground, particularly near the sea.'

Erinus

*Erinus alpinus L.

. **2** . **4** . . . . . .

Limestone walls and vertical rocks; rare.

2. On the stonework covering a well by the road E. of Black Head. On a vertical rock-face by the old road at Black Head (some distance above the modern road). **4.** On the wall of Renmore barracks.

First record: Praeger (1939). 'Above the main road near Black Head (Mrs Verschoyle).'

Chaenorhinum

[***Chaenorhinum minus** (L.) Lange]

. . (*3*) (*4*) . . . . . .

Railway-lines; probably extinct.

(*3*). *Tubber*, 1896 (CH2). (*4*). *E. of Galway* (ITB). *Moycullen*, undated (Phillips, 1924).

First record: CH2 (1898). 'Along the railway...at Tubber, near the Clare border, 1896.'

This species never got a very firm foothold in the extreme west, and has probably been extinct there since the systematic spraying of railway-lines with weed-killers began around 1939. There is in the *Atlas* a record for 1960 for the L. Bunny square, but it is not precisely localized and there are no details as to habitat. A railway runs through the square in question for less than a kilometre in the S.E. corner, but we think it best to omit the record pending confirmation.

Cymbalaria

*Cymbalaria muralis Gaertn. (*Linaria cymbalaria*)

. **2** . **45** . **78** **A** .

Limestone or mortared walls; more rarely on limestone pavement or shingle beaches. Occasional.

2. Locally frequent on pavement on Inisheer and Inishmaan; rarely on walls on Inishmore. **4–5.** Common in Galway city and its neighbourhood. *Spiddal* (Atlas). **7.** Walls near Clifden, on the N. side of the estuary. On a ruined house in Letterfrack. **8.** Walls in the N.E. part of Oughterard. Near the Hill of Doon. By the county-boundary bridge at Cong.

First record: Sim (1859). 'Common on old wall-tops in and about Galway.'

If it were not generally agreed that this species is endemic to southern Europe its presence on limestone pavement on Aran and on a shingle-beach E. of Galway would tempt one to consider it a native. It is remarkable that, although generally speaking calcicole, it has not been recorded for the Burren.

Scrophularia   Figwort   *Fothram*

Scrophularia auriculata L. (*S. aquatica*)

. 2 3 4 5 6 7 8   A B

Roadsides, waste ground, walls, hedges and wood-margins. Locally frequent in Connemara and near Galway; rather rare elsewhere.

2. Waste ground at Ballyvaughan harbour. Inishmore. **3**. Wood-margin at Dromore. On the parapet of the railway-bridge S. of Gort. **4**. Frequent around Galway and thence northwards. **5**. Frequent in the east; not seen in the west. **6**. Roadside 2 km E. of Carna. **7**. Frequent and locally abundant in the west; rather rare in the east. **8**. Occasional.

First record: More (1876). Inishbofin; in one place only, east of the village.

S. nodosa L.

1 2 3 4 5 6 7 8   A .

Hedges, roadsides, wood-margins and rocky ground. Locally frequent on the limestone; occasional elsewhere.

1–2. Frequent. **3**. Frequent between Corofin and Gort; rare elsewhere. **4**. Frequent E. of the Moycullen–Oughterard road; rare elsewhere. **5**. Occasional near the coast. **6**. Roadside 3 km W. of Cashel. **7**. Frequent on the Errislannan peninsula; rare elsewhere. **8**. Glann. Clonbur, and on the road to Ferry Br. By the river above Maam.

First record: Hart (1875). Aran Islands.

In parts of Connemara this species grows very luxuriantly and develops winged stems, which can cause it to be confused with the preceding species. It can always be distinguished, however, by its acute leaves and the very narrow membranous border to the sepals.

Mimulus

*Mimulus guttatus DC   Monkey-flower

. . . . 5 . (7) .   . .

Streamsides; very rare.

5. Established in fair quantity by the stream running into the E. end of L. Formoyle. **(7)**. *Stream N. of Ballynakill L.* (Praeger, 1934c).

First record: Praeger (1934c). 'Stream by Shanbollard, near Moyard (M. C. Knowles).'

M. luteus L. was recorded by Halliday *et al.* (1967) as collected in 1965 near the W. end of the 'Narrow Lake' at the S.W. corner of L. Mask, but his specimen in **TCD** indicates by the short but distinct pubescence of the inflorescence and the relatively short pedicels that it is really the hybrid *M. guttatus* × *luteus*. What seems to be another hybrid, probably involving *M. cupreus* L., is plentiful in a roadside ditch near Liscannor; it has large red patches on the corolla and sparse, but relatively long hairs on the pedicels.

Digitalis

Digitalis purpurea L.   Foxglove   *Méaracan phúca*
1 [2] . **4 5 6 7 8**   [A] .

Roadside banks, heathy grassland, scrub-margin, streamsides and disturbed ground. Frequent on siliceous soils; absent from the limestone.

1. Frequent, especially near the coast. **4**. By a lane leading to L. Corrib from the Oughterard road, 3 km N.W. of Galway. **5**. Frequent. **6**. Frequent in the east, occasional in the west. **7**. Occasional; mainly in the east and not on Inishbofin. **8**. Abundant.

First record: Wade (1802). '. . . rather scarce throughout Connemara, though here and there to be met with.'

Although rather commoner than Wade implies, the foxglove is scarce in much of the exposed coastal areas of Connemara and in the blanket-bog areas. The *Atlas* records for Inishmore and Burren village (district 2) are almost certainly errors.

Veronica   Speedwell   *Seamar chré*

‡**Veronica hederifolia** L.
. **2 3 4** . . . **8**   A .

Hedges and at the base of walls; rather rare.

2. Frequent at the base of walls on Aran. **3**. Ballinderreen. **4**. Hedges at Burnthouse. **8**. Near Oughterard.

First record: Hart (1875). Aran Islands.

A rare plant in western Ireland; its frequency on Aran is unusual.

‡**V. agrestis** L.
1 2 3 (*4*) (*5*) **6 7** .   A B

Cultivated or disturbed ground; rather rare, and obviously diminishing.

1. Garden weed 2 km S. of Lickeen L. **2**. Inishmore; rare. Formerly on the other two islands. **3**. On a disused track 3 km S.W. of L. Bunny. (*4*). *L. Atalia*; *Kilbeg ferry*; *Oranmore* (P. cat.). (*5*). *Gentian Hill* (P. cat.). **6**. Disturbed ground 3½ km S.S.W. of Carraroe. **7**. Doonloughan. Belleek. Inishbofin.

First record: Hart (1875). Aran Islands.

This species, which has declined everywhere in Ireland since 1900, seems to have disappeared from the intensively cultivated region around the head of Galway Bay, and now survives only on islands, mountain farms or remote peninsulas.

‡**V. polita** Fries
. (*2*) **3 4** (*5*) . **7** .   (*A*) B

Cultivated ground; very rare.

(*2*). *Ballyvaughan and Black Head*, 1899 (ITB). *Inishmore* (Praeger, 1895). **3**. 4 km W.S.W. of Kinvara, 1966 (JJM). **4**. N. side of Dunkellin estuary, 1964 (DW). (*5*). *Inveran*, 1895 (ITB). *Gentian Hill*, 1899 (**DBN**). **7**. One plant on Inishbofin, 1967 (DW).

First record: More (1876). Inishbofin; cultivated land and roadsides.

Nowhere seen in any quantity, and obviously decreasing rapidly. Both this and *V.*

agrestis seem to have been ousted by the aggressive *V. persica*, which was introduced to the region probably around 1870.

*V. persica Poir.

1 2 3 4 5 6 7 (*8*)   **A B**

Cultivated ground and roadsides. Very frequent in districts 3 and 4; occasional to frequent elsewhere, except in N.E. Connemara, where it seems to be rare, the only record for district 8 being from the *Atlas* (Leenane, 1957).

First record: Praeger (1895*b*). Inishmore.

*V. filiformis Sm.

. . . **4 5 6 7 8** . .

Grassy roadsides and damp waste ground. Occasional in Connemara; unknown elsewhere.

4. By the road from Oughterard to the quay. **5.** Waste ground by the former railway station at Oughterard. **6.** Roadside 1 km S.W. of Canal Br. **7.** Roadside bank 2 km E. of Clifden, and on a by-road S. of the town. **8.** Abundant on roadsides at Glann; also S.E. of Ferry Br. and immediately W. of Maam Cross.

First record: Atlas (1962).

This well-known weed of garden lawns is slowly but steadily spreading into native vegetation in various parts of Ireland. Many of the records, however, both in the *Atlas* and the literature merely refer to its presence in lawns.

V. arvensis L.

(*1*) **2 3 4 5 6 7 8**   **A B**

Roadsides, walls and semi-stable sand-dunes. Rare in district 1; widespread elsewhere, though nowhere really common. The only record from district 1 is from the *Atlas* (Cliffs of Moher, 1958).

First record: Hart (1875). Aran Islands.

V. serpyllifolia L.

1 2 3 4 5 6 7 8   **A B**

Grassy roadsides; less often in damp pastures or on rocky ground. Very frequent to abundant in Connemara; frequent elsewhere.

First record: Hart (1875). Aran Islands.

V. officinalis L.

1 2 3 4 5 6 7 8   **A** .

Roadsides, heathy grassland and limestone pavement. Very frequent to abundant throughout, though unexpectedly absent from Inishbofin.

First record: Wright (1871). Aran Islands.

V. chamaedrys L.

1 2 3 4 5 6 7 8   **A B**

Roadsides, scrub, wood-margins and grassland of nearly all types; abundant throughout.

First record: Wright (1871). Aran Islands.

V. montana L.

1 2 3 . . (*6*) . **8** [A] .

Woods and shady hedges; occasional.

1. Woods at Moy Ho. Beside the river at Ennistymon. In the ravine on the Aille R. above Roadford. **2**. Wood near Clooncoose. **3**. Garryland wood. Wood N. of Dromore. (*6*). *Thicket near Recess*, 1956 (DW). **8**. Hedge near the Hill of Doon. Woods N.W. of Clonbur and by the S. shore of L. Mask. *N.W. of Oughterard* (P. cat.).

First record: Wade (1802). 'Ballynahinch, Cunnamara.'

The earlier records for this species are most unsatisfactory. Although it has not been seen recently in the Ballynahinch woods it might well occur there; but Wade (1804) says that around Ballynahinch 'it is common in rather exposed situations', and Babington (1836) recorded it 'on the mountain behind the inn at Maam'. Both these statements run entirely counter to our experience: that it grows only in sheltered and shady situations, and on a soil which provides at least a little base. Even Praeger (FWI) raises doubts when he gives its distribution in the west of Ireland as 'chiefly in hilly country'; the data presented in the *Atlas* do not bear this out. We suspect that some of these records must have been based on slightly abnormal specimens of *V. chamaedrys*. On the same grounds we reject the *Atlas* records for the Clifden area and Fanore (both are based on field records). The *Atlas* record for Aran, however, seems to arise from an editorial error.

V. scutellata L.

1 2 3 4 5 6 7 8 . **B**

Ditches, marshes, lake-shores and wet meadows. Widespread, but frequent only in a few areas.

1. Occasional. **2**. By the turlough at Turlough village, and at L. Aleenaun. **3**–**4**. Frequent. **5**. W. and N.W. of Rosscahill. 4 km W. of Galway. **6**. Occasional. **7**. Occasional in the extreme west and by the N. coast; not seen elsewhere. **8**. Frequent.

First record: ITB (1901). H16.

V. anagallis-aquatica L.

1 2 3 4 5 6 7 8 **A** .

Ditches, streamsides, marshes and lake-shores. Frequent on the low-lying limestone; rather rare elsewhere.

1. In the R. Fergus at Killinaboy. Marsh near Elmvale Ho. **2**. L. Luick. L. Aleenaun. Carran turlough. Aran. **3**–**4**. Frequent. **5**. Very luxuriant in drains in the marsh 2 km N. of Killaguile Ho. **6**. Near Ardmore Point. **7**. Frequent on the Slyne Head peninsula. By the lakelet on Inishlackan. **8**. Frequent by the millpond at Cong. Ditches at Glann.

First record: Druce (1911). 'Near Galway.'

V. catenata Penn. (*V. aquatica*).

1 2 3 4 . . **7 8** **A B**

Ditches, turloughs and marshes. Frequent on the limestone; elsewhere only in the extreme west of Connemara.

1. Ditch near O'Dea's Castle. **2**–**4**. Frequent. **7**. Frequent in the marsh by the larger lake on Omey I. Inishbofin. **8**. Abundant by the millpond at Cong.

First record: Atlas (1962).

V. beccabunga L.   Brooklime

1 2 3 4 5 6 7 8 A B

Ditches, marshes, lake-shores and small pools.

Frequent throughout, except in district 2, where suitable habitats are rare; it is recorded, however, from two places in the valley of the Caher R., and from all three Aran Islands.

First record: Wright (1871). Aran Islands.

Hebe

This genus constitutes the shrubs usually known to gardeners as 'Veronica'. As they stand up well to salt-spray they are frequently cultivated in the coastal districts of Ireland, and have become locally naturalized on a small scale. In our region 3 species can be regarded as at least on the verge of naturalization:

Hebe salicifolia (Forst.) Penn., with narrow, willow-like leaves and small white flowers, sometimes tinged with pink; this has seeded itself on roadsides at Cashel;

H. × franciscana (Eastw.) Sout., of which there are some flourishing bushes on the walls of the castle at Killeany, though too high for certain identification (they might be *H. speciosa*);

H. elliptica (Forst.) Penn., which is naturalized S.E. of Hag's Head.

Limosella

Limosella aquatica L.

(*1*) **2 3** . . . [7] .   . .

Small pools, turloughs and lake-margins; rare.

(*1*). By the margin of Inchiquin L. (Levinge, 1893). **2**. In small, temporary pools in solution-hollows in the pavement at Poulsallagh. *Fisherstreet* (ITB). **3**. Fairly frequent around Coole L. Caherglassaun turlough. By the E. shore of L. Atedaun. *Tirneevin turlough, and the 'racecourse' turlough W. of Gort* (ITB).

First record: Levinge (1893). Inchiquin L. The record by Wade (1802) for Ballynahinch is, by general consent, dismissed as an error.

The only other known station in Ireland for this plant is in Co. Fermanagh.

Euphrasia   Eyebright   *Roisnin radhairc*

This attractive but notoriously difficult genus is so abundant over much of Connemara and the Burren that a few hints on identification may be useful. These notes, however, apply only to the genus as found in the Flora region.

Two species stand out by virtue of the long glandular hairs on the upper part of the plant; they are easily visible to the naked eye as a greyish down. *E. rostkoviana* is normally a tall, erect plant with large flowers; it has usually 2 or 3 slender, erect branches from near the base; *E. anglica* is more straggling, with often flexuous stems and slightly smaller flowers.

E. arctica also has glandular hairs on the lower side of the upper leaves, but they are very short and can be seen only when the plant is examined with a lens against the light. It is robust, with large leaves and flowers and usually a few long, spreading branches.

The remaining species have few glandular hairs or none (though they all have curved hairs without a glandular tip). *E. pseudokerneri* is in other respects similar to *E. arctica*, and grades into it; its flowers are even bigger. *E. tetraquetra* also grades into *E. arctica*, but in its typical form is recognized by its dense, 4-angled spike of flowers, like a church tower. *E. confusa* is very low-growing, with the stem either prostrate or extremely short. *E. scottica* is very slender, erect and sparingly branched, with small, narrow, blunt-toothed leaves and small flowers. The very rare *E. frigida* is rather like it, but with broader and more sharply toothed leaves.

All species so far mentioned have white flowers, only slightly tinged with purple, if at all. In *E. micrantha* and *E. nemorosa* the flowers are nearly always distinctly mauve. The former has small flowers and usually purple tinged stem and leaves. *E. nemorosa* is stouter and usually very bushy, with medium-sized flowers.

E. salisburgensis, one of the 'specialities' of the Burren, is very different from all the others. It has small, white flowers and very jaggedly toothed leaves much narrower than in other species. It is usually very bushy, and often copper-coloured all over. If in doubt examine the capsule; in all other species its upper margin is fringed with hairs, but in *E. salisburgensis* it is hairless.

Hybrids between different species undoubtedly occur, but they are so difficult to name with certainty that even experts usually qualify their determinations with a query. Hybrids of *E. salisburgensis* with other species are sterile, and occur, therefore, only as isolated individuals. Other hybrids are fertile, and one often finds what look like hybrid swarms. This is especially true of *E. micrantha* × *nemorosa*, *E. arctica* × *pseudokerneri* and *E. arctica* × *tetraquetra*.

Euphrasia salisburgensis Funck

. 2 3 4 5 . 7 8 A .

Shallow soil over limestone pavement; also on calcareous sands. Always, it would seem, parasitic on *Thymus praecox*. Abundant in the Burren and locally in W. Connemara; rare elsewhere.

2–3. Very frequent to abundant throughout; widespread on Aran, but less abundant than on the mainland. 4. Gravel-pit near Rinville. Limestone hill N. of Menlough. By L. Corrib, W. of Carrowmoreknock. *Common along the coast between Oranmore and Galway* (FWI). 5. Gentian Hill. 7. Dog's Bay and the lower slopes of Errisbeg. Abundant on the Slyne Head peninsula. 8. Pavement 2½ km N.W. of Cong.

First record: CH1 (1866), as *E. cuprea* Jord. 'Great Arran Island'. Oliver (1852) had earlier observed it on Aran and rightly guessed its identity, but deferred to Babington's erroneous naming of it as *E. gracilis* Fr. (now considered to be a synonym of *E. micrantha*). The first record under the correct name is by Townsend (1897).

This species ranges widely over the mountains of C. and S. Europe, and occurs at lower levels in Gotland (S.E. Sweden). Except for one unsatisfactory record for N. England, never confirmed, the nearest known station to Ireland is in the Vosges, 1200 km distant. In Ireland its headquarters is in the Burren, but it ranges in smaller quantity and discontinuously along the western limestones from Co. Limerick to Co. Donegal. Irish specimens differ slightly from those from the Continent, and have been distinguished as var. *hibernica* Pugsley.

E. micrantha Reichenb.

1 2 3 4 5 6 7 8 A .

Moorland, bogs and heaths; rarely in peaty pockets on limestone pavement or on tussocks in fens. Common in Connemara; rather rare elsewhere.

1. Bank by Derrymore Br. **2.** Summit of Cappanawalla. Inishmore (rare). **3.** Ballyogan and Rinroe bogs. On tussocks in a fen near Owenbristy. **4.** Knolls on the W. side of Ballycuirke L. Heathy ground S. of Carrowmoreknock. On tussocks in the fen at Coolagh. **5–8.** Very frequent, but not recorded for Inishbofin.

First record: Praeger (1902*b*), as *E. gracilis*.

Always, it would seem, associated with *Calluna vulgaris*.

E. scottica Wettst.

. **2 3 4 5 6 7 8** . B

Schoenus-fens; also occasionally on bog or moorland, usually in flushed areas. Frequent in the limestone lowlands; occasional elsewhere.

2. In the marsh S. of Feenagh. **3–4.** Very frequent in fens. **5.** Boggy ground 1½ km S. of Rosscahill. **6.** On Cashel Hill at *c.* 180 m (600 ft). By the outlet of Inver L. **7.** Abundant on the bog around L. Nalawney. Heathy slope N. of L. Fee. By the E. side of Derryclare L. Bog half-way between Clifden and Toombeola. Inishbofin. **8.** Occasional in the E., from the cliffs above L. Nadirkmore to near Oughterard.

First record: Praeger (1920*b*). Ross L.

E. frigida Pugsl.

. . . . . . **7** . . B

Mountain cliffs and exposed islands; very rare.

7. Fairly frequent on cliff-ledges on Bengower and more sparingly on Muckanaght (det. Yeo). Rocky grassland on Inishshark (det. Warburg).

First record: Webb & Hodgson (1968). Inishshark.

Normally a mountain plant; its occurrence at low level on Inishshark is very unusual.

E. confusa Pugsl.

. **2** . . . . **7 8** A B

Short grassland in exposed situations; very local.

2. Abundant at Rine Point. Frequent on Aran. **7.** Stabilized dunes at Dog's Bay and Doonloughan. *Renvyle* (Duncan, 1956). Inishbofin. **8.** At *c.* 300 m (1000 ft) on a north-east-facing slope above L. Nadirkmore. Sandy flats at Killateeaun.

First record: Duncan (1956). Renvyle.

As is usual in W. Ireland, plants are very dwarf and usually unbranched.

E. nemorosa (Pers.) Wallr. (inc. *E. curta*)

1 2 3 4 5 6 7 8 A .

In a wide variety of communities, but commonest in heathy pastures. Frequent on the limestone and in W. Connemara; rather rare elsewhere.

1. Heathy ground 2½ km S. of Fisherstreet. **2–3.** Frequent; abundant on Aran. **4.** Rinville. Very frequent in the area N. of Moycullen. **5.** Gentian Hill. **6.** Hillside above Cashel. S.E. corner of Gorumna I. **7.** Slyne Head peninsula. Belleek. Omey I. By Aughrusbeg L. **8.** Shore of L. Corrib, in several places. Hillsides N. of Maam and S.W. of Leenane.

First record: Oliver (1852). Near Roundstone.

E. tetraquetra (Bréb.) Arrond. (*E. occidentalis*)

1 2 3 4 5 6 7 .   A B

Cliff-tops, stabilized dunes and grassland near the sea; rarely inland on limestone. Frequent on exposed coasts; rather rare elsewhere.

1. Cliffs of Moher. **2.** Abundant on the dunes at Fanore and E. of Ballyvaughan. Poulsallagh. W. of New Quay. Aran. **3.** W. side of Kinvara Bay. Near Owenbristy. **4.** N. side of the Kilcolgan estuary. **5.** Costelloe Br., and on the E. side of Cashla Bay, near its mouth. **6.** Abundant by the 'coral strand' near Carraroe. S.W. of Kilbrickan. Furnace I. **7.** Frequent to abundant on most of the western dunes, from Dog's Bay to Claddaghduff and Omey I. Cliff-tops on Inishbofin and Inishshark.

First record: Praeger (1920*b*). Fanore.

E. arctica Rostr. (*E. borealis*, *E. brevipila*)

1 2 3 4 5 6 7 8   A B

Roadsides and pastures; common except in the south-west.

1. Meadows N. of Drumcullaun L. Near Elmvale Ho. Roadside W. of Kilnamona. **2.** Occasional, mainly in the east; rare on Aran. **3.** Frequent. **4–8.** Very frequent to abundant.

First record: Townsend (1897). 'Roundstone, 1852; *D. Oliver.*'

The commonest species in the region as a whole, as it is over most of Ireland. It is especially characteristic of grassland subject to some disturbance, as by field-gates and on roadside verges.

E. pseudokerneri Pugsl.

1 2 . . . .

Grassland overlying limestone pavement. Locally frequent in the W. part of the Burren; not recorded elsewhere.

1. On the limestone outcrop at Cahercloghane. **2.** Locally abundant in an arc extending from Poulsallagh through Formoyle, Feenagh, the Corkscrew Hill and Glensleade Castle to Carran.

First record: Census Catalogue (1972). 'Common in the Burren.'

A striking plant, of the general habit of *E. arctica* (with which it intergrades, doubtless by hybridization), but normally distinct in its larger flowers with very deeply divided lobes to the corolla-lip, and without glandular hairs on the leaves. It has been traced northwards from the Burren through Sligo to the limestones of S.E. Donegal. According to P. F. Yeo, it differs slightly from the *E. pseudokerneri* of the English chalk, but is closer to it than to any other described species.

E. rostkoviana Hayne

1 2 3 4 . 6 7 8 . .

Meadows, and rough grassland among bushes. Rather local.

1. Kilfenora. **2.** Frequent, but not seen on Aran. **3.** Near the 'race-course' turlough, W. of Gort. **4.** In two places by L. Corrib E. of Oughterard. **6.** Cashel. S.E. corner of Gorumna I. **7.** 3 km E. and 3 km S.W. of Clifden. Near the W. end of Ballynahinch L. N.W. of Moyard. Near Gowlaun school. **8.** Frequent.

First record: Druce (1909). Ballyvaughan.

E. anglica Pugsl.

. 2 3 4 5 6 7 8   A .

Heathy grassland; occasional almost throughout.

2. Hillside 4 km N.N.E. of Glencolumbkille Ho. Inishmore. **3.** 1½ km E. of Killinaboy. **4.** In two places N. of Ross L. **5.** By the old railway-track at Rosscahill and Oughterard; also on the hill S. of Oughterard station. **6.** At the outlet of Inver L. **7.** Roadsides near Derryclare L. and Ballynahinch L. By the wooded ravine N. of Dog's Bay. Doonloughan. **8.** Riversides near Maam. By L. Corrib, opposite the Hill of Doon. Maumturk Mts., above Ilion West.

First record: Webb (1980*a*).

Melampyrum

Melampyrum pratense L.   Cow-wheat

1 2 . . 5 6 7 8   A .

Woods, stony lake-shores, limestone pavement and blanket-bog. Locally abundant in parts of Connemara; rare elsewhere.

1. Pavement N.W. of Ballycullinan L. **2.** Pavement near the E. end of L. Aleenaun. On rocky outcrops in the marsh at Feenagh. Near Formoyle. Aran (very rare). **5.** Blanket-bog 3 km N.N.E. of Screeb, and also 2¼ km S.W. of Oughterard. **6.** Occasional, mainly in the north. **7.** Very frequent in woodland on the larger lake-islands; also in Ballynahinch wood. Occasional on lake-shores from L. Athola and Cloonagat L. southwards to L. Nalawney. Blanket-bog at Salruck and near Dawros Br. **8.** Locally abundant round the N.W. part of L. Corrib, both in woods and on the stony lake-shore; also on some of the islands. A few plants on blanket-bog N.W. of Maam.

First record: More (1860). 'Upon an island called "Canova" in L. Corrib.' Wade (1802), however, recorded *M. sylvaticum* at Ballynahinch, obviously in error for this species.

The ecological determinants which govern its distribution are not easy to grasp. It seems to be equally at home on limestone pavement and acid peat, in deep shade and in full exposure; yet it is far from ubiquitous. By the shore of L. Corrib, at and around Glann, it occurs in a prostrate form not seen elsewhere. Usually the flowers are bright yellow, but plants with much paler flowers occur here and there.

Odontites

Odontites verna (Bell.) Dum. (*O. rubra*)

1 2 3 4 5 6 7 8   A B

Roadsides and disturbed or heavily grazed grassland. Common over most of the limestone; occasional to frequent elsewhere.

1. Frequent. **2.** Very frequent in the east, but not seen much further west than Ballyvaughan; frequent, however, on Aran. **3–4.** Very frequent to abundant. **5.** Occasional by the coast. **6.** Occasional; mainly in the south and east. **7.** Frequent in the extreme west; very rare elsewhere. **8.** Near the middle of the Dooros peninsula. By the north shore of the 'Narrow Lake'.

First record: Ogilby (1845). Inishmore: 'everywhere a perfect weed'.

By the criteria adopted in *Crit. Suppl.* most plants are referable to subsp. *serotina* (Dum.) Corb., though one specimen in **TCD** from Aughrusbeg L. seems referable to subsp. *verna*, which is also recorded in the *Crit. Suppl.* for Ballyvaughan. But, if the additional criteria given in *Flora Europaea* are taken into account, a large number of plants are transitional between the two subspecies. We still have much to learn about the variation within this species.

Parentucellia

Parentucellia viscosa (L.) Caruel (*Bartsia viscosa*)

. . . . . **6 7** . . **B**

Heathy, rocky or sandy grassland; rare, and only in S. and W. Connemara.

6. Heathy ground on Annaghvaan I., 1971. Ardmore Point, 1976, 1981. By a lake N. of Carraroe, 1964. Roadside at Kylesa, 1967. Mweenish I., 1967. **7**. In several places at and near Dog's Bay, 1909–72. On the N. side of the Slyne Head peninsula, 1974–5. In two places on Inishbofin, 1911–67.

First record: Praeger (1909*b*). 'In a potato-field a quarter of a mile east of Dog's Bay.'

The fact that this conspicuous and unmistakable plant, long known from the south-west and from Donegal, was not recorded from Connemara until 1909 makes it fairly certain that it is a recent arrival in our region; and corroboration is given by its subsequent discovery in other counties where it was previously unknown (Sligo, Londonderry). Its distribution and habitats suggest, however, that its immigration took place by natural means.

Pedicularis

Pedicularis palustris L.　　Red rattle

1 2 3 4 5 6 7 8　(*A*) **B**

Marshes and other wet places; frequent but rather local.

1. Occasional. **2**. Very abundant in the marsh S.S.E. of Feenagh. *Inishmore* (Wright, 1871; Praeger, 1895*b*); not seen since and perhaps extinct. **3–5**. Frequent. **6**. By a pool near Glendollagh L. **7–8**. Frequent.

First record: Wright (1871). Aran Islands.

P. sylvatica L.　　Lousewort

1 2 3 4 5 6 7 8　(*A*) **B**

Heaths, moors and peaty pockets on limestone pavement. Abundant throughout Connemara; frequent elsewhere, except on Aran, where it has not been seen since 1869.

First record: Wright (1871). Aran Islands. First confirmed record: More (1876). Inishbofin.

Recorded from all three Aran Islands by Hart (1875), but not seen since; probably a long-term casual, introduced with turf.

Subsp. *hibernica* Webb (with hairy calyx) predominates in Connemara, and subsp. *sylvatica* in the Burren, but there are a few exceptions.

Rhinanthus

Rhinanthus minor L.   Yellow rattle

1 2 3 4 5 6 7 8   A B

Grassland of all types; abundant almost throughout.

1–4. Abundant. **5.** Frequent, mainly by the coast or the N.E. border. **6–7.** Abundant.
8. Rare in the south; frequent elsewhere.

First record: Hart (1875). Aran Islands.

Both subsp. *minor* and subsp. *stenophyllus* (Schur) occur in the region; the former
appears to be somewhat commoner, but data are not adequate for any confident
generalizations. Some plants are intermediate, and approximate therefore to the
aestival variant, named var. *elatior* Schur in *Flora Europaea*.

In the Burren there is a very distinctive variant (although it intergrades with more
typical plants) distinguished by its slender habit and very narrow and rather short
leaves. It is strongly reminiscent of subsp. *calcareus* (Wilm.) Warb., from the chalk
of southern England, but British botanists decline to admit it to this subspecies,
without specifying the characters in which it differs. It requires further study.

Lathraea

Lathraea squamaria L.   Toothwort

. 2 3 . . . . (8) . .

Woods; very rare.

2. Wood E. of Clooncoose, 1976, 1980 (JW). **3.** Garryland wood, 1977 (M. Mitchell).
(8). *Grounds of Ashford Castle*, before 1898 (cited in CH2).

First record: CH2 (1898). 'The Wilderness, Ashford, Cong; *Miss M. F. Jackson*.'

OROBANCHACEAE

Orobanche   Broomrape

Orobanche hederae Duby

. 2 3 4 . . . 8   A .

Parasitic on ivy growing over limestone rocks; very local.

2. Locally frequent on Aran. **3.** On walls and banks 4 km N.W. of Corofin. **4.** On
a wall about 1½ km N.W. of Moycullen. By Aughnanure Castle. **8.** Frequent just W.
of Cong.

First record: Oliver (1852). Inishmore.

We cannot ascertain the basis of the *Atlas* record for the Fanore region.

*O. minor Sm.

. 2 3 4 5 6 . .   A .

Pastures and disturbed ground. Frequent on Aran; rather rare elsewhere. Parasitic
mainly, though not exclusively, on clovers and other Leguminosae.

2. Frequent on Inishmore and Inishmaan. Turlough village. *Dunes E. of Ballyvaughan*

(Praeger, 1934c). *Corkscrew Hill* (Atlas). **3**. Near Boston. **4**. Field-margin at Kilcaimin. Railway embankment E. of Galway. *Oranmore*, 1952 (**DBN**). **5**. Wet pasture near Spiddal. **6**. Mweenish I.

First record: Praeger (1934c). Dunes E. of Ballyvaughan; also Inishmore. The record for Aran by Mackay (1806) is certainly an error for *O. hederae*.

O. alba Willd. (*O. rubra*)

. **2** . **4** . . . . **A** .

Limestone pavement, short grassland and fixed dunes; very local (Plate 3). Always, it would seem, parasitic on *Thymus praecox*.

2. Frequent, but local. Rare on Inishmore. **4**. S. of Carrowmoreknock.

First record: Graham (1840). 'Upon turf on a mound of limestone gravel, by the side of L. Corrib.' This station is probably close to or identical with that given for district 4, above.

LENTIBULARIACEAE

Utricularia   Bladderwort

The members of this genus are (except for the difficulty with *U. vulgaris* and *U. australis*) easily identified by their leaves, which is just as well as they are mostly shy to flower. *U. minor* is least so; on the other hand very few botanists have seen *U. intermedia* in flower. This last is distinguished by the fact that its bladders are on separate, leafless branches, not mixed in with the leaves. In *U. minor* the leaves are seldom more than 6 mm long; in *U. vulgaris* and *U. australis* they are 3–4 times that size.

Utricularia vulgaris (in the wide sense)

. . **3 4** . . (**7**) .   . .

Fens, drains and ditches; rare.

3. By the S. shore of L. George. Templebannagh bog. 3 km W. of Ardrahan. **4**. Coolagh fen. Occasional S. of Carrowmoreknock. *Ballycuirke L.* (Praeger, 1905). *River at Terryland* (ITB and Praeger, 1934c). (**7**). *Doonloughan* (Praeger, 1905). *Ballynahinch* (Wade, 1802).

First record: Wade (1802). Ballynahinch.

These records include *U. vulgaris* L. in the narrow sense and *U. australis* R. Br. (*U. neglecta*, *U. major*). They are almost impossible to separate except when flowering, and as many specimens have been recorded in a flowerless state they must remain ambiguous. The only certain records for the segregates appear to be *U. vulgaris* for Carrowmoreknock (**TCD**) and *U. australis* for Templebannagh bog (ICP, p. 230). Praeger at different times recorded both from the river at Terryland, but it is not clear whether he believed that both species grew there together, and there are no specimens.

U. minor L.

1 . **3 4 5 6 7 8** . **B**

Bog-pools and drains. Frequent in Connemara; local elsewhere.

1. Abundant at the W. end of L. Goller. Bog 1 km S. of Lickeen L. **3**. Frequent in fens and ditches in the area S. and E. of Mullaghmore. **4**. Occasional around Carrowmoreknock. **5–8**. Frequent throughout.

First record: Wade (1802). S. of Ballynahinch.

U. intermedia Hayne

(*1*) . **3 4 5 6 7 8** . **B**

Bog-pools, drains, lake-margins and fens. Frequent in Connemara; local elsewhere.

(*1*). *Lisdoonvarna*, 1900 (ITB). **3**. Frequent in reed-beds, fens and ditches in the area S. and E. of Mullaghmore. **4**. In several places around Carrowmoreknock. *Terryland and Menlough*, 1899 (ITB). **5–8**. Frequent throughout.

First record: Babington (1836). Ditches in deep bogs, especially near Oughterard.

Pinguicula   Butterwort   *Liath uisce*

Pinguicula vulgaris L.

1 2 3 4 5 6 7 8 **A B**

Bogs, marshes, fens, lake-shores, wet meadows and wet rocks. Very frequent throughout most of Connemara; rather rare elsewhere.

1. Frequent near the summit of Slieve Elva. **2**. Occasional in the north. Aran (rare). **3**. Occasional in the area between Corofin and L. Bunny. Fen at Carrowgarriff. **4**. Frequent in the north; rare in the south. **5–8**. Very frequent, except in the S. part of district 6, where it is rather rare.

First record: Wade (1802). 'Plentiful throughout Cunnamara.'

P. grandiflora Lam.

1 2 . . . .

Flushes and seepage-lines on cliffs and hills; rare.

1. In fair quantity on a wet shale cliff 100 m E. of the spa at Lisdoonvarna. **2**. In a diffuse flush at about 140 m (450 ft) on the S.E. face of Cappanawalla.

First record: Praeger (1903*b*). Lisdoonvarna.

The Lisdoonvarna station was at first regarded with some suspicion as a possible escape from a garden at the top of the cliff. The discovery of the Cappanawalla station by J. Heslop-Harrison in 1949 showed that these suspicions were unjustified.

For several years around 1970, a small proportion of the plants in this latter station had flowers of a very pale lilac colour, but searches in 1974 and 1975 failed to rediscover them. Close to this station plants have been seen which appear to be referable to the hybrid *P. × scullyi* (*P. grandiflora × vulgaris*).

P. grandiflora is regarded by many as the most beautiful plant in the Irish flora. It is common in Kerry and W. Cork, but does not occur in Britain. On the Continent it ranges from N. Spain through the Pyrenees and W. Alps to the Jura. Although its leaves are distinctly larger than those of *P. vulgaris*, there is a considerable overlap in size, and it is unsafe to name non-flowering specimens.

P. lusitanica L.

(*1*) **2 3 4 5 6 7 8** . **B**

Bogs, lake-shores and other wet places, usually in semi-open vegetation. Widespread in Connemara; rare elsewhere.

(*1*). *Lisdoonvarna*, 1879 (**BEL**). **2**. In a *Schoenus*-flush above Black Head. **3**. Temple-bannagh bog. Fen-margin by the N. end of L. Bunny. **4**. Bog 1 km S. of Carrow-moreknock; also a little further to the north. By an inlet of L. Corrib just south of Oughterard quay. **5–8**. Occasional to frequent throughout, except in the southern parts of districts 6 and 7, where it is rather rare.

First record: Wade (1802). Ballynahinch.

An inconspicuous plant, easily overlooked, but even when not in flower it can be distinguished by the greyish colour of the leaves, in contrast to the bright, yellowish green of the other two species.

VERBENACEAE

Verbena

‡**Verbena officinalis** L.   Vervain

 . **2** **3** (*4*) . . . **8** . .

Roadsides and waste places. Occasional on the limestone, but tending to decrease.

2. Roadside 1½ km N.W. of Bealaclugga. Waste ground near Corcomroe Abbey. *S. of Ballyvaughan* (Corry, 1880). **3**. Occasional, mainly in the south. (*4*). *Kilcolgan, L. Atalia, Kilbeg ferry*, 1899 (ITB). *Ross L.*, 1906 (**DBN**). **8**. Frequent in several places along the N. shore of L. Corrib, from Cong to Carrick.

First record: Corry (1880). 'By the road leading from the foot of the Corkscrew into Ballyvaughan.'

LABIATAE

Among the plants of this family which have been recorded from the region but only as casuals mention may be made of *Clinopodium vulgare* L., *Nepeta cataria* L. (catmint) and *Marrubium vulgare* L. (white horehound). The first, which is not effectively naturalized anywhere in Ireland, was seen as a single plant which persisted near Oughterard for several years but has now vanished. *Nepeta cataria* was recorded from several stations in districts 2, 3, 4 and 8 in the nineteenth century, but with no evidence of persistence. There is one recent record: Inishmaan, 1971 (JW, specimen in **DBN**), but it could not be found there in 1976. *Marrubium vulgare* was recorded as a transitory casual from several places on the limestone, but it has not been seen in the region for over 50 years.

Mentha   Mint

***Mentha suaveolens** Ehrh. (*M. rotundifolia*)

 . **2** (*3*) (*4*) . **6** (*7*) **8** A .

Roadsides and waste ground; rare, and seen recently only in Connemara and on Aran.

2. In two places on Inishmore. (*3*). *L. Bunny area* (Atlas). (*4*). *Between Galway and Oranmore* (More, 1872). **6**. Roadsides S. of Kilkieran and E. of Carna. (*7*). *Roundstone* (CH1). *Near Clifden* (Praeger, 1934c). **8**. Roadsides N. of Maam Cross and near the Hill of Doon.

First record: CH1 (1866). 'Roundstone, Connemara (the late W. MacAlla).'

The most widely naturalized of the garden mints, though now cultivated less widely than formerly. It is fairly easily recognized by its hairy, wrinkled, almost round leaves. Several other cultivated mints have been noted in the Flora region by ruined cottages or in abandoned gardens, where, although they can persist for many years, they can hardly be regarded as naturalized. These include *M.* × *gentilis* L. (*M. arvensis* × *spicata*), *M. longifolia* (L.) Huds., and hybrids of *M. suaveolens* with other garden species.

*M. spicata L. (*M. viridis*)   Spearmint

. 2 . . . 6 7 .   . .

Roadsides; rare.

2. Damp roadside at Burren village, 1975. **6**. Roadside at Cashel, 1977. **7**. S. of Moyard, 1973.

First record: Scannell & McClintock (1974). S. of Moyard.

Fairly easily recognized by its almost hairless, narrowish leaves with a very short stalk or none.

‡M. × **piperita** L. (*M. aquatica* × *spicata*)   Peppermint

. (2) . . . . 7 .   . .

Damp roadsides and waste places; rare.

(*2*). *Ballyvaughan* (Druce, 1909). *Fanore area* (Atlas). **7**. Roadside N.W. of Moyard, 1968 (MS, **DBN**). *Streamstown* (Baily, 1833). *Clifden* (Druce, 1911).

First record: Baily (1833). 'Banks of rivulets at Streamstown.'

Like the preceding, but with a sweeter, less pungent scent, and with stalks to the lower leaves at least 6 mm long. It is usually an escape from cultivation, but the possibility of spontaneous hybrids between *M. aquatica* and escaped *M. spicata* cannot be excluded.

M. aquatica L.   Water mint

1 2 3 4 5 6 7 8   A B

Marshes, ditches, lake-shores and other wet places. Frequent to abundant in many districts, but rather local.

1. Very frequent. **2**. Frequent in suitable habitats. On all the Aran Islands. **3–4**. Abundant. **5**. Frequent except in the east. **6**. Rather rare; seen only in the south-east. **7**. Abundant in the south and west; rather rare elsewhere. **8**. Rare in the west; very frequent elsewhere.

First record: Wright (1871). Aran Islands.

M. arvensis L.

. 2 3 4 5 6 7 8   A .

Roadsides, turloughs, lake-shores and cultivated ground. Frequent on the limestone; rare elsewhere.

2–4. Frequent, especially around turloughs and in cultivated fields. **5**. Wood-margin at Spiddal. **6**. Lake-shore near Carraroe. **7**. Inishlackan. **8**. Shore of L. Corrib N.W. of Oughterard. Waste ground by the mill-pond at Cong. Roadside at Glann.

First record: Hart (1875). Aran Islands.

M. × **verticillata** L. (*M. aquatica* × *arvensis*) (*M. sativa*)

. **2 3** . . **6** (*7*) **8** . .

By turloughs and in wet grassland; occasional.

2. L. Aleenaun. Turlough S.W. of Ballyvaughan. At the N. end of the Carran turlough. **3.** Frequent. **6.** In a marsh behind the 'coral strand' near Carraroe. (*7*). *Roundstone* (Praeger, 1906). **8.** Shore of L. Corrib N.W. of Oughterard, and opposite the Hill of Doon. S. of Cornamona. Drain by the shore of L. Mask N.E. of Killateeaun.

First record: Praeger (1906). Roundstone.

Usually in close proximity to both parents, but in the stations in district 8 listed above *M. arvensis* was not seen in the immediate vicinity. The only reliable character for distinguishing the hybrid from *M. arvensis* is the longer calyx, with narrower, sharply pointed teeth.

*M. requienii Benth.

. . . . . . **7 8** . .

Damp rocks and banks; very rare.

7. On a rock-face by the road 1 km W.N.W. of Clifden. **8.** On a clay bank and rocks between the road and the shore at the N. end of Leenane village.

First record: Scannell & McClintock (1974). Leenane.

A creeping plant with tiny leaves and flowers, sparingly naturalized from gardens in Ireland and Britain. It is native to the islands of the W. Mediterranean.

Lycopus

Lycopus europaeus L.   Gipsywort

1 (*2*) **3 4 5 6 7 8** (*A*) .

Marshes, lake-margins, river-banks and ditches. Frequent in Connemara, though irregularly distributed; rather rare elsewhere.

1. By Drumcullaun L. (*2*). *Inishmore* (Balfour, 1853; Hart, 1875). **3.** Pond-margin 3 km N.E. of Ruan. **4.** Stream near Aughnanure Castle. Ballycuirke L. Gortachalla L. Shore of L. Corrib S. of Oughterard. **5.** Occasional in the area S.E. of Costelloe. Streamside near Formoyle L. **6.** Abundant throughout, chiefly in marshes and by streams. **7.** Lake-shore at Callow. Marsh at Belleek. Damp wood N.W. of Dog's Bay. *Cregduff L.*, 1953 (**DBN**). **8.** By a small lake 1 km N. of Maam Cross. Occasional around the N.W. part of L. Corrib, and on one of its islands. Abundant and luxuriant on the limestone shore of L. Mask.

First record: Oliver (1851). Near Roundstone.

Origanum

Origanum vulgare L.   Wild marjoram

. **2 3 4** . . . **8** [A] .

Roadsides and grassy banks. Locally abundant in the Burren; rare elsewhere.

2. Very frequent and locally abundant in the north, as at Abbey Hill, Bealaclugga, New Quay, Formoyle, Ballyvaughan and thence to Corkscrew Hill. Recorded once from Aran (1890), but probably only as a casual. **3.** Frequent throughout, and locally abundant in the north, especially around Kinvara. **4**. Near the shore of Ross L. *E. of Galway* (ITB). **8**. Frequent W. of Cong.

First record: CH1 (1866). Garryland wood.

Thymus

Thymus praecox Opiz (*T. drucei, T. serpyllum, T. chamaedrys*)   Wild thyme

1 2 3 4 5 6 7 8   A B

Limestone pavement, dry grassland, roadsides, sand-dunes and mountain rocks and cliffs. Abundant almost throughout, but commonest on limestone or near the sea. Rather scarce in the south and east parts of district 1, and in parts of central Connemara; elsewhere very frequent to abundant.

First record: Wade (1802). 'At the foot of Lettery mountain.'

Records of *T. chamaedrys* (*T. pulegioides*) and *T. serpyllum* in the literature are based on taxonomic misunderstandings.

Calamintha

Calamintha sylvatica (Bromf.) (*C. ascendens*)   Calamint

. **2** . **4** . . [7] **8   A** .

Limestone pavement, roadsides and gravelly soil. Occasional near Oughterard and Cong; very rare or extinct elsewhere.

2. *Glenquin*, 1880, and *Muckinish Bay*, 1895 (CH2). Inishmore, 1972, in small quantity; much commoner 100 years earlier. **4**. Occasional in the north, from Rosscahill and Oughterard eastwards. **8**. Frequent on limestone pavement and roadsides N.W. of Cong.

First record: Wade (1802). 'In great quantity at Cong.'

Known at Cong since 1801 and at Aughnanure Castle since 1840. In view of its persistence here it is not easy to account for its apparent extinction in the Burren and its great decline in Aran. On balance, however, it would appear that it is likely to be native in some parts of the Flora region.

The *Atlas* record for the Letterfrack area requires confirmation.

Glechoma

Glechoma hederacea L.   Ground ivy

1 2 3 4 (5) **6** . (*8*)   **A** .

Woods, scrub, hedges, limestone pavement and grassland. Very frequent in the Burren; rare elsewhere.

1. Occasional along the N.E. margin. **2–3**. Very frequent over most of the Burren, but rare in Aran. **4**. By Ballycuirke L. On two islands in the S.W. part of L. Corrib. (5). *Gentian Hill* (P. cat.). **6**. Scrub at Flannery Br. (*8*). *N.W. of Oughterard* (P. cat.). *Leenane*, 1955 (DW).

First record: Hart (1875). Aran Islands.

Common over most of Ireland, but thinning out very strikingly near the west coast.

Scutellaria   Skull-cap

Scutellaria galericulata L.

1 . . . (*5*) **6** (*7*) **8** . .

Lake-shores and marshes; rarely in wet places by the sea. Rather rare.

1. Stony shore on N. side of L. Goller. Frequent in marshy ground on the S.W. margin of L. Raha. (*5*). *Inveran* (ITB). **6**. Wet ground by the shore 3 km S.S.W. of Carraroe. (*7*). *Ditch at the base of Errisbeg*, 1953 (**DBN**). *Shore at Renvyle* (Wade, 1802). **8**. Occasional in marshes and on stony shores around the N.W. part of L. Corrib, and on some of the islands. Between Clonbur and L. Mask.

First record: Wade (1802). 'Galway Bay, Bilberry Island, Lough Corib, Renvi shore. . . .'

S. minor Huds.

1 . . **4 5 6 7 8** . **B**

Marshes, bog-margins, river-banks and lake-shores. Occasional to frequent in Connemara; rare elsewhere.

1. Bog by Drumcullaun L. **4**. Bog by L. Corrib 2½ km W. of Carrowmoreknock. **5**. Bog S.E. of Costelloe. Stony shore of Slieveaneena L. Bog-drain 5 km W.N.W. of Galway. **6**. Frequent. **7**. E. end of Ballynahinch L. Frequent near the coast from Roundstone northwards to Ballinaboy. Inishbofin. **8**. Frequent.

First record: Wade (1802). 'Found with [*S. galericulata*] in Cunnamara, but much more common.'

Prunella

Prunella vulgaris L.   Self-heal

1 2 3 4 5 6 7 8   A B

Roadsides, pastures, heaths and wood-margins; frequent to abundant throughout.

First record: Wright (1871). Aran Islands.

Some plants on limestone or blown calcareous sand are unusually robust and show simultaneous opening of most of the flowers in the spike, giving the plant a handsome, orchid-like appearance.

Stachys   Woundwort

Stachys palustris L.

1 2 3 4 5 6 7 8   A B

Marshes, drains, damp grassland, roadsides, and as a weed of cultivated ground; frequent.

Very evenly distributed through the region, though somewhat scarcer on the Clare shales and in central Connemara than elsewhere.

First record: Wright (1871). Aran Islands.

S. sylvatica L.

1 2 3 4 . 6 7 8 A B

Woods, scrub, hedges and roadsides. Occasional, and locally frequent.

1. Abundant in woods at Ennistymon. Woods at O'Dea's Castle. River-bank near Drumcullaun L. By the spa at Lisdoonvarna. **2.** Occasional. **3–4.** Frequent. **6.** Occasional near the coast; also by the river at Oughterard. **7.** Inishbofin. *Near Clifden* (Atlas). **8.** Frequent around Cong and Clonbur. Waste ground at Oughterard, and in scrub 4 km to the W. By the Bealnabrack R. above Maam.

First record: Wright (1871). Aran Islands.

The hybrid between this species and the preceding (*S. × ambigua* Sm.) has been recorded from Aran, from Elmvale Ho. and from 5 km W. of Spiddal.

‡S. arvensis (L.) L.

. (2) (3) (4) 5 6 (7) . (A) (B)

Cultivated ground; formerly occasional; now very rare.

(2). *Aran Islands* (Hart, 1875; Praeger, 1895*b*). **(3).** *S. of Kinvara*, 1900 (ITB). **(4).** *Parkmore*, 1899 (ITB). **5.** Oughterard, 1979 (D. Lambert). *Rosscahill*, 1955 (Mrs Gough). **6.** Occasional in fields at the S.E. corner of Gorumna I., and on the mainland opposite, 1964–6 (DW). **(7).** *Clifden*, 1898 (ITB). *Inishbofin* (More, 1876; Praeger, 1911).

First record: Wade (1802). 'Common about Ballynahinch.'

Although it is still fairly frequent in S.E. Ireland this species has almost disappeared from the west. It is curious that, although it is normally regarded as distinctly calcifuge, the first three stations cited above are on limestone.

Galeopsis

‡Galeopsis tetrahit L.   Hempnettle

1 2 . 4 5 6 (7) 8 A (B)

Cultivated ground. Very rare on the limestone; occasional and locally frequent elsewhere.

1. 5 km E.N.E. of Ennistymon. **2.** Aran; formerly widespread, now very rare. **4.** Field near L. Corrib, 5 km E.S.E. of Oughterard. **5.** W.N.W. of Galway. S.E. of Costelloe. N.E. of Screeb. **6.** Very frequent in the south-east, rare elsewhere. **(7)** *Inishbofin* (Praeger, 1911). **8.** Glann, rare. Abundant in a potato-field on the N. shore of L. Nafooey. *Leenane*, 1957 (McClintock).

First record: Hart (1875). Aran Islands.

Like many weeds of cultivation formerly widespread, this survives now mainly in remote areas with old-fashioned agricultural methods.

Lamium

‡Lamium purpureum L.   Red dead-nettle   *Neanntóg chaoch*

1 2 3 4 5 6 7 8 A B

Cultivated fields and gardens; more rarely on roadsides or in waste places. Widespread throughout most of the region, but nowhere really common. It does best on light soils, and is therefore commonest in districts 3 and 4 and the coastal fringe of 7; it is rare on the peats and clays of districts 1, 5 and 8.

First record: Wright (1871). Aran Islands.

‡**L. hybridum** Vill.

. **2 3 4** . . **7** .   **A B**

Cultivated fields and gardens; rather rare, and mainly around Galway Bay.

2. By a lane N.E. of Ballyvaughan. Inishmaan. **3**. Frequent in the north; not seen in the centre or south. **4**. Occasional by the coast. **7**. Frequent on Inishbofin.

First record: Praeger (1911). Inishbofin; in several places.

Often confused with the preceding species, and perhaps commoner than these records indicate. It can be distinguished by its deeply and irregularly toothed leaves and its smaller flowers.

L. molucellifolium Fries (*L. intermedium*) and **L. amplexicaule* L., which are allied to the two preceding species, have been recorded on a few occasions in the Flora region, but neither has been seen recently. They were never much more than casual in the west, and now seem to be extinct there.

L. album L.   White dead-nettle

. **2** ?**3** . . . .

Roadsides and waste ground; very local.

2. In some abundance around Ballyvaughan harbour, 1967 (DW).

It is not clear whether the record by Praeger (1934c) for 'near Crusheen' is in the Flora region or not. We cannot discover the basis for the *Atlas* record for 93/25 (S. of Corofin), which is probably, though not certainly, in the Flora region.

This species is well naturalized in the counties N. and N.W. of Dublin, but is rare elsewhere in Ireland.

Teucrium

Teucrium scordium L.

. . **3** . . . .

Turloughs; very local.

3. Frequent in and around several of the turloughs, as at Tirneevin, Castle L., N. of Boston and S.W. of Mullaghmore; also in the turlough-like marsh on the county boundary S. of the ruins at Kilmacduagh.

First record: Levinge (1892). 'At the edge of a small lake near Glanquin.'

Elsewhere in Ireland known only from along the lower and middle course of the R. Shannon, where it is locally plentiful. Very local in Britain also, being almost confined to the fens of E. Anglia.

T. scorodonia L.   Wood sage

1 2 3 4 5 6 7 8   A B

Limestone pavement, woods, scrub, roadsides and rocky ground. Very frequent to abundant through most of the region.

1. Not common; seen only in scrub by L. Goller and S. of Lickeen L., and on a limestone outcrop near Elmvale Ho. **2–3.** Abundant. **4–5.** Frequent. **6–7.** Very frequent. **8.** Abundant in the east, but rare in the west.

First record: Hart (1875). Aran Islands.

Especially characteristic of the 'shattered' type of Burren pavement, which it shares with *Rosa pimpinellifolia*, *Prunus spinosa*, and often little else. Although it has elsewhere the reputation of at least a calcifuge tendency, in the region of this Flora it seems quite indifferent to the base-content of the soil. It demands good drainage, however, and is absent from heavy clays or wet peats.

Ajuga

Ajuga reptans L.   Bugle

1 2 3 4 5 6 7 8   A .

Woods, scrub, river-banks and shady roadsides. Occasional over most of the region, and frequent in the north-east.

1–3. Occasional, and locally frequent. **4.** Frequent around Ross L. and northwards; not seen S. of Moycullen. **5.** Scrub 2½ km S.S.W. of Oughterard. Roadside 4 km W.N.W. of Galway. **6.** Woods of Cashel House Hotel; perhaps introduced. **7.** Scrub E. of Clifden; abundant in woods at W. end of Ballynahinch L.; shady rocks E. of Kylemore; ravine by W. shore of L. Fee. **8.** Frequent.

First record: Foot (1864). 'By no means common. I only observed it at the Corkscrew Hill.'

A. pyramidalis L.

. 2 . . . . 7 .   A .

Thin soil overlying basic rocks; very rare.

2. Occasional over a small area at Poulsallagh. Occasional on Inishmore; recently discovered on Inishmaan (1971; JW). The record for the Slieve Elva region (93/17) in the first edition of the *Atlas* was an error; it was deleted in the second edition. **7.** *Bunowen Hill* (Praeger, 1907); repeatedly searched for here by DW and others, 1945–70, without result. It was discovered, however, in 1968 by Miss E. Bullard a few hundred metres to the north-west (on the other side of the road) on a low reef of basaltic rocks, and has been seen there several times since. A further station was discovered by MS in 1977, *c.* 2½ km further north, near Truska L.

First record: Moore (1854). 'Near Kilronan...sparingly.'

Not known elsewhere in Ireland; an odd distribution for a boreal-montane plant, all the stations being close to sea-level. As it is a biennial it varies considerably in abundance from year to year.

The hybrid *A. pyramidalis* × *reptans* has been reported from Ballyvaughan by Druce (1924), who obtained it from P. B. O'Kelly. We believe that the credentials of this plant must be treated with considerable reserve, if only because Ballyvaughan is a full 15 km from the Poulsallagh station of *A. pyramidalis*.

Fig. 16. *Ajuga pyramidalis* at Poulsallagh.

PLANTAGINACEAE

Plantago Plantain

Plantago coronopus L.

1 2 3 4 5 6 7 8 A B

In a wide variety of habitats near the sea; rare inland. Abundant in S. and W. Connemara, occasional to frequent elsewhere.

1. Frequent. **2.** Occasional. **3.** Frequent. **4–5.** Occasional. **6–7.** Abundant, except in the north part of 7, where it is only occasional. **8.** Rather rare.

First record: Babington (1836). Shore at Leenane.

Frequent on roadsides as much as 5 km from the sea along the road from Toombeola to Ballinaboy in district 7.

P. maritima L.

1 2 3 4 5 6 7 8 A B

Salt-marshes, and cliffs, rocks and banks near the sea; also inland on limestone pavement and on mountain cliffs.

1. Frequent on the coast. **2–4.** Very frequent. **5.** Occasional on the coast. **6–7.** Abundant on the coast; also on cliffs on many of the higher mountains. On rocks near the summit of Errisbeg. **8.** Occasional on the shore. On rocks and cliffs on the Maamtrasna plateau and the Maumturk range; also in the N.E. corrie of Bunnacunneen.

First record: Wade (1802). 'Lettery Hill, Ballynahinch.' Like many records for Benlettery this probably refers really to Bengower.

P. lanceolata L.   *Slánlus*

1 2 3 4 5 6 7 8 A B

Meadows, pastures, rough grazing, sand-dunes, roadsides, etc. Abundant throughout.

First record: Hart (1875). Aran Islands.

‡**P. major** L.

1 2 3 4 5 6 7 8 A B

Roadsides and other similar disturbed or trampled habitats; more rarely in turloughs, pastures and rough grazing. Common throughout.

First record: Wright (1871). Aran Islands.

The natural habitat of this species (if indeed it is native in the region) is obscure. It has not been observed on lake-shores or on stony sea-shores, and only a few isolated plants have been seen on limestone pavement.

Littorella

Littorella uniflora (L.) Asch. (*L. lacustris*)

1 2 3 4 5 6 7 8 A B

Lake-shores and turloughs; rarely in rivers, fens or small pools. Occasional to frequent throughout.

1. Drumcullaun L. Inchiquin L. **2.** L. Luick. In small pools on the pavement on Inishmore. **3.** Frequent in lakes and turloughs in the south part; much scarcer in the north, but seen in a fen N. of Kinvara. **4.** Ross L. Shore of L. Corrib at Burnthouse, and on an island nearby. E. shore of L. Corrib, 3 km above Galway. **5.** Occasional. **6–7.** Frequent. **8.** Abundant.

First record: Wade (1802). Marshy spots near Ballynahinch.

Although it has a number of stations on the limestone, this species seems more at home in acid waters. In L. Corrib it is most abundant in the N.W. part, which lies on siliceous rocks.

Among other submerged aquatics which resemble it superficially, it can be distinguished from *Isoetes* and *Lobelia* by the fact that it grows in mats, not as isolated rosettes, and from *Eriocaulon* by its brown roots, without transverse striations.

CHENOPODIACEAE

Chenopodium Goosefoot

‡**Chenopodium album** L.

. **2 3 4 5 6 7 8** **A B**

Cultivated ground; more rarely on sea-shores or in waste places. Local, and in several districts rare.

2. Among sugar-beet near Corcomroe Abbey. In barley at Finavarra. Potato-fields on Aran. **3.** Coole. Roadside S.W. of Ardrahan. S. of L. Fingall. **4.** Sand-pit by Gortachalla L. Rinville. Ballynacloghy. N. shore of Kilcolgan estuary. **5.** 5 km E.S.E. of Costelloe. 3 km N. of Gentian Hill. **6.** Carna. N. end of Furnace I. By Lettermullan Br. **7.** Omey I. Inishbofin (rare). **8.** Clonbur.

First record: Wright (1871). Aran Islands.

In most parts of Ireland this species is rarer than formerly, tending to be replaced by *Atriplex patula*. It does not, however, seem to have ever been very common in Connemara or the Burren.

C. rubrum L.

. **2 3** (*4*) . . . . **A** .

Around lakes and turloughs; rare and often transitory.

2. Portcowrugh, Inishmore. **3.** Locally abundant, at least in some years, around Coole L. In small quantity at Caherglassaun turlough. Tirneevin turlough. (*4*). *Ballyloughan*, 1899 (ITB).

First record: Praeger (1895*b*). 'In great abundance on the muddy margins of the brackish lake at Port Cowrugh.'

A capricious species, which, thanks to the power of dormancy of its seeds, comes and goes, and may re-appear after a long interval. It seems, however, to be reasonably well established on Aran and in a limited area near Gort, and although it is a great opportunist for colonizing man-made habitats it is probably native here.

A few plants of **Chenopodium bonus-henricus* L. were seen by a spring near Glencolumbkille Ho. (district 2) in 1965, but it is probably to be regarded as a casual.

Beta

Beta vulgaris, subsp. **maritima** (L.) Thellung (*B. maritima*)   Sea beet

1 2 3 4 5 . 7 . A B

Rocky, muddy and gravelly sea-shores, sea-cliffs and edges of salt-marshes. Frequent around Galway Bay; rather rare elsewhere.

1. Cliffs of Moher. **2.** Frequent, especially on the N. coast. Aran. **3.** Between Kinvara Bay and Aughinish. **4.** Near Kilcolgan. Parkmore. Oranmore, and thence to Galway. **5.** Gentian Hill. Derryloughaun. Spiddal. **7.** Inishbofin and Inishshark. *High I.*, 1831 (**DBN**).

First record: Graham (1840). 'On the shore near Galway, very sparingly.'

The distribution is capricious, some of the stations being on very exposed coasts and others in sheltered inlets. Its absence from most of S. and W. Connemara is hard to explain. The *Atlas* records for the Ballyconneely, Clifden and Kingstown Bay areas are open to some suspicion and require confirmation.

Halimione

Halimione portulacoides (L.) Aellen (*Atriplex portulacoides*)   Sea purslane

. 2 . (4) 5 . . .   . .

Rocky or stony shores; rare.

2. In crevices of limestone pavement just above high-water mark on Rine Point. Abundant over 100 m of shore at the extreme northerly point of Muckinish Bay. **(4).** *Tawin I.* (Praeger, 1939). **5.** Sparingly on the shore by Gentian Hill, 1965. A few plants E. of Spiddal, 1979.

First record: CH2 (1898). 'Rhyn Point near Ballyvaughan, 1895; *P. B. O'Kelly.*'

These are the only W. coast stations, apart from two in Co. Kerry which have not been recently confirmed. The late date for most of the records of this conspicuous plant suggests that it is a recent arrival in our region (cf. *Atriplex laciniata*).

Atriplex

Atriplex littoralis L.

. 2 . . . . . .   (A) .

Shingle beaches; very rare.

2. A few plants W. of New Quay, among abundant *A. glabriuscula*, 1969 (**TCD**); searched for in 1970 and several times since, but not found. *Inishmore*, 1869 (Hart, 1875); not seen here since.

First record: Hart (1875). Inishmore.

The inconstancy of this species in its mainland station (where it is backed by a specimen) gives credence to the Aran record. It is possible that like many species of shingly or sandy shores it may disappear for many years and then re-appear. There are no other records for the W. coast, and even on the E. coast it is very local.

A. patula L.

1 2 3 4 5 6 7 8   A B

Cultivated fields, waste places and roadsides. Occasional, and locally frequent on the limestone.

1. Occasional. **2**. Frequent around Ballyvaughan and on Aran. **3**. Frequent. **4**. Very frequent. **5**. Rubbish-dump at Gentian Hill. Tilled field at Glentrasna. **6**. Occasional near Carraroe. Furnace I. **7**. Fairly frequent by the coast, from Inishlackan to Cleggan and Inishbofin; not seen elsewhere. **8**. 2 km N.W. of Oughterard. N. of L. Nafooey. By the Bealanabrack R. above Maam.

First record: Hart (1875), as *A. angustifolia*. Aran Islands.

A. hastata L.

There has been so much confusion between this species and *A. glabriuscula* that it would be unsafe to present any detailed records for it. All we can say is that it appears to be rare in the region; that it probably grows only at the back of shingle-beaches and in waste places nearby (there are no records for cultivated land); and that most, if not all, of the plants are erect. We have tentative records for Inishmore and Oranmore, but even these require confirmation.

A. glabriuscula Edm. (*A. babingtonii*)

1 2 3 4 5 6 7 .   A B

Sandy and stony sea-shores, and on the margins of salt-marshes. Abundant in W. and S.W. Connemara; occasional and locally frequent elsewhere.

1. N. and S. of Lehinch. **2**. Abundant around New Quay. Rine Point. Aran. **3**. Occasional. **4**. Very frequent eastwards from Galway. Abundant on the N. shore of the Kilcolgan estuary. **5**. Occasional. **6–7**. Very frequent to abundant.

First record: More (1876). Inishbofin: common about the harbour etc.

The species is here interpreted in a relatively wide sense. Many of our specimens have been annotated by Tascherau as 'A. prostrata group', implying, if we understand him correctly, sometimes inadequacy of the specimen, but often a variable amount of hybridization from *A. longipes*. We have, however, no firm record of the latter species itself.

A. laciniata L. (*A. maritima, A. sabulosa, A. rosea*)

. 2 . . . . 7 .   A .

Sandy sea-shores; very local.

2. Aran; first noted by JW on Inishmaan (a single plant) in 1971; in considerable quantity on all the islands in 1976, but since then tending to diminish. **7**. In fair abundance on the beach at the W. end of the promontory separating Dog's Bay from Gorteen Bay, and in smaller quantity at the S. end of Dog's Bay. First reported here by B. E. Smythies in 1972; confirmed several times up to 1980, when it showed some diminution, but was still fairly plentiful.

First record: Webb (1980a).

Up to 1900 this species was known in Ireland only from the E. coast N. of Dublin. A station in Co. Wexford was added in 1901 (and several others reported more recently), and there is a specimen in **TCD** from Co. Waterford collected in 1944. In 1950 it had reached W. Cork, and during the next 20 years it evidently turned the corner and reached Connemara. There are as yet no records for Co. Kerry, but it would be surprising if a thorough search did not reveal its presence there.

Salicornia   Glasswort

This is a very difficult genus, but there is now fair agreement among those who have studied it that there are five or six species in Ireland. Of these, *S. europaea*, is by far the commonest, and it is the only one which we have observed. Dr I. K. Ferguson, however, who made a study of the genus in Ireland in 1961–4, has set out in an unpublished thesis records of two other species from the Flora region. In all their stations they grow in company with *S. europaea*, and it needs patience and an expert eye to pick them out.

S. europaea L.

1 2 3 4 5 6 7 .   A .

Salt-marshes; fairly frequent, but local.

1. N. of Lehinch. **2**. In several places on the N. coast, from Rine Point to New Quay. Inishmore. **3**. Aughinish, and half-way between here and Kinvara Bay. **4**. Clarinbridge estuary. In several places around Kilcaimin and Tawin I. Oranmore. **5**. Gentian Hill. E. of Spiddal. **6**. Very frequent. **7**. Fairly frequent in the south and west; rare in the north.

First record: Hart (1875). Inishmore.

S. ramosissima Woods

. **2** . **4** . **6 7** .   . .

Upper parts of salt-marshes and well-drained sandy pans; occasional.

2. New Quay and Finavarra. **4**. Clarinbridge estuary. Tawin I. Kilcaimin. **6**. Mweenish I. **7**. Roundstone.

First record: Census Catalogue (1972). H16.

These records have been assigned to *S. ramosissima*, as the plants come closer to it than to any other described species, but they differ in some respects from English specimens, and Dr Ferguson thinks that they may perhaps deserve separation as a distinct species.

S. fragilis Ball & Tutin

. **2** . **4** . . . .   . .

2. New Quay and Finavarra. **4**. Tawin I. Kilcaimin.

First record: Webb (1982).

Distinguished from the other two species by its long, smoothly cylindrical and scarcely 'knobbly' flowering spikes.

Suaeda

Suaeda maritima (L.) Dum.

1 2 3 4 5 6 7 .   A .

Stony, rocky, muddy, and more rarely sandy shores; also on limestone pavement on Aran, on cliff-tops or around brackish pools.

1. Near Liscannor. **2**. Frequent and locally abundant. **3**. Very frequent W. of Kinvara; not seen N. of it. **4**. Occasional. **5**. Occasional, mainly in the east. **6**. N. end of Furnace I. E. shore of Kilkieran Bay, W. of Rosmuck. S.E. of Carna. **7**. Beside the 'coral strand' at Ballyconneely. Gravelly shore of Clifden estuary. S. shore of Ballynakill harbour. Sandy shore N.E. of Aughrusbeg L.

First record: Wright (1871). Aran Islands.

Salsola

Salsola kali L. Saltwort

1 2 . . (*5*) **6 7** . **A B**

Sandy shores; occasional.

1. Near O'Brien's Br. **2**. Frequent by the dunes N.E. of Ballyvaughan. Aran (rare). (*5*). '*Galway*' (*probably at the Claddagh*) (P. cat.). **6**. S.E. of Carna. **7**. Dog's Bay. Shore near Aillebrack L. Frequent on the beaches N.E. of Gowlaun. Inishbofin (rare).

First record: Wright (1871). Aran Islands.

POLYGONACEAE

Polygonum

It is not easy to give any hints in non-technical terms for the identification of species in this genus, especially as many of them vary considerably according to the environment. The first 3 differ from the others in having inconspicuous flowers arranged singly and not massed into spikes. *P. hydropiper* can be recognized by chewing part of a leaf (cautiously!); after about 10 seconds it yields a hot, peppery taste.

Polygonum aviculare L. (*P. heterophyllum*)

1 2 3 4 5 6 7 8 **A B**

Roadsides, cultivated and waste ground, farmyards, sandy beaches, etc. Frequent in most districts.

Fairly evenly distributed, but rather scarce in districts 1 and 2. Much less common throughout the region than in eastern Ireland.

First record: Hart (1875). Aran Islands.

P. arenastrum Bor. (*P. aequale*)

1 2 . **4 5 6 7** . **A B**

Roadsides, tracks, paths, quays and stony ground. Occasional; perhaps frequent. Records for this species are clearly incomplete; only 11 stations have been noted. It is probably overlooked fairly often, but is certainly rarer than the preceding species.

First record: Webb (1982).

P. oxyspermum, subsp. **raii** (Bab.) Webb & Chater (*P. raii*)

(*1*) **2** . (*4*) **5 6 7** . **A B**

Beaches of sand or fine shingle; local and rather rare.

(*1*). *Lehinch* (FWI). **2**. Frequent at Rine Point. In small quantity on most of the sandy beaches of all three Aran Islands. (*4*). *Near Ardfry Ho.* (P. cat.). **5**. 5 km W. of Spiddal. **6**. N. end of Furnace I. Mweenish I. 1 km S.E. of Carna. **7**. Gorteen Bay; occasional in 1971 but abundant in 1979. Beaches N. of Aughrusbeg L. and W. of Cleggan. Beaches N.E. of Gowlaun. Near Bunowen. Inishbofin (rare).

First record: Graham (1840), as *P. roberti*. 'On the shore near Galway, very sparingly.'

A diminishing species, but holding its own better in the west of Ireland than in the east.

P. hydropiper L.   Water-pepper   *Glúineach the*

1 2 3 4 5 6 7 8   (*A*) **B**

Marshes, ditches, river-banks and turloughs; more rarely on roadsides or cultivated ground. Frequent over most of Connemara; local elsewhere.

1. Occasional. **2**. In the Carran turlough. **3**. By Cleggan L. Abundant by Caherglassaun L., around the Coole turlough, and by the Kiltartan R. **4**. Occasional in the north half. **5**. Frequent. **6**. Very frequent. **7**. Occasional. **8**. Frequent.

First record: Hart (1875). Aran Islands.

Although this plant is distinctly calcifuge in its general distribution, it occurs in some abundance in some calcareous habitats, especially in district 3. It was formerly widespread on Aran, but cannot now be found there; this is probably best explained by supposing that it was introduced with turf.

P. minus Huds.

. 2 3 4 . . . .   A .

Turloughs and lake-shores; very local, and only on the limestone.

2. In small quantity in L. Aleenaun, Carran turlough. Inishmore (rare). **3**. Abundant in the lower zones of the Tirneevin turlough, and in great abundance around the turlough at Coole. By the R. Fergus near Killinaboy, and in a marsh beside the river N. of Ennis. **4**. By the N. shore of Ross L.

First record: Praeger (1932*a*). Tirneevin turlough.

Most plants have much-branched, more or less prostrate stems with small leaves, but taller, less freely branched plants with leaves up to 75 mm long are sometimes seen. It is these which have been mistaken for *P. mite*.

[P. mite Schrank]

Recorded for the Tirneevin turlough by Praeger (1939), but certainly in error for tall plants of *P. minus*. Praeger's concept of *P. mite* had, unfortunately, been formed by specimens wrongly identified with this name by A. Bennett. A recent survey of all herbarium material from Ireland has shown that all specimens labelled *P. mite* are really referable to *P. minus*, *P. persicaria* or *P. hydropiper*, except for those collected since 1950 from the shores of L. Neagh, where *P. mite* is abundant; it seems, however, to be a recent introduction.

P. persicaria L.   Redshank, Spotted persicaria   *Glúineach dhearg*

1 2 3 4 5 6 7 8   A B

Cultivated ground, roadsides, disturbed grassland, waste places and turloughs. Frequent in Connemara and the Galway area; occasional elsewhere.

First record: Hart (1875). Aran Islands.

Commonest in districts 4 and 6; scarcest in 1, 2 and 3. In some of the turloughs it is extremely polymorphic, pink- and white-flowered and erect and decumbent plants being seen side by side.

P. lapathifolium L. (incl. *P. nodosum*)

. . . *(4)* . **6 7 8** . .

Cultivated or disturbed ground; rare.

(4). *Cultivated ground at Oranmore*, 1899 (**DBN**). *Rinville* (P. cat.). **6**. Disturbed ground 3½ km S.E. of Carna, 1974. **7**. Farmyard N.E. of Clifden, 1980. **8**. Potato-field at Glann, 1969. *Islands at S. end of L. Mask*, 1932 (**DBN**).

First record: ITB (1901). H16. On what authority this record rests we cannot discover; it was not noted for H16 in Praeger's catalogues.

Perhaps little more than casual. It differs from *P. persicaria* in its coarser growth, flowers of a greenish or dull, dirty pink colour, and in having small yellow glands on the inflorescence.

P. amphibium L.

1 2 3 4 5 6 7 8   A B

Turloughs, lake-margins, marshes and wet grassland. Very frequent and locally abundant in district 3 and parts of 7; occasional elsewhere. **1**. Occasional. **2**. Marsh near Turlough village. Aran. **3**. Very frequent throughout, and abundant in the extreme south. **4**. Occasional. **5–6**. Occasional near the coast. **7**. Frequent on the Slyne Head peninsula; occasional elsewhere near the coast. **8**. Shore of L. Mask, near White I.

First record: Wright (1871). 'Wet places to the west of Inishmore.'

*P. polystachyum Meissn.

1 . 3 4 5 6 7 8   . .

Roadsides; occasionally on river-banks. Fairly frequent, and increasing.

1. In great abundance, in some places far from houses, on several by-roads E. of Ennistymon. Near Kilfenora, both on the Ennistymon and the Corofin roads. **3**. 5 km W. of Kinvara. S.E. of Muckanagh L. 1½ km N.W. of Killinaboy. S. of Tubber. **4–5**. In several places on both sides of the Galway–Moycullen road. **6**. Roadside E. of Recess, and by Oorid L. Run wild below the Zetland Arms Hotel, Cashel. **7**. Near Moyard. **8**. W. of Leenane. By the ruins of the Recess hotel.

First record: Simpson (1936). Recess.

In all cases a garden outcast, but very persistent once established and soon spreading into large clumps. It can be distinguished from the other coarse perennials in this family which have been (somewhat unwisely) cultivated in gardens by its combination of large, feathery panicles of white flowers with lance-shaped leaves about 2½ times as long as wide.

**P. amplexicaule* Don, another garden outcast, has been noted on roadsides near Oranmore and between Galway and Moycullen, but it is much rarer and does not spread so fast. It is a less robust plant, and the flowers are bright red.

*P. campanulatum Hook.

. . . . 5 . 7 8 . .

Roadsides and woods; rare but increasing.

5. By the river at the W. end of Oughterard. **7.** Occasional in the north-west, from Ardbear to Cleggan and Letterfrack. **8.** Well naturalized in a wood and on the adjoining roadside between Ferry Br. and Clonbur.

First record: Scannell & McClintock (1974). Ardbear Bay and Renvyle.

A handsome plant, with pink flowers and very distinctive leaves, of which the underside is boldly veined and covered with a felt of whitish or buff-coloured hairs.

Fallopia

‡**Fallopia convolvulus** (L.) Löve (*Bilderdykia convolvulus, Polygonum convolvulus*) Black bindweed

1 2 3 4 5 6 7 8   A B

Cultivated ground; more rarely in waste places. Occasional, and locally frequent.

1. S. of Lickeen L. **2.** Near Corcomroe Abbey. Occasional on Aran. **3–4.** Frequent. **5.** N.N.W. of Screeb. 5 km N.W. of Galway. **6.** Fairly frequent. **7.** Belleek. Inishbofin (rare). **8.** 2 km N. of Oughterard. Waste ground at W. end of Cong.

First record: Hart (1875). Aran Islands.

Reynoutria

*Reynoutria japonica Houtt. (*Polygonum cuspidatum*)

1 2 3 4 5 6 7 8   . .

Roadsides, waste ground and abandoned gardens; occasional and locally frequent. Thinly scattered over the whole region except the islands, and tending to increase.

First record: McClintock (1960): Maam Br.

Of all the coarse herbaceous garden plants of this family this is the species which most readily becomes naturalized from a small piece of rhizome, which rapidly increases to form a good-sized thicket. Like the others it appears not to set seed in the British Isles.

*R. sachalinensis (Schmidt) Nakai (*Polygonum sachalinense*)

. . . . . . 7 ?8 . .

There are extensive thickets of this plant in some of the gardens and other enclosures in the village of Ballyconneely and near the end of the Errislannan peninsula (district 7), whence it has spread to the roadside verges in small quantity. In a few places N. of Maam (district 8) there is a similar plant, but it is scarcely naturalized in truly wild ground. Dr A. Conolly is of the opinion, however, that this latter is a large-leaved variant of *R. japonica*.

Oxyria

Oxyria digyna (L.) Hill

. . . . . . 7 8 . .

Mountain cliffs; rare.

7. On the N. face of Muckanaght and on the N.W. face of Bengower. At *c*. 360 m (1200 ft) by a stream on the N. face of Doughruagh. **8.** North-facing cliffs above L. Nadirkmore.

First record: Wade (1802). 'Lettery Mountain, Ballynahinch.' Like many records of montane plants from Benlettery, this almost certainly refers to Bengower.

Rumex   Dock   *Copóg*

†Rumex conglomeratus Murr.

1 2 3 4 5 6 7 8   A B

Marshes and wet ground, especially if trodden or disturbed. Occasional to frequent, and abundant in a few places.

Commonest in districts 3, 4 and 7; rarest in 2 (where it is almost confined to the margins of turloughs) and 8.

First record: Wright (1871). Aran Islands.

R. sanguineus L.

1 2 3 4 5 6 7 8   A .

Hedgerows, roadsides, wood-margins and pastures. Rather rare in Connemara and the more exposed parts of the Burren; frequent and locally abundant elsewhere.

1. Very frequent in places, but local, and rather rare near the coast. **2.** Fairly frequent. **3.** Abundant. **4.** Very frequent. **5.** Occasional along the coast, and frequent around Spiddal. Near the waterfall at Oughterard. **6.** Rough grazing near Carna. **7.** Occasional on roadsides near Clifden, and frequent around Ballynahinch and Letterfrack. **8.** By the Pigeon-hole at Cong. Roadside near Ebor Hall. *N.W. of Oughterard* (P. cat.).

First record: Praeger (1895*b*). Roadsides on Inishmore.

‡R. obtusifolius L.

1 2 3 4 5 6 7 8   A B

Roadsides, pastures, waste ground and sand-dunes; more rarely as a weed of cultivated ground. Very frequent throughout.

First record: Babington (1836). Maam.

R. crispus L.

1 2 3 4 5 6 7 8   A B

Stony sea-shores, sand-dunes, roadsides, pastures and cultivated ground. Abundant throughout, except in district 8, where it is only occasional.

First record: Babington (1836). Maam.

R. hydrolapathum Huds.

1 . 3 . . . 7 . . .

Marshes, drains and river-banks. Frequent in the area N. of Ennis; very rare elsewhere.

1. Marsh beside L. Raha. **3**. Very frequent in the Dromore area and by the small lakes N. of Ennis. Abundant by the R. Fergus above Ennis. **7**. Drains near Doonloughan, and by the lake at Bunowen.

First record: Corry (1880). 'Above the old bridge over the Fergus [near Ennis].'

A rare plant in Ireland, except in the basins of the rivers Shannon, Erne and Barrow.

R. acetosa L.   Wild sorrel   *Samhadh*

1 2 3 4 5 6 7 8   A B

In a very wide variety of habitats, ranging from roadsides and heathy grassland to sand-dunes and mountain-cliffs. Abundant almost everywhere, the only area in which it is relatively scarce being that covered by clayey calcareous drift at the head of Galway Bay.

First record: Hart (1875). Aran Islands.

An interesting variant, described by Rechinger (1961) as *R. hibernicus*, is found in some abundance on fixed dunes in districts 6 and 7, from Mweenish I. all round the W. coast to near Salruck. It has also been seen in the extreme S.W. of district 2 (opposite Crab I.) and perhaps on Inishmore. It differs from typical *R. acetosa* in its dwarfer habit, shorter basal leaves, narrower cauline leaves, narrow and dense inflorescence, and in the papillose-puberulent surface of the stems, petioles and sometimes the leaves. The distinctive characters are preserved to a variable extent in cultivation. Dr J. Akeroyd, who has recently made a close study of it, regards it as a distinct ecotype, probably deserving subspecific status; but intermediates are usually to be found where the dunes pass over into normal grassland. Outside our region it is known from other parts of the W. and S. coasts of Ireland, and from the Western Isles of Scotland. A plant with some of these characters, but not all, was collected in 1976 by JW from beside disused lead-mines near Carran (**DBN**).

R. acetosella L.   Sheep's sorrel   *Samhadh caorach*

1 (2) 3 . 5 6 7 8   (*A*) B

Bog-margins, peaty banks, roadsides, heathy pastures and outcrops of siliceous rock. Abundant on acid, and especially on peaty soils; unknown on the limestone.

1. Very frequent. (**2**). *Inishmore*; reported by Wright and Hart in 1867–9, but without details as to habitat or abundance. Its occurrence is best explained by supposing that it was a casual brought in with turf from Connemara. **3**. At the edge of a partly cut-away bog on the E. margin of the Ballyogan Loughs. **5–8**. Abundant.

First record: Wright (1871). Inishmore.

Most of the plants from our region are 'angiocarpic', i.e. with sepals adherent to the ripe fruit. Some botanists distinguish these as *R. angiocarpus* Murb., but as there is no agreement on other characters constantly correlated with this, it seems best to regard the angiocarpous condition as, at best, the basis of varietal distinction.

*R. scutatus L., a sorrel-like member of this genus, occasionally cultivated in herb-gardens, was naturalized on rocks and walls around Lisdoonvarna at least from 1893 to 1906 (**DBN**), and perhaps later. It has not, however, been seen in recent years.

EUPHORBIACEAE

Euphorbia   Spurge

*Euphorbia helioscopia L.

. 2 3 4 (5) 6 7 (8)   A B

Cultivated or disturbed ground. Frequent on the limestone; rather rare elsewhere.

2. Frequent near the coast and occasional on Aran. **3.** Very frequent. **4.** Occasional. (5). *Near Tonabrocky* (Atlas). **6.** Carraroe. Furnace I. S.E. corner of Gorumna I. Kilbrickan. **7.** Doonloughan. Inishbofin. *Near Clifden* (Atlas). *Roundstone* (P. cat.). (8). *N.W. of Oughterard* (P. cat.).

First record: Hart (1875). Aran Islands.

E. paralias L.

. 2 . . (5) . . .   A .

Sand-dunes; very local.

2. Abundant on the dunes at Fanore and N.E. of Ballyvaughan; present also on most of the dunes on Aran. (5). *Near Galway; also at Inveran* (ITB).

First record: Mackay (1806). 'Very plentiful on the south isles of Arran.'

Mainly an east-coast plant in Ireland; the stations cited above are the only ones known between N. Kerry and N. Donegal.

E. portlandica L.

. 2 . . . . 7 .   A .

Sand-dunes, shingle-beaches and maritime rocks; very local.

2. In small quantity on the dunes at Fanore. Locally frequent on shingle-beaches on Aran, spreading onto gravelly roadsides near the beaches. **7.** In sand-filled crevices in rocks just N. of the entrance to Kingstown Bay. *Slyne Head peninsula* (ITB).

First record: More (1872). 'Great and Middle Aran; H. C. Hart.'

It is easy to confuse this with the preceding unless the two are seen growing together. The best distinction is in the tip of the leaves; in *E. paralias* they are blunt or minutely notched, but in *E. portlandica* most of them end in a very small, sharp point.

‡E. peplus L.

. 2 3 4 5 6 7 (8)   A B

Cultivated or disturbed ground. Fairly frequent on the limestone; rare elsewhere.

2. Behind the dunes N.E. of Ballyvaughan. 1 km E. of New Quay. Near Glensleade Castle. Frequent on Aran. **3.** Very frequent. **4.** Occasional. **5.** 4 km S.E. of Costelloe. **6.** S.E. corner of Gorumna I. Near Carraroe. **7.** Doonloughan. Belleek. Inishbofin. **8.** N. of Oughterard.

First record: Wright (1871). Aran Islands.

The distribution is very similar to that of *E. helioscopia*, but the two species are seldom seen growing close together.

†**E. exigua** L.

. **2 3 4** . . . . . .

Rocky or stony ground; rare.

2. Abundant in a lane leading down to Aughinish Bay, N. of Abbey Hill, 1978. *Ballyvaughan* (ITB). **3**. Limestone pavement S.W. of Mullaghmore, 1958, 1962. Crusheen railway station, 1969. *Garryland, and S. of Kinvara* (P. cat.). **4**. Roadside at Carrowmoreknock, 1965. *Parkmore*; *shore at Oranmore*; *Kilbeg ferry* (P. cat.).

First record: ITB (1901). Ballyvaughan, Ross L., Kilbeg ferry.

Very capricious in its appearances and disappearances, and perhaps only casual in some of its stations.

Mercurialis

Mercurialis perennis L.   Dog's mercury

. **2 3** . . . .

Scrub, roadsides and rocky ground. Occasional in two limited areas of the Burren; unknown elsewhere.

2. Frequent at the steep defile, sometimes known as 'Khyber pass', S. of Deelin Beg, and extending thence discontinuously, at the base of walls and in small patches of scrub or crevices in partly overgrown pavement for about 1 km westwards and 2 km northwards. **3**. In several places in the area between Corofin and Ennis, as near Ballycullinan L., on roadsides W. of L. Cleggan and in scrub S. of Ruan.

First record: CH1 (1866). 'In the Kieber Pass, south of Bell Harbour in Burren, Clare; *Mr. F. J. Foot.*'

This plant, so common in woodlands almost throughout Britain, is in Ireland, apart from the Burren localities given above, confined to about a dozen stations, all in or close to estate woods, and obviously introduced. This naturally suggests the possibility that it may be introduced in the Burren also, for it is certainly unexpected that a woodland species common in Britain should be native here but nowhere else in Ireland. None of the Burren stations is, however, suggestive of introduction. The estate woods at Dromore might conceivably have served as a source for the southern area, but although it comes close to Dromore it has not been seen in the Dromore woods themselves; and in the area around Deelin Beg there is no house or estate where one can imagine imported trees being planted. We therefore accept it as probably native, with a slight doubt. For further discussion of its status in Ireland see Boatman (1956), Lambe, Mitchell & O'Connell (1978) and Webb (1978).

ULMACEAE

Ulmus   Elm   *Leamhán*

Ulmus glabra Huds. (*U. montana*)   Wych-elm

1 2 3 4 5 . **7 8** . .

Woods, hedges and scrub; widespread but rather rare.

1. Wood at Elmvale Ho. Hedges near O'Dea's Castle. **2**. In scrub at Formoyle and S.W. of Ballyvaughan. Clooncoose wood. Hedges W. of Turlough village. **3**. Woods at Coole and Garryland; also S.E. of Ruan. In the wood W. of the turloughs S.W.

of Mullaghmore. **4**. Hedge 3 km N.E. of Galway. **5**. Hedge half-way between Galway and Moycullen. Occasional by the coast, from Spiddal to near Galway. **7**. Woods at Kylemore and Letterfrack. **8**. Occasional in hedges around the N.W. part of L. Corrib. One tree by the waterfall W. of L. Nafooey.

First record: Praeger (1934c). Garryland.

Undoubtedly native in the Burren, and probably in some of its stations in N. Connemara. Elsewhere it is very difficult to distinguish between planted trees, those which are self-sown from planted trees, and those which are truly native. Nowhere does it occur in any great quantity.

U. procera Salisb. (*U. campestris*), the English elm, has been planted in woods and hedges in a few places, mainly near or in the grounds of the larger houses (Coole, Merlin Park, Clifden Ho., near Corofin). It tends to spread by suckering but nowhere can it be regarded as effectively naturalized.

CANNABACEAE

Humulus

*Humulus lupulus L.   Hop

. **2 3** . . . **7** . **A** .

Hedges and scrub; rare.

2. Inishmore; formerly frequent, now very rare. **3**. Roadside 1½ km N.N.W. of Ennis. In scrub by the ruined church S.W. of Ruan. *Abundant on old walls about Corofin* (O'Kelly, 1903). **7**. In hedges between Kylemore and Letterfrack.

First record: Mackay (1806). 'In the south isles of Arran, among the lime-stone rocks.'

URTICACEAE

Soleirolia

*Soleirolia soleirolii (Req.) Dandy (*Helxine soleirolii*)

. . . **4 5 6 7 8** . .

On and at the base of damp walls and banks. Rare, but tending to increase.

4. On a wall by the canal in Galway. *Merlin Park, c.* 1933 (Praeger, 1934c). **5**. In two or three places around Spiddal. **6**. Cashel. **7**. In some quantity on walls in the N.W. part of Clifden. **8**. Opposite the hotel at Leenane and on the quay.

First record: Praeger (1934c). 'Abundantly by a spring in the wilder part of Merlin Park.'

In all cases close to the garden from which it escaped, but effectively naturalized.

Urtica   Nettle   *Neanntóg*

‡**Urtica dioica** L.

1 2 3 4 5 6 7 8 A B

In a wide variety of habitats, but mostly close to roads, human settlements or places frequented by cattle or sheep. Abundant throughout, except in the wilder parts of Connemara.

First record: Wright (1871). Aran Islands.

Possibly native in damp woods in a few places, but in most of its stations an obvious follower of man or his beasts.

***U. urens** L.

. 2 3 4 5 6 7 . A B

Cultivated ground and farmyards; widespread and abundant in a few places, but rare over most of the region.

2. Fairly frequent on Aran; not seen on the mainland. **3.** Abundant as a garden weed at Kinvara. Near the edge of Garryland wood. **4.** Occasional. **5.** On the shore between Galway and Salthill. Farmyard 5 km S.S.W. of Costelloe. **6.** Farmyard near Carna. **7.** Waste ground, trodden by cattle, at Doonloughan. Inishbofin.

First record: Wright (1871). Inishmore: 'Only met with in the immediate neighbourhood of Kilronan.'

Formerly common over most of Ireland, it is now local, and has for the most part retreated to the regions of relatively primitive farming.

Parietaria

Parietaria diffusa Mert. & Koch (*P. ramiflora*, *P. officinalis*)   Pellitory

1 2 3 4 5 . . 8 A .

Walls, boulder- and shingle-beaches and limestone pavement. Frequent on the limestone; not seen elsewhere.

1. Walls of O'Dea's Castle and the adjacent ruined church. **2.** Limestone pavement near the sea at Black Head, and in great abundance at the drift-line above the stony beach 1–1½ km to the south. Frequent on Aran, on shingle-beaches, on pavement, and at the base of walls. **3.** Stony track near the sea at Aughinish. Among rocks by the W. shore of Kinvara Bay. On the old bridge at Kilcolgan estuary. **4.** Frequent on walls in and around Galway and Oranmore; also near Aughnanure Castle. Shingle-beaches on Tawin I. and E. of L. Atalia. **5.** On walls in the S.W. part of Galway city. On the boulder-beach at Gentian Hill. **8.** On a wall by the river at Cong.

First record: Wright (1871). Aran Islands.

In most of Ireland this species is seen only on old mortared walls. Its great abundance on stony beaches all round Galway Bay is remarkable, and gives clear evidence of its native status. South of Black Head it forms, with *Armeria maritima* and *Saxifraga rosacea*, a dense, hummocky community just above the storm-beach.

MYRICACEAE

Myrica

Myrica gale L.   Bog myrtle   *Railleóg*

1 . 3 4 5 6 7 8   . B

Lake-shores, bogs and fens. Abundant in Connemara; absent from most of the Burren; occasional elsewhere.

1. Frequent around L. Goller and Lickeen L. **3.** Frequent in and around the fens from L. Briskeen southwards to Barefield; not seen N. of Gort. **4.** Frequent from Galway northwards. **5–8.** Abundant.

First record: Babington (1836). Bogs between Clifden and Roundstone.

BETULACEAE

Betula   Birch   *Beith*

Betula pendula Roth (*B. verrucosa, B. alba*)

. . . 4 . . . 8   . .

Rocky ground and woods; rare.

4. On limestone pavement S. and E. of Ross L. **8.** In woods on the S. shore of L. Mask. In estate woods at Ashford Castle and Gortdrishagh, but here probably derived from planted trees.

First record: ITB (1901). Clonbur.

Grossly over-recorded in Ireland, from confusion with glabrous forms of *B. pubescens* (see below); it is relatively rare as a native and is found mainly on the margins of lowland bogs or by limestone lakes. Elsewhere it is usually naturalized from planted trees. In the first two stations cited above it is probably native, but is accompanied here by hybrids with *B. pubescens*, which bear on their young shoots a few pale raised glands but also some hairs. Their leaves are closer in shape to those of *B. pubescens*.

B. pubescens Ehrh.

1 2 3 4 5 6 7 8   . .

Woods, copses and rocky ground. Very frequent on siliceous soils where there is some protection from wind and grazing; local on the limestone.

1. By a lakelet S. of Lickeen L. **2.** Frequent in the woods at Poulavallan and the Glen of Clab; also in copses in the Caher R. valley. On the S.W. slope of Gleninagh Mt. **3.** Woods at the base of Slieve Carran, and S.W. of Mullaghmore. W. of Coole. Near the cross-roads N.N.E. of Tulla, and 2½ km S.S.E. of Tulla. **4.** Occasional around Ross L. S. of Carrowmoreknock. **5–8.** Abundant in suitable habitats, especially in district 8; a constant ingredient of the scrub or woodland on lake-islands in districts 6 and 7.

First record: CH2 (1898). 'Islands in L. Inagh and mountains above it; *N.C.*'

Throughout the region, and especially in the Burren, trees with almost or completely glabrous young shoots are to be found. They have often been recorded as *B. pendula*, but in all other characters they conform with *B. pubescens*, and may tentatively be referred to its subsp. *carpatica* (Willd.) Asch. & Graebner. Everywhere, however, they grade into typically pubescent forms.

Alnus   Alder   *Fearnóg*

Alnus glutinosa (L.) Gaertn.

1 2 3 4 5 6 7 8 . .

Lake-shores, river-banks, damp woods and copses, and on marshy flats. Common near L. Corrib and L. Mask; rather local elsewhere.

1. Hedges near Moy Ho. (perhaps planted). A few trees by the stream at O'Dea's Castle. **2.** Sparingly in a fen 1½ km N. of Carran. **3.** Several trees by the river W. of Kiltartan cross-roads. **4.** Abundant at the W. end of Ross L. W. shore of Ballycuirke L. At several places near L. Corrib, E. of Moycullen. **5.** Occasional by the coast, and by the old railway-track on the N.E. margin; also in small quantity by lakes and in ditches in the northern part of the blanket-bog between Rosscahill and Screeb. **6.** Frequent in copses E. of Recess. A few trees by the stream draining Inver L. **7.** Occasional, including a few lake-islands. **8.** Very common on the shores of L. Corrib and L. Mask, and on many of the islands; also along several of the streams flowing into the lakes.

First record: ITB (1901). H16.

CORYLACEAE

Corylus

Corylus avellana L.   Hazel   *Coll*

1 2 3 4 5 6 7 8 A .

Dominant in scrub over wide areas; less often as a minor constituent of mixed woodland or as isolated bushes on cliffs. Very abundant on the more sheltered parts of the limestone; frequent elsewhere.

The only areas in which the hazel has not been seen are the coastal strip of district 1 (where there is virtually no woody vegetation), the drift-covered country at the head of Galway Bay, and the most exposed parts of S. and W. Connemara, including Inishbofin. Although most abundant on limestone, it grows freely on all types of rock.

First record: Hart (1875). Aran Islands.

The distribution of the hazel seems to be governed largely by the amount of shelter afforded, because in places exposed to strong winds the buds on the leading shoots are killed in the course of the winter; the whole shoot then dies and growth has to be resumed from the lower part of the plant. A small amount of shelter is immediately marked by an increase in the height of the bushes. It suffers also from grazing, but less severely than do most other woody plants; undamaged seedlings have been seen on moderately grazed grassland on the margin of scrub.

FAGACEAE

Small coppiced trees of *Castanea sativa*, the Spanish chestnut, have been noted in mainly native scrub or woodland in districts 3 (S. of L. Bunny), 7 (Ballynahinch) and 8 (W. of Glann), but it is only rarely that this tree sets viable seed in western Ireland, and there is no evidence that it is in any real sense naturalized.

Quercus   Oak   *Dair*

Quercus petraea (Matt.) Liebl. (*Q. sessiliflora*)

1 . 3 . 5 6 7 8 . .

Woods, scrub and rocky ground. Very rare on the limestone; fairly frequent elsewhere.

First record: CH2 (1898). Island in L. Inagh, 1897.

It is to be presumed that a substantial part of Connemara was at one time covered with woods or scrub in which this species was dominant, though the upper mountain slopes, the coastal strip and the wetter bogs probably carried no woodland. Some fair-sized fragments of this woodland survive, as by the Owenboliska R. (district 5), near Glendollagh (district 6), W. of Ballynahinch (district 7) and in several places in district 8. All of them, however, have been subjected to at least some felling and replanting, and in some of them hybrids with *Q. robur* are almost as common as pure *Q. petraea*. Smaller fragments, mostly of the nature of scrub, which have been subject to little interference except light grazing and sporadic felling, exist by mountain streams or under the shelter of rocky bluffs. It is only on the larger lake-islands that the oak-woods may be seen in something like their original state; for details see Webb & Glanville (1962) and Roden (1979).

On the limestone *Q. petraea* has been seen only in the western part of district 3: in scrub near the Ballyeighter Loughs, by the turloughs S.W. of Mullaghmore, and *c*. 1 km N.E. of Corofin. In all cases it occurs in small quantity, and is accompanied by *Q. robur* and hybrids.

†Q. robur L.

1 . 345 . 78 . .

Woods, hedges and scrub; local.

1. On and around Clifden Hill. Occasional in hedges near O'Dea's Castle. **3.** Abundant in the woods at Coole. Occasional around Dromore. Sparingly, with *Q. petraea*, in small copses from Corofin north-eastwards to the Mullaghmore turloughs and Ballyeighter Loughs. **4.** On limestone pavement E. of Ross L. By the R. Corrib above Galway. **5.** Moycullen wood. In scrub S. of Oughterard. **7.** In scrub by the lake at Renvyle Hotel. **8.** By the Pigeon-hole at Cong, and self-sown in various parts of the Ashford Castle woods. On islands in L. Mask partly covered by calcareous drift. One sapling on a bluff at the Maumeen gap at 450 m (1500 ft).

First record: Tansley (1908). Garryland wood.

Only trees which appear to be self-sown have been listed above. Although this species may well be native around Galway or on the eastern fringe of the Burren, there is no station that is not under some suspicion as being derived ultimately from planted trees.

Fagus

*Fagus sylvatica L.   Beech   *Fea*

1 . 3 4 5 6 7 8   . .

Woods and plantations; frequently planted in all but the most exposed areas, and regenerating in many of them.

Regeneration is most obvious in districts 4 and 8 and in the eastern parts of districts 3 and 5, but even in W. Connemara it has been observed between Ballynahinch and Toombeola, at Letterfrack, and on an island in L. Fadda. Young trees were also seen on islands in L. Corrib (district 8), to which the seed must have been transported from the mainland by birds.

First record: Tansley (1908). Garryland wood. The context implies that self-sown as well as planted trees were observed.

SALICACEAE

Salix   Willow, Sally   *Saileach*

It is difficult to present satisfactory records for this genus on account of the frequency with which its species and hybrids are planted in hedges. *S. atrocinerea*, *S. caprea*, *S. aurita*, *S. repens* and *S. herbacea* are undoubtedly native and are rarely if ever planted, so that their distribution presents no difficulty. *S. pentandra* and *S. viminalis* are certainly introduced, the former for hedging and the latter for basket-making, but both appear to be occasionally naturalized; this is also true of *S. alba* and *S. fragilis*, which are planted for ornament. *S. purpurea*, though seen mainly in hedges, has some claim to be considered as a possible native. *S.* × *smithiana* is seen only in hedges, and is not naturalized.

Willow-hedges are usually made from cuttings of a single clone, and are therefore entirely of one sex, with no chance of reproduction by seed unless (as might rarely happen) a planting of the other sex has been made nearby; we have, however, nowhere observed this. Such plants have, in our opinion, no place in a Flora, though a curious convention in the past has decreed that hedges of willow should be recorded while those of Lawson's cypress or *Lonicera nitida* are ignored. A difficulty arises, however, when one finds a single plant of a willow in a hedge mainly composed of some other shrub; it might be a seedling, or it might derive from a cutting put in to fill a gap.

The situation is further obscured by the fact that recorders for the *Atlas* were instructed to record all plants in field-hedges, whether they showed any sign of naturalization or not. We have had, therefore, to ignore the *Atlas* records for all introduced species, and we are unable to give convincing first records except for the native species.

In the closing years of last century an attempt was made to establish a basket-making industry at Letterfrack and Moyard, and plantings were made of *S. triandra* and perhaps some other species. The enterprise was not a success, as the plants soon fell victims to a severe fungal attack; today no traces of the introduced species remain, except for a single bush which has been identified as a *triandra*-hybrid.

Willows are notorious for the ease with which they can hybridize, and for the difficulty in naming the hybrids with confidence. *S. atrocinerea* × *aurita* is probably fairly common in Connemara, as both parents frequently grow together. *S.*

aurita × *repens* has been recorded from near Tully Cross and from the Errislannan peninsula, and may well occur elsewhere in W. and C. Connemara. *S.* × *smithiana* and the very similar *S.* × *sericans* (hybrids of *S. viminalis* with *S. atrocinerea* and *S. caprea* respectively) are planted in hedges in districts 1 and 7, but are not naturalized. *S. fragilis* × *pentandra* (*S.* × *meyerana*) has been recorded from Maam and Letterfrack, but as almost certainly having been planted.

*S. pentandra L.

Frequently planted in hedges in districts 1, 4, 7 and 8, and sparingly naturalized by the salt-lake S. of Clifden and S. of Clonbur.

*S. fragilis L.

Frequently planted in hedges and occasionally in copses in districts 1, 4, 5, 6 and 7; naturalized here and there in Galway city, mainly by the river.

*S. alba L.

Occasionally planted in woods, demesnes and hedges, and occasionally naturalized in fens or hedges or on river-banks. It has been seen at the S. end of L. Bunny, by the Kiltartan R., on the W. bank of the R. Corrib above Galway and in a hedge at Glann.

S. purpurea L.

1 . 3 . . . (7) 8 . .

Hedges and copses; rare.

1. In a hedge S. of Lickeen L., and in another half-way between Kilfenora and Ennistymon. *By Inchiquin L.* (Praeger, 1905). **3.** Drains by the R. Fergus, N. of Dromore. (**7**). *Near Clifden* (Praeger, 1934*c*). **8.** In scrub 5 km W. of Cong.

First record: Praeger (1905). Inchiquin L.

Possibly native in district 3; very doubtfully elsewhere. It is planted in hedges, especially around Lisdoonvarna.

S. atrocinerea Brot. (*S. cinerea*)

1 2 3 4 5 6 7 8   A B

Woods, copses, hedges, river-banks, lake-shores and islands. Frequent throughout and locally abundant.

Scarcer in districts 2 and 3 than elsewhere. It is more tolerant of wind-exposure than any native tree of the region, except perhaps *Sorbus aucuparia*.

First record: More (1876). Inishbofin.

Many botanists prefer to treat this as a subspecies of *S. cinerea*, and it is certainly true that the most reliable character for separating them – the presence in *S. atrocinerea* of rust-coloured hairs on the underside of the leaves – is difficult to observe in many specimens from Connemara.

S. caprea L.

1 2 3 4 5 . 7 8   A .

Scrub, hedges, woods and limestone pavement. Occasional and locally frequent on the limestone; rare elsewhere.

1. Hedge W. of Kilnamona. **2–3.** Occasional. **4.** Frequent in the north; rare elsewhere.

5. Frequent along the N.E. margin; rare elsewhere. **7.** In a copse near Letterfrack. **8.** Occasional.

First record: Hart (1875). Inishmore (but doubtfully native here).

S. aurita L.

1 2 3 4 5 6 7 8 . .

Bogs, moorland, heathy ground, streamsides, hedges and scrub. Occasional on the limestone; very frequent to abundant elsewhere.

1. Frequent and locally abundant. **2–3.** Occasional. **4.** Frequent in the north; not seen in the south. **5–8.** Abundant.

First record: Babington (1836). Maam.

*S. viminalis L.

1 2 3 . **5 6** . . **A** .

Widely planted for basket-making, and occasionally naturalized in marshes or by streams.

1. By a stream at Derrymore Br. By Garvillaun L. **2.** In a marsh at the W. end of Inishmore. **3.** Marsh S.E. of Ruan. **5.** Streamside W. of Inveran. **6.** Streamside near Carna.

S. repens L.

1 2 3 4 5 6 7 8 **A B**

Heathy or rocky ground (especially on mountains), limestone pavement and stony lake-shores; more rarely on sand-dunes. Frequent in most districts and locally abundant.

1. Frequent. **2.** Very frequent. **3.** Occasional. **4.** Frequent in the north; not seen elsewhere. **5.** Occasional, mainly near the coast. **6.** Very frequent. **7.** Abundant in the north and west; rarer elsewhere. **8.** Occasional.

First record: Wade (1802), as *S. fusca*. 'Common on the turfy and heathy mountains, Cunnamara.' He also mentions *S. depressa* 'with the former; silvery on both sides of the leaf'.

We have interpreted this species in a wide sense, as we can find no reliable guide to the taxonomy of its variants. The characters given by Warburg (1962) for subsp. *argentea* (Sm.) Camus (heterophylly, leaves silky on upper surface, and with female catkins on leafy shoots) differ widely from those used by Rechinger in *Flora Europaea* to distinguish *S. arenaria* L. (thick, dark, hairy twigs, stipules often present, leaves with toothed, revolute margin and numerous veins, capsule hairy, etc.). Most of our specimens from the region seem to some slightly closer to *S. arenaria, sensu* Rechinger than to *S. repens, sensu stricto*, but there is a sad lack of correlation between the supposedly diagnostic characters, and we are in no position to divide our material into two taxa at any level. All specimens are sparsely to fairly densely hairy on the upper side of the leaves, but none can be described as silky or silvery. The species has been noted on only two dune-systems in our region (Slyne Head peninsula and N.E. of Gowlaun), and the plants from these stations are, if anything, less hairy than those from rocky ground.

S. herbacea L.

. . . . . . **7 8** . .

Mountain cliffs, and rock-crevices on mountain summits; occasional.

7. Frequent on the Twelve Pins. In the corrie on the S.W. side of Benchoona. Sparingly on the summit of Garraun. **8**. Abundant on the rocky ridge between Bencorragh and Benbeg. Very luxuriant on the cliffs S.W. of L. Nadirkmore. Frequent on the Maumturk range N. of the Maumeen gap. Devilsmother, just S. of the summit. Leckavrea.

First record: Mackay (1806). 'On Lettery mountain and other mountains in Connemara.'

Populus   Poplar

Populus tremula L.   Aspen

. **2 3 4 5 6 7 8** . **B**

Woods, scrub and hedges; occasional.

2. Occasional in scrub in the middle part of the Caher R. valley. Two large trees at the entrance to Poulavallan; probably planted, but now spreading by suckers. **3**. Hedges 3 km N. of Ruan. **4**. In scrub 6 km E.S.E. of Oughterard. By the Oughterard–Galway road S. of Ballycuirke L. **5**. Near Glentrasna. **6**. On a bank by the sea W. of Cashel. **7**. Hedge on the N. side of Clifden estuary. Near the W. end of Ballynakill L. Inishbofin (rare). **8**. Frequent and locally abundant on several of the islands in L. Corrib.

First record: More (1876). Inishbofin: 'sparingly on rocky banks at east end of the harbour'.   ˙

Perhaps planted in some of its stations, but native in most of them.

*P. × **canescens** (Ait.) Sm. (*P. alba* ×*tremula*)   Grey poplar

1 . **3** . **5** . . . . . .

Hedges and thickets; rare.

1. Suckering freely near Clifden Ho., Corofin. **3**. A few trees near the S. end of L. Bunny. Several trees by the Kiltartan R. **5**. By the Oughterard road *c*. 8 km from Galway. In a hedge near Gentian Hill.

First record: Atlas (1962).

Always planted, but readily naturalized on account of its freely suckering habit.

**P.* × *canadensis* Moench (*P. deltoidea* × *nigra*) is occasionally planted on field-boundaries, but much less commonly, especially in Connemara, than in most other parts of Ireland. It does not sucker, and cannot set seed (all trees being males), so in no sense can it be regarded as naturalized.

MONOCOTYLEDONS

ORCHIDACEAE

Hammarbya

Hammarbya paludosa (L.) Kuntze (*Malaxis paludosa*)

(*1*) . . . . ?6 . **8** . .

Wet bogs; very rare, or perhaps overlooked.

(*1*). *Near Lisdoonvarna*, before 1866 (CH1). **8**. One plant by a flush above the lane on the W. side of Devilsmother, 1966 (GH). *Below Carnseefin* (Praeger & Carr, 1895).

First record: CH1 (1866). 'In a marshy place near Lisdoonvarna, on the coal measures; *Rev. T. Warren.*'

Perhaps more widespread; it is a very inconspicuous plant. We have received a verbal report of its occurrence E. of Cashel (district 6), but confirmation is required.

Neottia

Neottia nidus-avis (L.) Rich.   Bird's-nest orchid

. . **3 4** . . . (*8*) . .

Woods and scrub; rare.

3. Garryland wood, 1980. In dense hazel-scrub 1 km N.E. of Killinaboy, 1976. **4**. Wood by the N. shore of Ross L., 1965. At the edge of hazel-scrub 3 km N.N.W. of Ross L., 1970. (The ITB record for 'Oughterard' may well apply to this station.) (*8*). *Ashford Castle*, 1895 (CH2).

First record: CH2 (1898). 'In the wilderness, Ashford, Cong, 1895; *Miss M. F. Jackson.*'

Listera

Listera cordata (L.) R.Br.

. **2** . . . . (*7*) **8** . .

Peaty ground on mountains or lake-islands; rare.

2. Near the summit of Gleninagh Mt., 1966 (JJM). (*7*). *Islands in L. Nambrackmore* (Tansley, 1908), *and in L. Anessaundoo*, 1957 (E. Glanville). **8**. In some abundance on bare peat on the summit of the hill (1307 ft) E. of Leckavrea Mt., 1965 (DW). Maumturk Mts., N. of the Maumeen gap, 1968 (DW). *Lusteenmore I. in L. Mask* (Praeger, 1909*a*). *Island in L. Corrib, N. of Inishdoorus* (Praeger, 1933).

First record: Hart (1883). Maumturk Mts.

L. ovata (L.) R.Br.   Twayblade

1 2 3 4 5 . (*7*) **8**   A .

Grassland, woods, marshy ground and limestone pavement. Common in the Burren; rather rare elsewhere.

1. By L. Garvillaun. Roadside by E. end of Lickeen L. Woods at Elmvale Ho. **2–3**. Very frequent. **4**. Frequent. **5**. By the old railway-track N.W. of Rosscahill. (7). *Bunowen Hill*, 1957 (DW). **8**. By the lake 1 km N. of Maam Cross. Marsh N.E. of Clonbur. Near Cong. On islands in L. Corrib.

First record: Foot (1864). 'Everywhere abundant, all through the district [Burren].'

Spiranthes

Spiranthes spiralis (L.) Chev.   Ladies' tresses

. 2 3 4 5 6 7 . A .

Dry grassland, usually on limestone or on stabilized calcareous dunes. Locally abundant, but very variable in its frequency of flowering from year to year.

2. Fanore dunes. Occasional on Aran. *Ballyvaughan* (Druce, 1911). **3**. In several places between Killinaboy and the N. end of L. Bunny. *Kilcolgan, Ballinderreen and S. of Kinvara* (P. cat.). **4**. By the N. shore of Gortachalla L., and near L. Corrib, W. of Carrowmoreknock. *Frequent E. and S.E. of Galway, from Menlough to Ardfry Ho.* (P. cat.). **5**. By the river-mouth at Spiddal. **6**. Stabilized dunes on Furnace I. Behind the 'coral strand' S.S.W. of Carraroe. Mweenish I. S.E. of Carna. **7**. Locally frequent along the coast, from near Roundstone to Salruck. *Grassland on W. slope of Errisbeg* (Tansley, 1908).

First record: Ogilby (1845). 'Diffused over the whole island [of Inishmore].'

Probably under-recorded, partly from its irregularity of flowering, and partly because it is rarely in flower before the end of August.

S. romanzoffiana Cham.   American ladies'-tresses

. . . . 8 . .

Stony lake-shores; very local (Plate 3).

8. Locally frequent on the shore of L. Corrib from near Currarevagh Ho. to a point S. of Clonbur, including the Dooros peninsula and Inishdoorus; it has not, however, been seen on the narrow arm of the lake that runs from the Hill of Doon up towards Maam. It has also been reported to us from Ballydoo L., N. of Cornamona (1960; Mrs W. M. Barron).

First record: Webb (1959). Shore of L. Corrib, S. of Clonbur.

Long known from S.W. and N.E. Ireland, this species was first found in our region by Miss K. Harding on the N. shore of L. Corrib, while doing mapping work for the *Atlas* in 1958. In 1964 DW saw it on the west shore, near its apparent southern limit on the lake, and in 1969 he spent two days with GH mapping its distribution round the lake. Outside Ireland its European stations are confined to a few places in W. Scotland, and also the edge of Dartmoor, where it was first seen in 1958. It is widespread in North America and the Aleutian Islands (though not in E. Asia, as is often stated). It is, accordingly, an amphi-Atlantic plant of strikingly asymmetrical distribution.

Like many orchids it varies greatly from year to year in the abundance of its flowering spikes; DW kept one colony under close observation for 7 years and found that the number of spikes varied from 21 to 110, without any constant tendency to increase or decrease. Its flowering period is precise and short (normally 20 July to 7 August in Connemara), and when not in flower it is quite remarkably inconspicuous. It seems probable, therefore, that further stations, both in Ireland and in Britain, remain to be discovered, and we do not think that there is any real evidence that the species is extending its range.

By L. Corrib it grows at a well-defined level, where it is subject to occasional flooding. Its most constant associates are *Myrica gale, Molinia caerulea, Mentha aquatica, Ranunculus flammula, Leontodon autumnalis, Carex panicea, C. serotina, Anagallis tenella* and *Salix repens.*

Epipactis   Helleborine

Epipactis palustris (L.) Crantz

. **2 3 4** . . **7** . . .

Fens and other wet, calcareous habitats; occasional.

2. Very abundant in the marsh 1½ km S.S.E. of Feenagh. In a flush on the Corkscrew Hill. *Ballyvaughan*, 1891 (**DBN**). **3**. A few plants in the fen at the N. end of L. Bunny; also between here and Boston. *Near Dromore* (Atlas). *Corofin*, 1905 (**DBN**). *Kilmacduagh marsh* (CH1). **4**. Coolagh fen. Roadside 3 km S. of Carrowmoreknock. Island in L. Corrib E. of Gortachalla L. *Rinville* (P. cat.). **7**. Meadow on the Slyne Head peninsula, N.W. of Truska L.

First record: CH1 (1866). 'Bog at Kilmacduagh; *A.G.M.*'

E. helleborine (L.) Crantz (*E. latifolia*)

1 2 3 4 . . . **8** . .

Scrub, woods and limestone pavement. Occasional on the limestone; very rare elsewhere.

1. Scrub near Elmvale Ho. **2–3**. Occasional, and locally frequent, but not on Aran and rare near the coast. **4**. Occasional S. and E. of Oughterard. **8**. On an island in L. Corrib N.W. of Oughterard. W. of Cong.

First record: CH1 (1866). Garryland.

E. atrorubens (Hoffm.) Schultes

1 2 3 4 . . . (*8*) [A] .

Crevices in limestone pavement, and in *Dryas–Empetrum* heaths. Rather frequent in the Burren above 150 m (500 ft), and occasional at lower levels; very rare elsewhere.

1. On a limestone outcrop near Elmvale Ho. **2**. Frequent on the northern mountains; also on Mullaghmore and other hills on the E. margin. **3**. Limestone pavement N. of Boston. By Shandangan L. *Kinvara*, 1900 (**DBN**). **4**. Hill N.E. of Menlough. Limestone pavement at Srue. (*8*). *In a rocky valley on the Clonbur side of Cong* (Marshall & Shoolbred, 1896).

First record: O'Mahony (1860), as *E. ovalis.* 'Near the summit of the highest limestone mountains in Burren.'

A few of the above records are subject to some doubt, as non-flowering plants are not always easy to distinguish from *E. helleborine.* The leaves are supposed to be evenly disposed around the stem in the latter, and 2-ranked in *E. atrorubens,* but many specimens are bafflingly intermediate. The possibility that some of these plants are hybrids is not to be dismissed, as even flowering specimens can be ambiguous: a plant in **TCD** from near Cong resembles *E. helleborine* in most characters but has a densely pubescent ovary.

The *Atlas* record for Aran is almost certainly an error (see Webb, 1980*a*).

Cephalanthera

Cephalanthera longifolia (L.) Fritisch

. (*2*) **3** . . . .

Rocky ground in the Burren, very rare.

(*2*). *Hilly woods near Ballyvaughan*, 1912 (Mrs O'Callaghan, **NMW**). **3**. Low ground near Mullaghmore, 1980 (Mrs Keane).

First record: Atlas (1962).

A very local plant in Ireland, though abundant in a few of its stations. The nearest to the Flora region are in W. Mayo (N.W. of Mweelrea) and S.E. Galway (near Woodford). In view of its rarity and attractiveness we think it advisable not to specify Mrs Keane's station very precisely.

Vines & Druce (1914) say that this species was found by Sherard at Ballynahinch, Co. Galway, and that the specimen is at Oxford. It was, however, collected at Ballynahinch, Co. Down, not in Connemara.

Anacamptis

Anacamptis pyramidalis (L.) Rich.   Pyramidal orchid

1 2 3 4 5 . **7** .   **A** .

Pastures, roadsides and sand-dunes. Frequent and locally abundant on dry calcareous soils; very rare elsewhere.

1. Frequent on the dunes N. of Lehinch, and occasional in pastures near Moy Ho. **2**. Very abundant on the Fanore dunes. Castletown. Dunes N.E. of Ballyvaughan. Locally abundant on Aran. **3**. Very frequent. **4**. Waste ground E. of Galway. Drift hillock at Kilcaimin. Near Kilcolgan, N. of the estuary. **5**. Frequent on Gentian Hill. In pastures 4 km W. of Spiddal. **7**. Frequent on fixed dunes on the Slyne Head peninsula, also on Omey I. and the mainland opposite.

First record: Foot (1864). 'In great abundance through Burren, from the sea-shore to the tops of the hills.'

Strictly calcicole; usually confined to dry soils, but it has been seen in a fen beside a small lake N. of Barefield.

Orchis

Orchis morio L.   Green-veined orchid

. . . . . **6 7** .   . .

Rocky knolls in heath or bog; rare.

6. On heathy hillocks behind the 'coral strand' S.S.W. of Carraroe. At the quay near Ardmore Point. **7**. On small granite knolls in blanket-bog 4 km N.N.W. of Roundstone. On thin peaty soil at Bunowen. *Errisbeg*, 1896 (**DBN**). *By the road between Clifden and Roundstone* (CH1).

First record: CH1 (1866). 'Roadside in one place, between Clifden and Roundstone; *Prof. Melville*.'

O. mascula (L.) L.   Early purple orchid

1 2 3 4 5 6 7 8   A .

Scrub, limestone pavement, rocky grassland and mountain cliffs. Frequent and locally abundant on the limestone; occasional elsewhere.

1. In some quantity on a roadside bank 5½ km E. of Inagh. Edge of hazel-copse 5 km N.W. of Ennis. **2–4.** Frequent throughout and locally abundant, especially near the W. coast. Local on Aran. **5.** Gentian Hill. **6.** Mweenish I. **7.** Dog's Bay peninsula. Frequent on the Slyne Head peninsula. Renvyle. Near the sea N. of the entrance to Streamstown Bay. **8.** Occasional on mountain cliffs, as on Lissoughter, at Maumeen and on the mountains N.W. of Cornamona. Edge of copse 4 km W. of Oughterard. Limestone pavement N.W. of Cong.

First record: Foot (1864). 'In great abundance through Burren.'

This plant makes a notable contribution to the floral display in the Burren in the latter half of May. Besides the normal deep reddish purple, plants with lilac-pink or pure-white flowers are not uncommon.

Dactylorhiza

Dactylorhiza fuchsii (Druce) Soó (*Dactylorchis fuchsii*)

1 2 3 4 5 6 7 8   A .

Grassland and scrub on basic, neutral or mildly acid soils. Abundant on the limestone; rather rare elsewhere, and mostly on base-rich drift or near the sea.

1. Occasional, mainly on the northern margin or near the coast. **2–4.** Abundant. **5.** Occasional on the N.E. margin and near the coast; also on drift near Knocka L. **6.** Meadow near Flannery Br. Near the outlet of Inver L. **7.** Very frequent in the extreme west, especially on sandy ground on the Slyne Head peninsula; also at Belleek, on Omey I. and near Cleggan. Frequent on lake-islands. **8.** By the shore of L. Mask 3 km N. of Ferry Br. By the shore of L. Corrib near Cong and near Oughterard; also on one of its islands.

First record: Druce (1909), as *Orchis maculata*, var. *okellyi*. 'Quite frequent over a large area of limestone country about Ballyvaughan.'

A variant with unspotted leaves and pure-white flowers (or occasionally slightly tinged with lilac), sometimes distinguished as subsp. *okellyi* (Druce) Soó, is frequent throughout the Burren, and has been seen near Galway city; the *Crit. Suppl.* indicates it also for S.W. Connemara. We do not find that it differs from the type in anything but the absence of anthocyanins. Variants with deep reddish purple flowers and shorter stems have been called subsp. *hebridensis* (Wilmott) Soó, but they too cannot be regarded as more than a variety at best. Plants answering more or less to this description have been seen at Dog's Bay and opposite Omey I.

D. maculata (L.) Soó (*Dactylorchis maculata*)

1 2 3 4 5 6 7 8   A B

Moors, bogs and acid grassland; rarely on peaty pockets on limestone pavement. Occasional on the limestone; abundant elsewhere.

1. Very frequent. **2.** Occasional throughout, and locally abundant on the high ground. Frequent on Inishmore. **3.** By a small lake E. of Killinaboy. On limestone pavement N.N.E. of Tulla. **4.** Occasional in the area W. of L. Corrib. **5–8.** Abundant.

First record: Praeger (1934c). H16. Earlier records do not distinguish between this and *D. fuchsii*.

D. incarnata (L.) Soó (*Dactylorchis incarnata*)

1 2 3 4 5 6 7 . **A** .

Fens, marshes, bogs and wet meadows, mainly, but not exclusively on moderately base-rich soils. Frequent on the limestone lowlands; rare elsewhere.

1. Marshes and wet meadows by Drumcullaun L., Inchiquin L. and Ballycullinan L. **2.** Inishmore (rare). **3.** Very frequent throughout, in fens and calcareous marshes. **4.** Very frequent from Galway northwards. **5.** Occasional in flushed areas of blanket-bog; also by L. Naneevin and L. Bofin. **6.** Marsh near Moyrus. **7.** Doonloughan, and near Bunowen Hill.

First record: ITB (1901). H16.

The above records are mostly referable to subsp. *pulchella* (Druce) Soó, with flowers of a fairly deep purple. Subsp. *incarnata*, with flowers of a relatively pale pink, occurs, but is rare. Subsp. *coccinea* (Pugsl.) Soó, with ruby-red flowers, has been recorded from Bunowen and near Letterfrack, but we have not seen it. It seems rather doubtful whether these colour-variants are worth subspecific rank. There is, however, a better differentiated and interesting variant with dark spots on both sides of the leaf, which is usually referred to subsp. *cruenta* (Müll.) Sell, but which differs in some respects from the plant to which the epithet was originally given by Müller. It is occasional to frequent in a limited area N. and N.W. of Corofin, extending from Rinroe to L. Bunny and the turloughs S.W. of Mullaghmore. In all its stations, however, it intergrades with normal subsp. *pulchella*.

D. majalis (Reichenb.) Hunt & Summerhayes (incl. *D. kerryensis* and *D. purpurella*)

1 2 3 4 5 6 7 8 **A B**

Marshes and wet meadows; more rarely on heaths, bogs or limestone pavement. Frequent in W. Connemara and on the Clare shales; rather rare elsewhere.

1. Frequent. **2.** Heathy ground on the hill N. of Mullaghmore. Poulsallagh. By the Caher R. Occasional on Aran. **3.** Rinroe bog. Peaty pocket on limestone pavement N.E. of Tulla. L. *Bunny* (Pugsley, 1935). **4.** Marsh on N. shore of Ross L. **5.** Coastal marsh E. of Inveran. On Gentian Hill, among the *Dryas*. **6–7.** Frequent. **8.** Halfway between Maam Cross and Recess, in a bog N. of the road. On an island in L. Corrib S. of Cornamona.

First record: Marshall (1899). Maam.

Most of the plants are referable to subsp. *occidentalis* (Pugsl.) Sell, but several of them show, to a variable extent, an approach to subsp. *purpurella* (T. & T. A. Steph.) Moore & Soó.

The *Atlas* records of this species for Black Head and Ballyvaughan require confirmation.

Hybrids with *D. maculata*, and more rarely with *D. fuchsii*, have been recorded from several stations in the Burren.

D. traunsteineri (Sauter) Soó

. . 3 4 . 6 7 .   . .

Fens and marshes; rare.

3. Fens by L. Bunny and L. George; also near Rinroe and on the E. edge of the big marsh at Kilmacduagh. **4.** Fen at Ballindooly L. **6.** Damp hollow in the dunes on Mweenish I. **7.** Marsh on the E. side of L. Bollard.

First record: Gough (1952). By a bog drain in the neighbourhood of L. Bunny.

We list this, in deference to the taxonomy adopted in *Flora Europaea*, as a distinct species, but recent intensive work by T. Curtis has convinced him (and us) that it should be reduced to the status of another subspecies of *D. majalis*. As we go to press, however, the new combination has not been published.

Ophrys

Ophrys apifera Huds.   Bee orchid

1 2 3 4 5 . 7 *(8)*   **A B**

Pastures, stony ground and the drier parts of fens. Occasional on the limestone and calcareous sands; not seen elsewhere.

1. Meadow at Ballynalackan Ho. **2.** In grassy hollows in the dunes at Fanore; abundant in some years. Poulsallagh. Ballyvaughan, 1884 (**DBN**). Occasional on Aran. **3.** Occasional in the south, from Dromore L. to L. Cullaun and Muckanagh L. *Garryland* (P. cat.). *Drumconora* (Corry, 1880). **4.** Occasional in the north and near Galway. **5.** Gentian Hill. On the old railway-track W. of Rosscahill. **7.** Inishbofin. *(8).* '*Frequent in the Cong district*' (FWI).

First record: Foot (1864). 'Decidedly rare in Burren, being confined, as far as I could see, to one spot at Acres village, near Ballyvaughan.'

O. insectifera L. (*O. muscifera*)   Fly orchid

. **2 3 4** . . . .

Limestone pavement, calcareous heaths and fens. Occasional in a limited area of the Burren; very rare elsewhere.

2–3. Occasional and locally frequent in a narrow strip of country stretching from near Killinaboy past Mullaghmore and its turloughs and the mountains W. of Tulla to the pavements N.E. of Tulla. Fens by L. Bunny, L. Cullaun and L. George. **4.** By the W. end of Ross L. On gravel ridges S. of Carrowmoreknock.

First record: CH1 (1866). Near Ross L.

Gymnadenia

Gymnadenia conopsea (L.) R.Br.   Fragrant orchid

. **2 3 4** *(5)* . **7** .   **A** .

Pastures, fens and damp, peaty places. Abundant on the limestone; very rare elsewhere.

2–4. Abundant throughout, except on Aran, where it is rather scarce. *(5). Gentian Hill* (Atlas). **7.** Abundant in a meadow N. of L. Anaserd. *Sea-sands near Bunowen* (FWI).

First record: Graham (1840). 'On limestone near Galway...a white-flowered variety.'

Most of the plants from the Flora region, and perhaps all, seem to be referable to subsp. *densiflora* (Wahlenb.) Camus *et al.*

A plant collected by F. H. Perring in 1968 from 3 km S.E. of Costelloe has been determined by P. D. Sell as *Dactylorhiza fuchsii* × *Gymnadenia conopsea*. The nearest known station for the latter parent is 27 km distant.

Neotinea

Neotinea maculata (Desf.) Stearn (*N. intacta*)

(*1*) 2 3 4 (*5*) . (*7*) . A .

Dry grassland on calcareous drift, or overlying limestone pavement; rarely in woodland. Occasional in the Burren; very rare elsewhere (Plate 3, Figure 8).

(*1*). *In two places S. of the Lehinch–Ennistymon road* (Praeger, 1901*a*). No details of habitat are available, but it is remarkable that these stations are on the shale and not on the limestone. **2**. Evenly but rather thinly dispersed over almost the whole district; rarest, perhaps in the centre, and seldom ascending above 100 m (330 ft). **3**. By the turloughs S.W. of Mullaghmore. Near the N. end of L. Bunny. By the cross-roads N.E. of Tulla. *Kinvara and 3 km W. of it; Fingall L.* (all **DBN**). **4**. On gravel-ridges S. of Carrowmoreknock. *Menlough, Ballycuirke L. and Ross L.* (all **DBN**). (*5*). *Gentian Hill* (Tomlinson, 1909). (*7*). *On the Dog's Bay peninsula, and on the roadside near the bay; 3 stations on the Slyne Head peninsula* (Praeger, 1906, 1907).

First record: Corry (1880). 'On a rocky plateau at the base of the Glanquin hills.'

Though by no means the most spectacular, this is one of the most interesting species in the flora of western Ireland, as its distribution is almost entirely Mediterranean, and not Mediterranean–Atlantic or Cantabrian–Pyrenean. Outside the Mediterranean region, strictly defined, it has been recorded only from Madeira and the Canaries, Portugal, S.C. France, one station in N. Spain, and (fairly recently) the Isle of Man. It is not known from Britain. In Ireland it extends northwards to L. Carra in E. Mayo and eastwards to Clonmacnoise, Offaly; there is also a station near Cork.

Even if allowance is made for Praeger's (and Mrs Praeger's) flair for spotting this inconspicuous plant, even when out of flower, it is curious that none of his Connemara records has received recent confirmation.

In the Burren it normally grows in open situations, but White & Doyle (1978) have recorded it in ash–hazel woodland S. of Glasgeivnagh Hill.

Some plants have spotted leaves and perianth spotted with dull pinkish-purple; others have unspotted leaves and a perianth variously described as greenish-white or straw coloured. Phillips (**DBN**, 1933) described plants from Menlough as having 'primrose-coloured flowers'.

Pseudorchis

(*Pseudorchis albida* (L.) A. & D. Löve) (*Leucorchis albida*)

(*1*) (*2*) . (*4*) (*5*) . . (*8*) . .

Heathy pastures; not seen recently, and apparently now very rare.

(*1*). *Ennistymon*, 1901 (**DBN**). *Lisdoonvarna* (Foot, 1864; **DBN** (Levinge), 1891). (*2*). *Above Black Head*, 1952 (C. D. Pigott). (*4*). *Ballycuirke L.*, 1952 (C. D. Pigott). (*5*). *A little west of Galway* (CH1). (*8*). *Benwee*, 1899 (ITB). *Rocky pastures at the foot of Mount Gable* (Marshall & Shoolbred, 1896).

First record: O'Mahony (1860). 'Frequent on high grounds in Burren, 1851.'

This is, to judge from the data published in the *Atlas*, a diminishing species both in Ireland and in Britain (apart from the west highlands), but even so it is curious that its disappearance from our region should be apparently so complete.

Coeloglossum

Coeloglossum viride (L.) Hartm.   Frog orchid

1 2 3 4 5 . 7 .   A (*B*)

Dry grassland. Fairly frequent in the Burren; rare elsewhere.

1. Roadside by the E. end of Lickeen L. On a bank near Kilnamona. **2.** Frequent, but rare on Aran. **3.** Occasional. **4.** Gravel knolls S. of Carrowmoreknock. **5.** Gentian Hill. **7.** Frequent on the N. side of the Slyne Head peninsula. *Dog's Bay* (Atlas). *Renvyle*, 1832 (**DBN**). *Inishbofin* (Praeger, 1911). Near the youth hostel N.W. of Salruck.

First record: Foot (1864). Ballyvaughan and Lisdoonvarna.

Platanthera   Butterfly orchid

Platanthera bifolia (L.) Rich.

1 2 3 4 5 6 7 8   A .

Damp or peaty grassland, bog-margins and cut-away bog. Widespread, but seldom in great abundance.

Some 45 distinct stations have been noted, well distributed over the whole region, except in district 7, where it appears to be confined to the western margin (though not on Inishbofin), and district 8, where it has been seen in only three places: two near Cornamona and the third N.W. of Maam Cross.

First record: Foot (1864). 'Widely spread over the whole district [Burren].'

P. chlorantha (Custer) Reichenb.

1 (*2*) (*3*) **4 5 6 7 8**   . (*B*)

Meadows, rough grazing and heathy ground. Rare in the Burren; occasional elsewhere.

1. Near the Cliffs of Moher. W. of Kilnamona. *Between Kilfenora and Lisdoonvarna* (Atlas). (*2*). *Deelin Beg*, 1954 (DW). (*3*). *Near Killinaboy* (Atlas). **4.** Near Ballycuirke L. Near Garrynagry. *Ardrahan*, 1900 (**DBN**). **5–7.** Occasional. **8.** Frequent around Maam and Cornamona; not seen elsewhere.

First record: Babington (1836). Maam.

The relationship between this species and the preceding with regard to ecology and distribution is as baffling here as elsewhere in Ireland. *P. chlorantha* may, perhaps, prefer drier ground, but in several places the two can be seen growing together; on the other hand each can occur in isolation, fairly remote from the nearest station of the other.

The *Atlas* records for Inishbofin for both these species are erroneous. *P. chlorantha* was recorded by Praeger in 1911, but has not been seen since. *P. bifolia* has never been recorded from the island.

IRIDACEAE

Iris

*Iris foetidissima L.

. 2 . . . . . 8 . .

River-banks and lake-shores; very rare.

2. By the Caher R., near the coast road. **8.** Shore of L. Corrib, just N. of the mouth of the Oughterard R.

First record: Atlas (1962).

Whatever may be thought of its status in eastern Ireland, there is no doubt that this species is, in the west, an escape from cultivation.

I. pseudacorus L.   Yellow flag   *Seileastram*

1 2 3 4 5 6 7 8   A B

Ditches, marshes and wet grassland. Very frequent to abundant on acid soils; occasional on the limestone.

1. Abundant. **2.** In a marsh near the church at Carran. By the pond at Burren village. *Near Glencolumbkille Ho.* (Atlas). Sparingly on all three Aran Islands. **3.** Frequent in the extreme south, from Ruan to Ennis. By L. George. Marsh N. of Ballinderreen. **4.** Frequent north of Galway; not seen round the head of Galway Bay. **5.** Occasional near the coast. Abundant at the E. end of L. Formoyle. **6.** Very frequent. **7.** Abundant. **8.** Frequent.

First record: Hart (1975). Inishmaan.

Sisyrinchium

*Sisyrinchium californicum (Ker-Gawler) Aiton

. . . . 8 . .

8. Naturalized in fair abundance on damp, stony and grassy ground by L. Corrib, 3 km N.W. of Oughterard, 1968–80. Obviously escaped from a nearby garden, but well established and spreading by seed. A small colony was seen in 1978 in a ditch at the W. end of Oughterard.

First record: Webb (1982).

Tritonia

*Tritonia × crocosmiflora (Lem.) Nicholson (*Crocosmia crocosmiflora*) Montbretia

1 2 3 4 5 6 7 8   A B

Roadsides, ditches, river-banks and waste places. Frequent in most of Connemara and around Galway city; less common elsewhere and not so completely naturalized.

1. S. of L. Mooghna. Near Lisdoonvarna. **2.** Roadside between Ballyvaughan and Black Head. Occasional on Inishmore. **3.** 6 km from Crusheen, on the road to Gort. **4.** Very abundant on waste ground $1\frac{1}{2}$ km E. of Galway; also in the city by the river. **5.** Frequent. **6.** Recess. Cashel. By a stream at Carna. **7–8.** Frequent.

First record: Druce (1911). Near Clifden.

A hybrid between two South African species, *T. aurea* and *T. pottsii*, cultivated in Irish gardens (under a variety of names!) for nearly a century.

AMARYLLIDACEAE

Narcissus

*Narcissus × medioluteus** Mill. (*N. biflorus*)

. 2 . . . . 7 . [A] .

Waste ground and margins of sand-dunes; rare.

2. Well naturalized on waste ground by Ballyvaughan harbour, 1967–79. Formerly on Aran, but long extinct. 7. On the edge of sand-dunes W. of Cleggan, 1976.

First record: Moore (1854). 'On sandy pasture ground between Kilronan and the sea.'

A hybrid between *N. poeticus* and *N. tazetta*, widely cultivated, and naturalized in much of W. and S. Europe.

LILIACEAE

Phormium

*Phormium tenax** J. R. & G. Forster   New Zealand flax

. . . . . 6 7 8 . .

Frequently planted as a windbreak and occasionally for ornament; more or less naturalized in a few places in Connemara.

6. A large thicket by the sea W. of Cashel. 7. By a stream near Gowlaun, not obviously planted, and accompanied by self-sown *Gunnera tinctoria*. An isolated plant in a hedge W. of Cleggan. 8. W. of Leenane, in pastures. On the bog 1½ km S.E. of Ferry Br.

First record: Webb (1982).

The above records all refer to sites where the species is not likely to have been planted. Whether it spreads by seed we do not know. It is frequent as a hedge plant, especially in the western part of district 7.

Hyacinthoides

Hyacinthoides non-scripta (L.) Chouard (*Endymion non-scriptus*)   Bluebell

1 2 3 4 5 6 7 8 [A] **B**

Woods and scrub; also in meadows and rocky pastures and on steep, rocky slopes, probably indicating in most cases the former existence of scrub. Locally frequent in the Burren; occasional elsewhere.

1. Woods at Moy Ho. and Elmvale Ho. Abundant in woods at Ennistymon. 2. Abundant in Poulavallan. Frequent in scrub 3 km W. of Castletown. Near Glencolumbkille Ho. 3. Frequent in the south. Woods at Coole. 4. S. of Ross L. 5. Frequent in meadows W. of Barna; also near Inveran. 6. Among gorse and heather by the sea at the N. end of Furnace I. Scrub at Flannery Br. *Carraroe* (Atlas). 7.

Locally frequent in the north near the coast, southwards to near Clifden. **8**. Abundant N. of Clonbur. Woods at Teernakill Br. and at Ashford Castle. Islands in L. Corrib and L. Shindilla.

First record: ITB (1901). H16.

The Atlas record for Aran requires confirmation.

Like many woodland plants the bluebell is able to survive in Connemara in exposed habitats near the sea.

*H. hispanica (Mill.) Rothm. (*Endymion hispanicus*)   Spanish bluebell

. **2** . . . . . **8** . .

Locally naturalized in woods or on waste ground.

2. Well naturalized around Ballyvaughan harbour; also sparingly in a lane 2 km to the east. **8**. In the woods at Ashford Castle, where it hybridizes with *H. non-scripta*.

First record: Webb (1982).

This is the bluebell of most gardens (as opposed to demesne woods), where it often becomes a tiresome weed. It differs from the native species mainly in the erect (not nodding) spike, in which the flowers are arranged evenly all round.

Narthecium

Narthecium ossifragum (L.) Huds.   Bog asphodel

1 . **3 4 5 6 7 8** . **B**

Bogs and wet heaths. Abundant in Connemara; frequent in suitable habitats elsewhere.

1. Frequent. **3**. Locally frequent on the remains of raised bogs (Rinroe, Templebannagh, Ballyogan). **4**. Bog by Ballindooly L. Frequent from Ballycuirke L. northwards. **5–8**. Abundant almost throughout.

First record: Babington (1836). Bogs between Roundstone and Clifden.

Allium   *Gairleóg*

†Allium babingtonii Borrer (*A. ampeloprasum*, var. *babingtonii*)

1 2 . . **5 6 7** . **A B**

Field-margins, roadsides, limestone pavement and waste ground near the sea. Fairly frequent in W. and S. Connemara and on Aran; rare elsewhere.

1. Roadsides near Liscannor and 1½ km S. of L. Goller. **2**. Roadside S. of Bealaclugga. Waste ground at Ballyvaughan harbour. Very frequent on Aran. **5**. Wood-margin near the bridge at Spiddal. **6**. At the base of walls on Furnace I. Roadside N.E. of Carraroe. Waste ground near Carna. Field-margins on Mweenish I. E. of Lettermullan village. **7**. By the harbour at Roundstone. Field-margins near Bunowen quay. Belleek. Waste ground near cottages at Claddaghduff. *Dog's Bay* (Druce, 1911). Inishbofin (rare).

First record: Borrer (1849). 'In Ireland it is probably a true native, Mr W. MacCalla having found it near Roundstone, Galway, and in the South Isles of Arran.' Several earlier authors had reported from Aran plants which may have been this species, but under erroneous or ambiguous names.

The status of this plant is hard to assess, in regard both to its taxonomic rank and to the question of whether it is native or alien. It is certainly closely related to *A.*

ampeloprasum (the leek), and often treated as a variety of it; nevertheless it has several distinctive characters. It is unusually seen fairly close to houses, but this is not always true, especially on Aran, and it seldom grows in *close* proximity to a house, nor is it ever cultivated. It is not used in cookery, and although it is held to be of medicinal value it is rarely used for this purpose today.

It is rather difficult to understand how a distinctive cultivar should have become dispersed only along the extreme Atlantic fringe of the British Isles (in Britain it is confined to Cornwall), entirely in regions of relatively primitive agriculture, and yet never to have been cultivated there within the last 150 years. For this reason we believe that it is at least arguable that it is an endemic species of N.W. Europe which has become to some extent commensal with man. Praeger (1932*b*) also considered it native.

A. ursinum L.   Wild garlic

1 2 3 4 5 . 7 8   A .

Woods and scrub, mainly on basic soils; rarely in crevices of limestone pavement. Local, but usually in some abundance where it occurs.

1. Locally abundant on Clifden Hill. Hedges 3 km W.S.W. of Corofin. **2.** Crevices in limestone pavement near the sea, just S. of Black Head. In scrub on Cappanawalla. Abundant in woods in Poulavallan and near Clooncoose. Very frequent on Inishmore; rare on Inishmaan. **3.** Locally abundant in the woods at Coole. **4.** In woods and scrub near the ruins of Menlough Castle. **5.** By a stream in a small patch of woodland S. of L. Bofin. *Near Spiddal and Moycullen* (Atlas). **7.** Woods W. of Letterfrack. In several places near Clifden. *By Derryclare L.* (ITB). **8.** Occasional in scrub N.W. of Clonbur.

First record: Nowers & Wells (1892). Inishmore: near L. Atalia.

*A. triquetrum L. was recorded by 3 different observers at Ballyvaughan, 1956–61, but has not been seen since. It is naturalized just outside our region at the head of Killary harbour. **A. vineale* L. was seen on Inishmaan in 1971 and 1974, but only in small quantity. The lack of earlier records and the fact that it was not found there in 1976 suggests that it was probably only casual.

DIOSCOREACEAE

Tamus

*Tamus communis L.   Black bryony

This species has been noted at various times from 1971 to 1979 in a hedge 4 km S.S.W. of Ballyvaughan, and, as it is a difficult weed to eradicate from gardens, it is likely to remain there for some time. How it got there is a mystery.

First record: Webb (1982).

JUNCACEAE

Juncus   Rush   *Luachair*

In identifying a rush the first thing to determine is what its leaves are like: they may be reduced to short brown sheaths; they may be green and cylindrical, very like the stems, or they may be flat or hair-like, somewhat like the leaves of grass. In the first category we have *J. inflexus*, *J. effusus* and *J. conglomeratus*. The first is distinct in its slender, tough, wiry, bluish-green stems, inside which the pith is frequently interrupted by air-spaces. In the other two the stems are softer, with continuous pith, but in *J. conglomeratus* they are rather strongly ridged (best observed immediately below the flowers), while in *J. effusus* the ridges are very low. Among species that have leaves like the stem, *J. maritimus* is distinct in both its maritime habitat and the spine-like tip to leaves and stem. The others, constituting the 'jointed rushes' (so called because the leaves are hollow, with transverse partitions) are not so easy to distinguish, but *J. articulatus* is usually less than 60 cm high, with blackish flowers; the other two are taller, *J. acutiflorus* with chestnut-brown and *J. subnodulosus* with pale buff flowers. Species of the third group, with grass-like leaves, cannot be distinguished here; reference must be made to a descriptive Flora.

Juncus bufonius L.

1 2 3 4 5 6 7 8 A B

Marshes, damp roadsides, lake-shores and wet, semi-open habitats of all kinds. Very frequent and locally abundant throughout.

First record: ITB (1901). H16.

Two other closely related plants, recently recognized as species, occur in our region, but their frequency and distribution is not yet known, as they differ from *J. bufonius* only in somewhat technical characters which require close examination. *J. foliosus* Desf. has been recorded from several places near the sea in district 7, from near Renvyle southwards to near Errisbeg, and also from near Roadford in district 1. *J. ranarius* Song. & Perr. (*J. ambiguus*) has been found near the shore S.W. of Ballyconneely (district 7) and on a sandy lake shore near Carna (district 6).

J. squarrosus L.

1 . . 4 5 6 7 8 . B

Moors, heaths and bog-margins. Occasional to abundant on acid soils; unknown on the limestone, except where overlain by acid bog.

1. Occasional. **4**. In small quantity on the bog S. of Carrowmoreknock. **5**. Very frequent. **6**. Occasional. **7**. Frequent. **8**. Abundant.

First record: More (1876). Inishbofin: 'hilly ground at west end of the island'.

J. gerardi Lois.

1 2 3 4 5 6 7 8 A B

Salt-marshes; more rarely on stony shores or on rocks near the sea. Very frequent to abundant in S. and W. Connemara; rare to occasional elsewhere.

1. N. and S. of Lehinch. Near Hag's Head. **2**. Maritime rocks at Poulsallagh. Limestone pavement at Rine Point. New Quay. By the brackish lakes between Ballyvaughan and Bealaclugga. Locally abundant on Aran. **3**. In several places between Aughinish and Kinvara. Kilcolgan estuary. **4–5**. Occasional. **6**. Abundant. **7**. Very frequent. **8**. Leenane.

First record: More (1876). Inishbofin: along the south-west shore.

***J. tenuis** Willd. (*J. macer*)

. [2] . . . . . 7 8 [A] .

Frequent in parts of Connemara, mainly around the N.W. arm of L. Corrib; unknown elsewhere.

[2]. Recorded in CH2 from a salt-marsh on Inishmore, but never confirmed; for

reasons for rejecting this record see Webb (1980*a*). **7.** Near Kylemore, 1962. Between Kylemore and Letterfrack, 1973. In fair quantity beside the by-road on the W. side of Ballynahinch L., and on forestry tracks nearby, 1967–80. **8.** One plant by the road at Glann, 1969. Frequent between Clonbur and Maam, southwards to the Dooros peninsula and the Hill of Doon, and northwards to near Srahnalong, on the S.W. arm of L. Mask; noted in 8 different stations, 1965–70.

First record: Halliday *et al.* (1967). Kylemore, Maam and Cornamona.

First noticed in 1962 near Kylemore. Its date of introduction is impossible to establish, but is unlikely to have been earlier than 1950. Up to about 1973 it appeared to be spreading, but recent observations suggest that it may have passed its maximum; its history in S. Co. Wicklow, where it was abundant in 1935–50 but is now relatively rare, is consistent with this.

J. inflexus L.

1 2 3 4 (*5*) . . . **A** .

Marshes and wet fields. Frequent on or near the limestone; unknown elsewhere.

1. Frequent along the boundary with district 2, and on the limestone inlier around Kilfenora. **2.** Occasional; rare on Aran. **3.** Very frequent and locally abundant. **4.** Frequent around Galway city. Kilcolgan estuary. (*5*). *Gentian Hill*, 1899 (P. cat.).

First record: ITB (1901). H16.

Essentially a plant of gley soils over calcareous boulder-clay; relatively rare, therefore, on the bare limestone of the Burren and Aran.

J. effusus L.

1 2 3 4 5 6 7 8 A B

Marshes, wet fields, bog-margins and roadsides. Abundant on acid soils; more local on the limestone.

1. Abundant. **2.** Very frequent, but rare on Aran. **3.** Occasional in the area E. of Corofin; not seen elsewhere. **4.** Very frequent in the north; not seen S. of Ballycuirke L. **5–8.** Abundant.

First record: Hart (1875). Aran (probably Inishmore).

Almost entirely replaced by *J. inflexus* in the country around Galway city and the head of Galway Bay.

J. conglomeratus L.

1 2 3 4 5 6 7 8 A B

Moorland, bog-margins, poor acid pastures, roadsides, etc. Very frequent on acid soils; rather rare on the limestone.

1. Very frequent. **2.** At the foot of cliffs on Gleninagh Mt. In a small marshy hollow $3\frac{1}{2}$ km N.E. of Glencolumbkille Ho. Roadside at the entrance to the Glen of Clab. One small colony on Inishmore. **3.** E. shore of Muckanagh L. By a small lake S.W. of Barefield. Marsh N.W. of Crusheen. **4.** Occasional to frequent, mainly in the north. **5–8.** Very frequent.

First record: More (1876). Inishbofin.

J. maritimus Lam.

1 2 3 4 5 6 7 8 . B

Salt-marshes. Abundant in S. Connemara; local elsewhere.

1. Near O'Brien's Br. **2**. Rine Point. By the brackish lake N.W. of Bealaclugga. **3**. Abundant by the inlet 5 km W. of Kinvara. **4**. Frequent by L. Atalia. *Oranmore*, 1899 (P. cat.). **5**. Gentian Hill. **6**. Abundant, especially in the east and south. **7**. Widespread, but usually in small quantity. **8**. Leenane.

First record: Oliver (1852). Near Roundstone.

Especially characteristic of the sinuous, brackish inlets S. of Screeb, where the fringing band of this species alone betrays the tidal nature of the water.

J. bulbosus L.

1 . **3 4 5 6 7 8** . B

Bog-pools, acid streams, lake-shores, bogs and acid marshes; also on calcareous marl in fens and turloughs. Abundant everywhere on acid peat; local, but in places abundant, in calcareous habitats.

1. Very frequent. **3**. Abundant on acid peat at Rinroe and Templebannagh; also on calcareous marl in the turloughs S.W. of Mullaghmore, and frequent in a calcareous fen S. of L. Fingall. Recorded also by ICP from L. Bunny and Travaun Loughs. **4**. Frequent in a wide variety of habitats between L. Corrib and the Moycullen–Oughterard road. **5–8**. Abundant throughout.

First record: More (1876). Inishbofin.

The occurrence of this calcifuge species in abundance in the calcareous fens and turloughs of the Burren is very striking. It occupies a low place in the zonation, where it is submerged for a large part of the year. In a dry summer, if exposed for long, it turns bright red and contrasts vividly with the otherwise bare white marl, offering an extraordinary spectacle, suggestive rather of the saline deserts of Central Asia than of the west of Ireland.

Nearly all plants from the region, as elsewhere in Ireland, have 6 stamens, but specimens have been collected from sandy flats at Killateeaun, on the shore of L. Mask, with only 3.

The var. *fluitans*, with long, limp, hair-like leaves and usually few flowers, occurs frequently in lakes, and is sometimes cast up in rope-like rolls after storms.

J. subnodulosus Schrank

1 2 3 4 5 . **7 8** A .

Fens, marshes and reed-swamps; also occasionally on blanket-bog. Locally abundant, but absent from wide areas.

1. Abundant in the marsh at the N. end of Inchiquin L. Occasional by Garvillaun L. **2**. Dominant in the large marsh S. of Feenagh. By L. Luick. Inishmore (rare). **3**. Very frequent and locally abundant. **4**. Frequent. **5**. Abundant in ditches by the Oughterard road 3 km E. of Maam Cross. Bog-pool N.E. of Tonabrocky. **7**. Frequent in the south, on blanket-bog and in flushes; not seen N. of Ardbear Br. **8**. Ditches by the road 3 km E. of Maam Cross. Locally abundant in a bog 1½ km S. of Teernakill Br.

First record: Babington (1836). 'Common in the bogs of Connemara.'

The frequent but very capricious appearance of this normally calcicole species on the acid peat of the Connemara blanket-bogs provides yet another ecological paradox of our region. Although it tends to favour flushes, where it may be accompanied by *Eriophorum latifolium* and *Carex dioica*, it can sometimes be seen on the unbroken surface of level bog, amid a purely calcifuge flora.

J. articulatus L.

1 2 3 4 5 6 7 8   A B

Wet places, especially on roadsides or other trodden or stony ground. Abundant throughout.

First record: More (1876). Inishbofin.

J. acutiflorus Hoffm.

1 [2] 3 4 5 6 7 8   [A] B

Acid marshes, bogs and wet meadows. Frequent on non-calcareous soils; very rare on the limestone.

1. Very frequent. **[2].** Recorded for Aran by earlier authors, but apparently in error for *J. articulatus*. **3.** On acid peat on Rinroe bog. **4.** Marsh near the W. shore of Ballycuirke L. **5.** Very frequent. **6.** Occasional around Screeb; not seen elsewhere. **7–8.** Frequent.

First record: More (1876). Inishbofin.

The hybrid with *J. articulatus* has been noted in districts 1, 2, 5 and 7; it doubtless occurs elsewhere. Being sterile, it may often be recognized at a glance by its clonal growth, giving rise to few, very solid stands in contrast to the frequent but small patches of the parent species.

*J. planifolius R.Br.

. . . . . 6 . .   . .

Lake-margins, streamsides and wet bogs. Well established in several places in the Cashel–Carna area; unknown elsewhere.

6. Locally abundant in a small area running from the coast N.E. of Mweenish I., through Glinsk to near Toombeola and to the road-junction E. of Cashel. Its eastern limit, as at present known, lies 5 km E. of Carna.

First record: Scannell (1973).

This species, which, on account of its flat, relatively wide leaves, looks more like a *Luzula* than a *Juncus*, was discovered by MS in July 1971. It is native to Australia, New Zealand and temperate South America. One can hardly suppose that it was transported to Ireland by natural means, but the machinery of its introduction remains completely obscure. (It is reported as naturalized also in Hawaii and Oregon.) Several species of *Juncus* from the southern hemisphere have been reported from time to time in Europe, but they have mostly come in with Australian wool – an unlikely import for a remote part of Connemara. Many plants of Chilean origin are grown in the gardens of western Ireland, but direct import seems extremely improbable. It does not seem likely that the mystery will ever be solved.

Luzula   Wood-rush

Luzula sylvatica (Huds.) Gaud.

1 2 . . 5 6 7 8   . B

Woods, scrub, river-banks and mountain slopes. Local, though abundant in places, on siliceous rocks; rare on the limestone.

1. Occasional in scrub or woodland, as on Clifden Hill, S. of L. Goller, near Roadford, and at Ennistymon. **2.** In scrub on the Corkscrew Hill. In small quantity on pavement near the summit of Gleninagh Mt. Frequent in Poulavallan. **5.** Frequent in woods at Moycullen. Among rocks on the shore of Slieveaneena L. **6.** Woods at Glendollagh. Abundant on islands in Athry and Oorid Loughs. In scrub at Flannery Br. **7.** Abundant on all lake-islands examined which carry a substantial growth of woodland. Abundant in woods W. of Ballynahinch. On cliffs and grassy slopes on Muckanaght. In the ravine on the S.W. side of L. Fee. Inishbofin (rare). **8.** Very frequent and locally abundant, extending on to the limestone in the grounds of Ashford Castle.

First record: Babington (1836). Mountains above Maam; Leenane.

Characteristic especially as the dominant species in the field-layer of most wooded islands, both in the N. part of L. Corrib and in the smaller lakes to the west.

L. campestris (L.) DC.

1 2 3 4 5 6 7 8   A B

Banks and dry grassland. Frequent in the Burren and near the coast; occasional elsewhere.

Commonest in districts 2, 3 and 7; rarest in 1 and 5.

First record: Hart (1875). Inishmore.

L. multiflora (Retz.) Lej.

1 (2) 3 4 5 6 7 8   (A) B

Bogs, heaths, marshes and acid fens. Abundant an acid soils; rare on the limestone, and only where peat has formed.

1. Abundant. (2). *Inishmore* (Praeger, 1895a); probably a temporary importation with peat from Connemara. **3.** Occasional in the south, in acid fen and on cut-away bog, from Barefield northwards to Templebannagh L. **4.** On bogs N.E. of Srue and S. of Carrowmoreknock. **5–8.** Abundant.

First record: More (1876). Inishbofin.

TYPHACEAE

Typha   Bulrush

Typha latifolia L.   *Coigeal na mban sidhe*

1 2 3 4 5 6 7 8   A .

Lakes, rivers, ditches, ponds and fens, chiefly in base-rich waters. Rather rare in Connemara; occasional to frequent elsewhere.

1. Frequent in the east; not seen near the coast. **2.** Very abundant in the large pond at Burren village. Occasional on Aran. **3.** Templebannagh L. Marsh N.W. of

Crusheen. L. Cleggan. By the R. Fergus above Ennis. **4**. Frequent. **5**. Marsh in the Claddagh, 1971. Fen by L. Naneevin. **6**. Pool E. of the road 7 km N. of Carna. Pool at the E. end of Oorid L. **7**. In several marshes on the Slyne Head peninsula. Ballinaboy. Kylemore. **8**. E. of Recess, in a small lake between the old and new roads to Maam Cross.

First record: ITB (1901). H16.

T. angustifolia L.

. . . . . . 7 . . . .

7. Known only from the N. shore of Ballynakill L., 1968, 1980 (DW). Not flowering when seen *in situ*, but the identity was confirmed by cultivation.

First record: Webb (1982).

A rare species in Ireland and, except for a large colony S. of Ennis, not previously recorded west of the Shannon.

Sparganium   Bur-reed

Sparganium erectum L. (*S. ramosum*)

1 2 3 4 5 6 7 8   . B

Ditches, marshes and streams; frequent in base-rich water, local elsewhere.

1. Very frequent. **2**. Pond at Burren village. **3**. Frequent in the extreme south; also in a fen near Ardrahan and by the estuary of the Dunkellin R. **4**. Very frequent. **5**. Abundant in a stream 2½ km E. of Barna. **6**. L. Skannive. **7**. Frequent near the W. coast, from near Roundstone to Ballynakill L. and Inishbofin. **8**. Occasional from near Maam northwards to L. Nafooey.

First record: Marshall & Shoolbred (1896). Frequent about Maam and Clonbur.

Of the three subspecies known from Ireland subsp. *erectum* and subsp. *neglectum* have been recorded from the Flora region.

S. emersum Rehm. (*S. simplex*)

. . 3 4 5 6 7 8   . B

3. Frequent. **4**. N. shore of the Dunkellin estuary, Ballyloughaun (P. cat.). **5**. At the S. end of L. Bofin. **6**. L. Keeraun. **7**. Abundant in the smaller lake on Omey I. In the stream flowing out of Church L., Inishbofin. *Renvyle*, 1832 (**DBN**). **8**. In the river above Maam Br. Drains on the N. bank of the river below Oughterard.

First record: ITB (1901). H16.

The Inishbofin record was overlooked by Webb & Hodgson (1968); there is a specimen in **TCD** (DW, 1967). It was growing alongside *S. erectum*.

S. angustifolium Michx. (*S. affine*)

. . . . 5 6 7 8   . B

In fairly deep water in lakes and pools; more rarely in streams. Locally frequent in base-poor waters; absent from those with more than a very small base-content.

5. Pool S.E. of Lettercraffroc L. Slieveaneena L. **6**. Frequent. **7**. L. Fadda. Cregduff L. Plentiful on Inishbofin. **8**. L. Nafooey and the Finny R. L. Nadirkmore.

First record: Wade (1802), as *S. natans*. 'Many places in Cunnamara.'

Characteristic especially of lakes of the mountain tarn type, containing little other vegetation. It can here be recognized at a distance by its long, ribbon-like floating leaves, only occasionally accompanied by flowers. It is perhaps commoner than the above records indicate, as it has often been misidentified as *S. minimum*, since the key character used by most Floras (leaves inflated at base or not) is very unreliable.

S. minimum Wallr.

1 . 3 4 5 6 7 .   . .

Ditches, drains, lake-margins and marshes; occasional.

1. Swamp at the W. end of L. Goller. **3.** Ditch by Templebannagh bog. Acid fen, transitional to bog, S.E. of Ballyogan Loughs. Marshy lake-margin E. of Tubber. *Garryland* (Druce, 1909). **4.** Occasional in drains in the area N. and N.E. of Ross L. *Menlough* (ITB). **5.** By the S.E. end of L. Bofin, and by the lake to the E. of it. **6.** Lake-margin on Lettermullan I. **7.** Marshy margin of Cregduff L. By the mouth of the Culfin R., N.E. of Gowlaun.

First record: ITB (1901). Menlough; Roundstone.

Some of the *Atlas* records are probably errors for *S. angustifolium*, and that for Inishbofin is certainly an error.

ARACEAE

Arum

Arum maculatum L.   Lords and ladies

1 2 3 4 5 . . 8 A .

Woods, hedges and scrub; rarely in grassland or on river-banks. Frequent on the limestone; rare elsewhere, and only where there is evidence of some base-enrichment of the soil.

1. N.W. of Ennis. W. of L. Raha. W. of Killinaboy. Abundant at Ennistymon. **2–4.** Frequent throughout. **5.** Moycullen wood. Near Gentian Hill. Copse 4 km W. of Oughterard, on slightly calcareous drift. **8.** Across the main road from the last station. Inchagoill and other islands in L. Corrib. Frequent around Cong and Clonbur, westwards to Ebor Hall.

First record: Hart (1875). Aran Islands.

LEMNACEAE

Lemna   Duckweed

Lemna trisulca L.

. (2) 3 4 . . . 8 . .

Marshes, lakes and rivers; very local.

(2). *Ballyvaughan area* (Atlas). **3.** Frequent in the south, northwards to Tubber and Corofin. In the lake at Coole. **4.** Coolagh fen. **8.** Marsh 1½ km N.E. of Clonbur.

First record: ITB (1901). Near Galway.

L. minor L.

1 2 3 4 5 6 7 (*8*) (*A*) .

Ditches and marshes; less often in lakes, ponds or rivers. Widely distributed, but nowhere very common, and rare in some districts.

1. Occasional. **2**. Carran turlough, at Castletown. *Aran* (Wright, 1871; Hart, 1875); not recorded since. **3–4**. Occasional. **5**. Marsh in the Claddagh. **6**. 3 km N.W. of Costelloe. Marsh between Carna and Mace Head. Mweenish I. **7**. Very frequent in the extreme west; rather rare elsewhere. (*8*). *N.W. of Oughterard* (P. cat.).

First record: Wright (1871). Aran Islands.

L. gibba L. was recorded by Praeger (1939) as having been found by Mrs Gough 'three quarters of a mile east of Kilfenora'. Mrs Gough was an accurate observer, and a mistake is unlikely, but we have searched the area without finding the plant or even a suitable habitat, and must conclude that it is extinct here.

L. polyrhiza L.

1 . **3** (*4*) . . . .

Ponds, ditches and backwaters of rivers; rare.

1. In a pool on the N. side of the road 2½ km E. of Kilfenora; varying considerably in quantity from year to year, but usually in fair abundance. **3**. Backwaters of the R. Fergus above Ennis. (*4*). *Ditches at Menlough*, 1899 (**DBN**).

First record: ITB (1901). Menlough.

ALISMATACEAE

Alisma

Alisma plantago-aquatica L.   Water plantain

1 2 3 4 5 . (*7*) **8** . .

Lake-margins, slow rivers, turloughs, drains and marshes. Frequent in suitable habitats on the limestone; rather rare elsewhere.

1. L. Raha. Lickeen L. Drumcullaun L. Drains near O'Brien's Br. **2**. Carran turlough at Castletown. Turlough W. of Turlough village. **3**. Very frequent, especially in the south and around Gort. **4**. Frequent. **5**. Occasional in drains 4 km W.N.W. of Galway. *Spiddal* (Atlas). (*7*). *Ballyconneely area* (Atlas). **8**. Abundant in the river at Maam. Occasional in drains by L. Corrib at Carrowgarriff, and also near Oughterard, and on one of the islands N.E. of Oughterard. Shore of L. Mask N. of Clonbur. Effluent stream from Coolin L.

First record: ITB (1901). H16.

Baldellia

Baldellia ranunculoides (L.) Parl.

. **2 3 4 5 6 7 8** **A** .

Lakes, marshes, ditches, streams and bog-drains. Frequent over most of the region, but rather local.

2. L. Luick, and in the turlough 3 km S. of it. Carran turlough. Locally frequent on

Fig. 17. *Baldellia ranunculoides* in a lake-margin near Carraroe.

Aran. **3**. Very frequent, especially in the south. **4**. Frequent. **5**. Marsh by the coast 5 km W. of Spiddal. **6**. Occasional. **7–8**. Frequent, but not on Inishbofin.

First record: Hart (1875). Inishmore: L. Atalia.

BUTOMACEAE

Butomus

Butomus umbellatus L.   Flowering rush

1 . . . .

Marshes and ditches; very rare.

1. Occasional in the marsh by L. Raha, 1967 (DW). *Ditches near Corofin* (Mackay, 1806). *Abundant in the R. Fergus, near Corofin* (O'Kelly, 1903).

First record: Mackay (1806). 'Ditches near Corofin, Clare, in great abundance.'

Perhaps introduced in some of its Irish stations, but certainly native by L. Raha. The stations recorded by Mackay and O'Kelly may lie partly in district 3, but district 1 seems more probable. It is hard to see why it cannot be found there now if it really occurred 'in great abundance'. The record by Wade (1804) for marshes by the R. Fergus lies just outside our region.

in district 2, but it is abundant in L. Luick and near Turlough village, and occurs also on Aran.

First record: Hart (1875). Inishmore: L. Atalia.

Stunted forms from shallow water may be difficult to separate from the following species; we find that the very long stipules of *P. natans* are usually a safe guide. The 'flexible joint' at the top of the petiole mentioned in most Floras is sometimes very difficult to see.

P. polygonifolius Pourr.

1 (2) 3 4 5 6 7 8   [A] B

Bog-drains, flushes, small lakes and streams; also semi-terrestrial in the wetter parts of bogs. Abundant in blanket-bog areas; occasional elsewhere.

1. Occasional. (2). *Carran turlough*, 1930 (BM). 3. Rinroe bog. Templebannagh bog. Marsh by L. Briskeen. Turlough S.W. of Mullaghmore. 4. Occasional in the north. 5–8. Abundant.

First record: More (1872). 'Plentiful in lakes and streams at Ballynahinch, Connemara.'

In Connemara, and especially in district 7, var. *pseudofluitans* Syme, with narrow, thin, more or less translucent floating leaves, is frequent in bog-pools. The Aran record is (*fide* Dandy) an error for *P. natans*.

P. coloratus Hornem.

. 2 3 4 . . 7 .   . .

Lakes, rivers, ditches and bog-drains. Locally frequent on the limestone; very rare elsewhere.

2. Frequent in the Caher R. In a small stream running into Carran turlough. 3. In the southernmost of the turloughs S.W. of Mullaghmore. Drains in Rinroe bog. Frequent in Muckanagh L., and in small lakes S.E. of L. Bunny. *Kilmacduagh* (CH1). 4. Ditches near Rinville. Bog-drains S. of Carrowmoreknock. 7. Lake and adjoining drains at Bunowen. In Truska L. and some of the other lakes on the Slyne Head peninsula. *Cregduff L.* (Praeger, 1906).

First record: CH1 (1866). Kilmacduagh; Roundstone.

All records (even those from bog-drains) are from stations on limestone, basalt or calcareous sands, except for Cregduff L., where there is no very obvious source of base. Praeger's specimen from here in **DBN** is authentic, but the plant has not been seen there since.

P. alpinus Balb.

1 . 3 . (5) . (7) .   . .

Lakes and pools; rare.

1. In the small lake W. of Lickeen L., 1969 (MS). 3. In a pool by the fen in Ballyogan Loughs, 1966 (MS). (5). *Near Galway* (CH1). (7). *Kylemore L.* (Hurst, 1902).

First record: CH1 (1866). 'Near Galway; *Prof. Melville.*'

The presence of this species in Connemara requires confirmation. Neither of the last two records is backed by a specimen.

P. gramineus L. (*P. heterophyllus*)

. (*2*) 3 (*4*) . . 7 **8** . .

Lakes and turloughs; local.

(*2*). *Carran* (Atlas). (There is also in the herbarium of University College, Aberystwyth, a specimen labelled 'Ballyvaughan, 13 July, 1914, W.S.') **3**. Frequent in lakes and turloughs in the area around Castle Lough, L. Bunny and Mullaghmore. Tullaghnafrankagh L. (*4*). *Gortachalla L.* (Praeger, 1906). **7**. Cregduff L. Larger lake on Omey I. Bunowen L., and several other lakes on the Slyne Head peninsula. Pollacappul L. **8**. Sandy bay on the S. shore of L. Mask.

First record: Linton & Linton (1886). Cregduff L. The record by Wade (1802) for 'in the water around Ballynahinch' needs confirmation.

P. lucens L.

1 . **3 4** . . . **8** . .

Locally frequent in the basins of the rivers Fergus and Corrib; apparently confined to calcareous waters.

1. R. Fergus, below Inchiquin L., and also in the lake. **3**. Frequent in the R. Fergus from L. Atedaun to Ennis; also in Ballyallia L. **4**. In the R. Corrib above Galway. S.W. shore of L. Corrib, E. of Ballycuirke L. *Ditch near Moycullen*, 1899 (**DBN**). **8**. Frequent along the N. shore of L. Corrib, and in the river at Cong.

First record: ITB (1901). Terryland R., 1899.

P. praelongus Wulf.

. . **3 4** . . **7 8** . .

Rivers and lakes; rather rare.

3. Kiltartan R. **4**. L. Corrib, E. of Moycullen. **7**. Bunowen L. *Doonloughan* (Sledge, 1942). *Ballynahinch* (CH2). **8**. L. Corrib (E. side of Dooros peninsula). *L. Mask, N. of Clonbur* (Marshall & Shoolbred, 1896).

First record: Marshall & Shoolbred (1896). They describe the situation as 'W. side of L. Mask, about 3 miles N. of Clonbur'. But at 3 miles (*c.* 5 km) N. of Clonbur we are in the middle of the lake. 3 miles N.W. of Clonbur takes us to the W. shore N. of Ferry Br., and this seems the most probable station.

P. perfoliatus L.

1 2 3 4 5 6 7 8 . .

Lakes and rivers; widespread and locally abundant.

1. Inchiquin L. **2**. L. Luick. **3**. Coole L. L. Atedaun. **4**. Abundant in the R. Corrib near the Cathedral, and also upstream. Ross L. **5**. L. Bofin. Canal in Galway city. **6**. Frequent. **7**. Frequent in the extreme west; also in Derrylea L. **8**. Frequent in L. Corrib and L. Mask. Lehanagh Loughs.

First record: ITB (1901). H16.

P. crispus L.

1 . **3 4 5** . . **8** . .

Lakes, rivers, turloughs and drains. Occasional on the limestone; very rare elsewhere.

1. Lickeen L. Ditch near O'Dea's Castle. **3**. Pond 3 km N.E. of Ruan. Frequent W. and N.W. of Gort. Lakelet E.N.E. of Kinvara. **4**. R. Corrib, just above Galway. L.

Corrib at Burnthouse. **5.** Canal in Galway city. *Spiddal* (Atlas). **8.** L. Corrib, at Ashford Castle.

First record: ITB (1901). H16.

P. berchtoldii Fieb.

1 . 3 4 . 6 7 8 . B

Lakes and rivers. Fairly frequent in Connemara; occasional elsewhere.

1. R. Fergus above Inchiquin L. **3.** L. Atedaun. Tirneevin turlough. **4.** Stream by Aughnanure Castle. In the canalized river running into Ballycuirke L. **6.** Lake near Tullywee Br. Pool between Glendollagh L. and the old railway embankment. In several lakes near Carna. **7.** Occasional, mainly in the north. **8.** Lehanagh Loughs.

First record: More (1876), as *P. pusillus*. Inishbofin: Church L.

P. pusillus L.

. . 3 4 . . 7 . . .

Lakes, turloughs and rivers; very local.

3. In the 'race-course' turlough W. of Gort. Coole L. **4.** Very frequent in the R. Corrib, in Galway city and upstream. **7.** In the larger lake on Omey I. Maumeen L., S.E. of Ballyconneely.

First record: Impossible to establish; see following note.

Hard to distinguish from the preceding species, and many earlier authors did not attempt the separation; moreover, nomenclatural ambiguities meant that even those authors who recognized two species used the name *P. pusillus* to indicate *P. berchtoldii*. The records given above are all based on specimens determined by the late J. E. Dandy, but many North American authors today are inclined to regard *P. berchtoldii* as only a subspecies of *P. pusillus*.

P. obtusifolius Mert. & Koch

. . . . . 6 (7) . . .

Lakes; very rare.

6. L. Truskan, 1970 (MS); also in a small lake 1 km W. of Carna, 1974 (MS). (**7**). *Lakelet* 1½ *km W. of Bunowen* (Praeger, 1939).

First record: Praeger (1939).

Rare in western Ireland; the nearest recorded stations are in S. Clare and N. Mayo.

P. friesii Rupr. was recorded from Roundstone by Linton & Linton (1866), as *P. mucronatus*, but no details are given. We have been unable to trace a specimen, and the record cannot be accepted without confirmation, as it is a very rare plant in Ireland.

P. pectinatus L.

1 2 3 4 . . 7 . A B

Lakes and estuarine waters. Fairly frequent on limestone or near the sea; not recorded elsewhere.

1. Inchiquin L. **2.** L. Luick. L. Murree. Abundant in the brackish lake on Inisheer; also at Portcowrugh (Inishmore). **3.** L. Atedaun. Kilcolgan estuary. *Travaun L.*, 1905 (**DBN**). **4.** L. Corrib, near Oughterard. Brackish pools E. of Galway. **7.** Doonloughan. Lake on Omey I. Inishbofin (L. Bofin). Aughrusbeg L.

First record: More (1872). 'Aran Islands; *H.C.H.*'

P. filiformis Pers.

. . . . **8** . .

Lake-margins; very rare.

8. In a sandy bay on the S. shore of L. Mask, near White I., 1965 (GH and DW).

First record: Marshall & Shoolbred (1896). 'A single fruiting spike...floating near the shore of L. Mask...about 3 miles N. of Clonbur.'

Hybrid Potamogetons

Hybrids are readily formed in this genus, and although they are nearly all sterile (*P. × zizii* alone is sometimes fertile) they reproduce readily by vegetative means, and are often found in some abundance in the absence of one, or even of both parents. Nine such hybrids have been reliably reported from Ireland, of which the following six occur in the region of this *Flora*.

P. × lanceolatus Sm. (*P. berchtoldii × coloratus*). **2**. Plentiful in the Caher R. over most of its course.

P. × nitens Weber (*P. gramineus × perfoliatus*). **1**. Small lake W. of Lickeen L. Mooghna L. *Ballyportry L.* (Praeger, 1905). **3**. Kiltartan R. (*4*). *Ross L.* (ITB). *Ballycuirke L.* (Praeger, 1906). **7**. Very frequent in the lakes near the west coast, from Cregduff L. northwards to Aughrusbeg L. **8**. S. shore of L. Mask.

(*P. × nerviger*) (*P. alpinus × lucens*). (*1*). R. Fergus, above Inchiquin L., 1948, 1952 (**DBN**).

P. × salicifolius Wolfg. (*P. lucens × perfoliatus*). **1**. R. Fergus above Ennis, 1972 (**TCD**).

P. × sparganifolius Fr. (*P. gramineus × natans*). **8**. In the Bealanabrack R. above Maam Br.

P. zizii (*P. gramineus × lucens*). **1**. R. Fergus, near Killinaboy. (*4*) *Ballycuirke L.*, 1906 (**DBN**). (*7*). *Aillebrack L.* (Praeger, 1906). **8**. In several places in the N.W. part of L. Corrib, and in the river at Cong.

RUPPIACEAE

Ruppia

Ruppia maritima L. (*R. rostellata*)

1 2 3 4 (*5*) **6 7** . A .

Salt-marshes and brackish pools and lakes. Local, but abundant in some of its stations.

1. Salt-marsh near O'Brien's Br. **2**. In a stream near the shore, N.E. of Kilronan. L. Murree. **3**. Abundant in a salt-marsh $3\frac{1}{2}$ km W. of Kinvara. **4**. Brackish pool near L. Atalia. *Oranmore and Kilcolgan* (ITB). (*5*). *Gentian Hill* (ITB). **6**. In the lake at the head of Ard Bay. Salt-marsh at Canower. **7**. Brackish lake at Cleggan. Inishbofin: abundant in L. Bofin.

First record: ITB (1901). Lehinch, Barna, Oranmore, Kilcolgan.

R. cirrhosa (Pet.) Grande

. . . . . **6** . . . . .

Rocky shore of a brackish lake; very rare.

6. In small quantity, with *R. maritima*, on the shore of the brackish lake at the head of Ard Bay.

First record: Scannell (1975).

This record, the only one for the west coast of Ireland, is subject to some slight doubt. Specimens from this station were sent to J. E. Dandy, who named some of them as *R. cirrhosa*, giving as the differentiating characters a long, spirally coiled fruiting peduncle, nearly symmetrical fruits and rounded (not acute) leaf-tips. One specimen, which showed the first two characters fairly plainly, was retained by Dandy and could not be found after his death. Those returned to **DBN** show a coiled fruiting peduncle, longer than is usual in *R. maritima*, but shorter than the length given in most Floras for *R. cirrhosa*, but the shape of the fruits varies widely on the same plant, and no difference in leaf-shape can be seen between the two species. It would seem, therefore, either that the two species are not really distinct, or that at the Ard Bay station there were a few plants of *R. cirrhosa*, and more numerous hybrids with *R. maritima*. Further collections have failed to yield any plants which are indubitably *R. cirrhosa*.

ZANNICHELLIACEAE

Zannichellia

Zannichellia palustris L.

1 2 3 . . . . . **A** .

Fresh or brackish water. Occasional in and near the Burren; not recorded from Connemara.

1. Salt-marsh near O'Brien's Br. E. side of Inchiquin L. **2**. Abundant in L. Aleenaun. In a pool near Kilronan. *Caher R.*, 1955 (**TCD**). **3**. Occasional in the north and centre.

First record: Nowers & Wells (1892). In a pool near Kilronan.

Similar in general appearance to *Potamogeton pectinatus* and *Ruppia maritima*, but even in the absence of fruit or flowers it can be distinguished from both by its opposite leaves.

ZOSTERACEAE

Zostera

Zostera marina L.

. **2** (*3*) **4** (*5*) **6 7** . **A B**

Sand or mud below, or sometimes just above low-water of spring tides. Occasional in S. and W. Connemara; rather rare elsewhere.

2. Beach E. of Ballyvaughan. Inishmore. (*3*). *Kinvara Bay* (P. cat.). **4**. Parkmore. *Oranmore and L. Atalia* (P. cat.). (*5*). *Gentian Hill* (P. cat.). **6**. Occasional. **7**. Dog's Bay. Occasional in the Clifden–Ballyconneely area. Beach E. of Gowlaun. Inishshark.

First record: Hart (1875). Inishmore.

The records cited are those of places where the plant has been seen growing, or has been washed up in sufficient quantity to make it certain that it came from a colony not far away.

Nearly 50 years ago this species was attacked by a disease which greatly reduced its abundance in Britain, and it has never made a complete recovery. In Ireland the populations in Co. Down and Co. Cork are known to have been similarly affected, but no observations have been published relating to the west coast. The number of old but unconfirmed stations suggests, however, that here too the abundance of the plant has been reduced.

Z. noltii Hornem. (*Z. nana*) was reported by Praeger & Carr (1895) from mud-flats W. of Gentian Hill on the strength of a verbal report from Prof. Johnson. No voucher specimen exists, and nobody seems to have seen the plant here since; in view of the rarity of the species and the difficulties of identification the record is best held over pending confirmation.

Z. angustifolia (Hornem.) Reichenb. is indicated for Inishmore in the *Atlas*; this is the result of an editorial error.

NAJADACEAE

Najas

Najas flexilis (Willd.) Rostk. & Schmidt

. . . . . 6 7 8 . .

Lakes. Occasional in Connemara; unknown elsewhere.

6. In a small lake W. of Ard Bay. **7.** In several lakes near the W. coast, from near Roundstone to Renvyle; also in Kylemore and Pollacappul loughs. **8.** Lehanagh Loughs.

First record: Oliver (1852). Cregduff L.

Often classed as a 'Hiberno-American' species, but in view of its fairly wide, though sparse, distribution in Europe, eastwards to Russia and southwards to Switzerland, this is scarcely justified. It is even more widely distributed in Europe as a post-glacial fossil. In Ireland it is confined to the extreme west, and, as it usually grows in fairly deep water, it was recorded almost entirely from fragments cast up on lake-shores. Recent exploration of the submerged flora by diving has revealed several new stations, and it seems probable that an extension of this technique will reveal several more.

ERIOCAULACEAE

Eriocaulon

Eriocaulon aquaticum (Hill) Druce (*E. septangulare*)   Pipewort

. . . 4 5 6 7 8 . B

Lakes; more rarely in drains or wet bogs. Very frequent throughout a large part of Connemara; unknown elsewhere.

4. Ballycuirke L. **5.** Frequent in the west; rather rare in the east. **6.** Abundant in the north and east; local in the south and west. **7.** Abundant in the south; more local in the north. **8.** Lakes N. of Maam Cross. Coolin L. In the small lake near L. Nadirkmore. Plentiful on the shore of L. Corrib opposite the Hill of Doon; also at Carrick.

First record: Wade (1802). 'Conspicuous in the majority of the loughs in Cunnamara.'

Although it ranges throughout Connemara, from Inishbofin to L. Corrib, and from Renvyle to Carraroe, it is much more abundant in some parts than others. This is partly because it tends to favour small lakes rather than large ones, the only large lake in which it grows in abundance being the N.W. arm of L. Corrib.

It can grow in water as deep as $1\frac{1}{2}$ m; on the other hand it is sometimes semi-terrestrial. It is conspicuously calcifuge, but, as with *Lobelia dortmanna*, this is to be interpreted in terms of the substratum rather than of the water, as at Carrick, on L. Corrib, although the substratum is derived from siliceous rocks, the water is slightly alkaline.

Fig. 18. *Eriocaulon aquaticum* (pipewort) near Maam Cross.

This species has its Irish headquarters in Connemara, but it is also found in smaller amounts in Kerry, Clare, Mayo and Donegal. It is unknown elsewhere in Europe except for a few stations in western Scotland, but is widespread in eastern North America. It is, therefore, a classical 'Hiberno-American' plant, or better, perhaps, an amphi-Atlantic plant of extremely asymmetrical distribution. It has been suggested on the basis of chromosome counts that the European plant is endemic and that the American one is specifically distinct, but this interpretation has not been generally accepted.

It is easily identified when not in flower by its pale green, sharply pointed leaves and its white, transversely banded roots.

CYPERACEAE

Eleocharis

Eleocharis acicularis (L.) Roem. & Schult.

(*1*) **2 3** . . . .

On mud or marl in lakes, pools or turloughs. Locally abundant in a few parts of the Burren.

(*1*). *Inchiquin L.* (CH2). **2**. L. Aleenaun. Carran turlough. **3**. Abundant in turloughs W. and N.W. of Gort and in Coole L. In a pool by the road near Ruan. L. Atedaun.

First record: CH2 (1898). Inchiquin L. The record by Wade (1802) for Connemara is generally agreed to be erroneous.

E. quinqueflora (Hartm.) Schwarz (*E. pauciflora*)

. **2 3 4 5 6 7 8**   . (*B*)

Lake-shores, marshes and fens. Occasional to locally abundant.

2. Frequent and luxuriant in a marsh S.S.E. of Feenagh. **3**. Shores of L. Bunny. Abundant in a fen 2 km E. of Killinaboy. Very luxuriant in the fen S. of L. Fingall. **4**. Abundant at the E. end of Ross L., on the shore of L. Corrib near Burnthouse, and by small lakes between these two stations. **5**. Frequent in a marsh by the Drimneen R. **6**. Shore of Athry L. **7**. Marshes on the Slyne Head peninsula. *Inishbofin* (Praeger, 1911). **8**. Frequent on the shore of L. Corrib at Glann and on the Dooros peninsula. Flushes by L. Nafooey and on the hillside above it. By the S. shore of L. Mask.

First record: Babington (1836). Connemara: 'forming a large part of the herbage on heathy ground'.

E. palustris (L.) Roem. & Schult.

1 2 3 4 5 6 7 8   A B

Marshes, turloughs, rivers and lake-margins. Widespread and locally abundant.

1. Frequent. **2**. In the turloughs at Carran and by Turlough village; also by L. Aleenaun and L. Luick. Locally abundant on Aran. **3–4**. Very frequent. **5**. Near Spiddal, and in a marsh 6 km to the west. In a small lake S.E. of Maam Cross. **6**. Frequent on the Carraroe peninsula. L. Nageeron, W. of Carna. **7–8**. Abundant.

First record: Babington (1836). Connemara bogs.

E. uniglumis (Link) Schult.

1 2 . 4 . 6 7 .ꞏ A B

Saline marshes, and by brackish lakes or streams near the sea. Occasional.

1. Salt-marsh near O'Brien's Br. **2**. By the brackish lakes N.W. of Bealaclugga and by the turlough E. of Ballyvaughan. **4**. Saline meadows at Oranmore. Brackish marsh E. of Galway. **6**. Abundant at the N. end of Furnace I. At the tip of the Carraroe peninsula. By a brackish lake S. of Moyrus. **7**. Occasional in the extreme west; also N.E. of Gowlaun.

First record: Webb & Hodgson (1968). Inishbofin.

Praeger (1939) reported this species as 'frequent in lakelets between Roundstone and Clifden, in up to three feet of water, often viviparous, with young spikelets bearing six-inch stems and roots in August'. We have failed to confirm this, and suspect that what he saw was an unusual form of *E. palustris* or *E. multicaulis*.

E. multicaulis (Sm.) Sm.

(*1*) . **3 4 5 6 7 8**   . **B**

Lake-shores, bogs and wet, peaty ground of all types; also locally on marl or calcareous mud around lakes or turloughs. Common throughout Connemara; local elsewhere, but abundant in places.

(*1*). *S. of Ennistymon* (Atlas). **3**. In the flood-zone of L. Bunny, L. George and Travaun L. In the upper zone of the turloughs S.W. of Mullaghmore, and in some of the *Carex elata* fens in this region. Templebannagh bog. **4**. Plentiful in bogs W. and S. of Carrowmoreknock; also on marl by Gortachalla L. **5–8**. Abundant throughout.

First record: Mackay (1806). 'Very common in Cunnamara.'

The abundance of this normally calcifuge species in the fens and around the lakes and turloughs of the Burren is one of the most striking paradoxes of their flora. It has also been seen on limestone on the S. shore of L. Mask.

Scirpus

Scirpus cespitosus L. (*Trichophorum cespitosum*)

1 . . 4 5 6 7 8 . .

Bogs, moorland and wet heath. Abundant throughout Connemara; rather rare elsewhere.

1. Occasional, and locally abundant, as in bogs near the Cliffs of Moher and S. of Lickeen L. **4**. Bog by Ballycuirke L. **5–8**. Abundant, but absent from Inishbofin.

First record: Hart (1883). N.E. slopes of the Twelve Pins.

S. fluitans L. (*Eleogiton fluitans*)

. . 3 4 5 6 7 8 . B

Drains, streams and pools. Very frequent in acid waters; very rare on the limestone.

3. In considerable quantity on calcareous marl in the fen S. of L. Fingall, 1970 (DW). Near L. Bunny, 1959 (ICP). **4**. Bog-drains S. of Carrowmoreknock. **5–8**. Frequent.

First record: Mackay (1836). 'Cunnamara.'

S. setaceus L. (*Isolepis setaceus*)

1 2 . 4 5 6 7 8 A B

Damp grassland, roadsides, marshes and drains. Very frequent in Connemara; rare elsewhere.

1. Near Ballynalackan Castle. By the Aille R., near Roadford. *Near L. Goller* (Atlas). **2**. Aran; rather rare. *Fanore, and near Rine Point* (Atlas). **4**. Occasional N. and E. of Moycullen. Rinville. *Menlough* (P. cat.). **5–8**. Very frequent.

First record: Babington (1836). Oughterard.

Atlas records for the Killinaboy and Mullaghmore areas require confirmation.

S. cernuus Vahl (*Isolepis cernua, Scirpus savii*)

. 2 3 4 5 6 7 8 A B

Lake-shores, marshes and other damp places. Frequent in most of Connemara; rare elsewhere.

2. Inishmore. *Fanore*, 1954 (DW). **3**. Near L. Bunny. **4**. N.E. of Moycullen. **5**. On the E. side of Cashla Bay, near the mouth. **6–8**. Frequent.

First record: Mackay (1836). 'First observed in Ireland by Mr. Shuttleworth in wet bogs near Rynvile.'

S. lacustris L. (*Schoenoplectus lacustris*)

1 2 3 4 5 6 7 8 A B

Lakes, rivers, fens and turloughs. Very frequent and locally abundant. Rather more abundant in basic than in acid waters, but well distributed throughout Connemara, except in district 8, where it is rather rare.

First record: Babington (1836). Lakes between Clifden and Galway.

Subsp. *tabernaemontani* (Gmel.) Syme, distinguished by its glaucous stems and usually 2 stigmas, occurs in all districts but 8; it is on Aran, but not on Inishbofin. It grows mostly in brackish waters, but also in lakelets on the Slyne Head peninsula. It is, through most of our region, very distinct from subsp. *lacustris*, but transitional forms have been seen on Aran and in some of the Burren turloughs.

S. maritimus L.

1 2 3 4 5 6 7 8 A B

Salt-marshes and margins of brackish lakes. Frequent in S. and W. Connemara; local elsewhere.

1. Salt-marsh near O'Brien's Br. **2**. By all the brackish lakes on the N. coast. Locally abundant on Aran. **3**. Estuary of the Dunkellin R. **4**. Occasional. **5–6**. Frequent. **7**. Very frequent. **8**. Drains in the salt-marsh at Leenane.

First record: Hart (1875). Aran Islands.

Blysmus

Blysmus rufus (Huds.) Link

1 . . . 5 6 7 . . .

Salt-marshes and ditches by the sea. Occasional in S.W. Connemara; rare elsewhere.

1. Near O'Brien's Br., on the N. side of the estuary. **5**. By the mouth of the Cashla R., near Costelloe. **6**. Sparingly in a salt-marsh S.S.W. of Carraroe. Frequent in ditches by the sea 3 km W. of Rosmuck. **7**. E. shore of Ballyconneely Bay.

First record: Wade (1804), as *Schoenus rufus*. 'In the marine marshes near Custrower Bay [N.E. of Roundstone].'

Eriophorum   Bog-cotton   *Ceannabhán*

Of the four species of this genus the first two, which are much the commonest, are distinguished by their long 'cotton', and by their perfectly smooth peduncles. *E. angustifolium* has usually 2–4 flower heads, drooping in fruit, narrow but flat leaves, and a far-creeping rhizome; *E. vaginatum* has a single, erect flower-head, hair-like leaves, of which the uppermost is expanded at the base into a swollen sheath, and grows in compact tufts. Of the others, with much shorter 'cotton' and minutely scabrid peduncles, *E. latifolium* grows in tufts and has fairly broad leaves, especially the uppermost; *E. gracile* has very narrow leaves and a far-creeping rhizome.

Eriophorum vaginatum L.

1 . 3 4 5 6 7 8 . B

Bogs and wet moorland. Common except on the limestone.

1. Very frequent. **3**. Sparingly on Ballyogan bog. **4**. Occasional on a bog by L. Corrib E. of Moycullen. **5**. Abundant. **6**. Very frequent. **7–8**. Abundant, but rare or extinct on Inishbofin, though present on Inishshark.

First record: Wade (1802). Near Ballynahinch.

E. angustifolium Honck.

1 2 3 4 5 6 7 8   A B

Bogs, drains, marshes, lake-margins and fens. Local on the limestone; abundant elsewhere.

1. Very frequent. **2.** In the marsh S. of Feenagh. In damp hollows on the S.W. slopes of Gleninagh Mt. and Slieve Carran. Two good colonies on Aran. **3.** Occasional in the fens; widespread, though usually not very abundant. **4.** Very frequent. **5–8.** Abundant.

First record: Wade (1802). Near Ballynahinch.

The occurrence of this species, generally regarded as calcifuge, in the Burren and Aran is hard to explain. It is accompanied mainly by calcicole species.

E. latifolium Hoppe

. 2 3 4 5 (6) 7 8   . .

Bogs, fens and flushes; occasional.

2. Abundant in the marsh S.S.E. of Feenagh. 'Frequent in curious swampy hollows below cliffs S.S.E. of Black Head, near the summit of the hills, *c.* 1954' (R. D. Meikle, *in litt.*). **3.** Fen at Owenbristy. **4.** N. of Ross L. In a damp hollow near Porridgetown. **5.** On sloping bog by the N. shore of L. Bofin, and in a flush 1 km to the N.W. (6). *N. shore of Oorid L.* (Praeger, 1933). **7.** A small colony *c.* 2 km W. of L. Nalawney. **8.** S.E. of Ardderry L., and between this and Maam Cross.

First record: More (1872). 'Bog on the north-west side of Urrisbeg mountain.'

E. gracile Roth

. . . . 5 6 7 8   . .

Very wet marshes and bogs, and in shallow water at lake-margins. Occasional in S. Connemara; unknown elsewhere.

5. Marsh by the Drimneen R., 1967–71; perhaps extinct now as the result of drainage. **6.** Very abundant among *Phragmites* in Callaherick L. In a marsh by the road E. of Cashel. In several small lakes near the road 5–6 km N.W. of Derryrush. **7.** S. shore of Cloonagat L. Abundant in the scraw bog W. of Cregduff L. Marsh W. of Errisbeg. **8.** In the small lake immediately N.W. of Maam Cross.

First record: Rose (1967). By Cregduff L.

The belated discovery of this species in such a well-botanized area is partly because it is very inconspicuous except during its rather short fruiting season (late June), and partly because it grows only in very wet places and can, from a distance, easily be mistaken for *E. latifolium*. The only other station known in Ireland is in Co. Westmeath; in Britain it is rare and diminishing.

Rhynchospora

Rhynchospora alba (L.) Vahl

1 . 3 4 5 6 7 8   . (B)

Wetter parts of *Sphagnum*-bogs, often in the pools. Abundant in S. and E. Connemara, but scarcer in the north-west; occasional in suitable habitats elsewhere.

1. By the W. end of L. Goller. By Drumcullaun L. **3.** Rinroe and Templebannagh bogs, and N.E. of Ballyogan Loughs. **4.** Bogs N. and N.E. of Ross L. By Ballycuirke

L. **5–6**. Abundant. **7**. Abundant in the centre; rather rare in the north, east and extreme west. **8**. Abundant.

First record: Wade (1802). 'In bogs near Ballynahinch.'

R. fusca (L.) Aiton fil.

. . 3 . 5 6 7 8 . .

In similar habitats to the last, but much less abundant and usually confined to the pools. Widespread in Connemara, but rather local; very rare elsewhere.

3. Templebannagh bog. **5**. Frequent. **6**. Very frequent. **7**. Frequent in the bogs between Clifden and Roundstone; not seen elsewhere. **8**. Frequent around Maam and the N.W. arm of L. Corrib; also by L. Nafooey and the S.W. corner of L. Mask. Abundant in a bog N. of L. Inagh.

First record: Mackay (1806). 'In several places between Ballynahinch and Oughterard.'

Schoenus

Schoenus nigricans L.

. 2 3 4 5 6 7 8   A B

Fens, lake-shores and by springs on the limestone; also very common on acid blanket-bog. Abundant except on the Clare shales and in the drier parts of the Burren.

First record: Mackay (1806). 'Very abundant on wet mountains in Connemara.'

The abundance of this normally base-loving species on acid blanket-bog in western Ireland (and Scotland) has been the subject of much speculation and some experiment, but there is no generally agreed explanation.

Cladium

Cladium mariscus (L.) Pohl

1 2 3 4 5 6 7 8   . .

Fens and lakes. Very frequent and locally abundant.

1. W. end of L. Goller. By Garvillaun L. **2**. L. Luick. A small clump on an island in the brackish lake N.W. of Bealaclugga. **3–4**. Very frequent in fens. **5**. In a lake N.N.E. of Screeb, and in the lake E. of L. Bofin. Frequent around Killaguile. **6**. Frequent, especially in the area S. of Recess. **7**. Very frequent throughout. **8**. Lakelet N.W. of Tullywee Br. Lake immediately N.W. of Maam Cross. In L. Corrib, at the end of the Dooros peninsula, and in L. Mask, N. of Clonbur.

First record: Wade (1802). 'In the small loughs between Ballynahinch and Oughterard.'

This species provides yet one more anomaly in acid–base relations in our area, for although it is normally calcicole or at least strongly base-demanding, it is very widespread in the highly oligotrophic lakes of Connemara. Here it is often accompanied by *Lobelia dortmanna*, a combination unknown in England or Wales and rare in Scotland.

A variant with narrow, light green (not glaucous) leaves which lack the usual serrated, cutting margin, has been collected from seepage-channels $1\frac{1}{2}$ km N.E. of Carna. It retains its character in cultivation.

Carex

Of the 51 species of *Carex* found in Ireland all but 7 occur in the region of this Flora. There is no point, therefore, in attempting a key, as it would merely duplicate those in descriptive Floras. The following hints may, however, prove helpful in narrowing down the list of species to be considered.

Leaves very narrow; few, if any, over $2\frac{1}{2}$ mm ($\frac{1}{10}$ in.) wide
  extensa (maritime)
  dioica, diandra, echinata, limosa, lasiocarpa (wet places)
  divulsa, remota, pilulifera, caryophyllea (fairly dry places)
  pulicaris (habitat variable)
Leaves very wide; many of them at least 8 mm ($\frac{1}{3}$ in.) wide
  acutiformis, riparia, pseudocyperus, pendula, strigosa
  Also some plants of *otrubae, laevigata*

Leaves more or less glaucous (bluish-greyish-green)
  bigelowii (mountain-tops)
  extensa (maritime)
  rostrata, nigra, acuta, elata, lasiocarpa (in water or wet marshes)
  flacca, panicea (dry to medium-damp places)

Leaves hairy
  hirta, pallescens

Fruits finely downy
  lasiocarpa (lakes and fens)
  hirta (dampish grassland)
  caryophyllea, pilulifera (dryish grassland or heath)

Forming large tussocks
  elata, paniculata

Carex dioica L.

. 2 3 4 5 6 7 8 . B

Bogs, marshes, fens and moorland flushes. Rare in the south; widespread in the north and frequent in N.E. Connemara.

2. Flush on the S.W. slope of Gleninagh Mt. Marsh S. of Feenagh. **3.** Rinroe bog. Cut-away bog at Ballyogan Loughs. **4.** Fen on N. side of Gortachalla L. Fen by small lake S. of Carrowmoreknock. **5.** Bog by E. end of L. Formoyle. Between L. Bofin and the road. Bog by Slieveaneena L. 7 km W. of Barna. **6.** Bog $2\frac{1}{2}$ km E.S.E. of Screeb. On the saddle 5 km S.W. of Maam Cross. Near the mouth of a small stream S.E. of Cashel. **7.** Bog near L. Nalawney. In a drain by L. Conga. Bog E. of Dawros Br. Inishbofin. **8.** Frequent.

First record: Mackay (1806). 'Very plentiful in bogs between Cahill and Ballynahinch.'

C. pulicaris L.   Flea sedge

1 2 3 4 5 6 7 8 A B

In a wide range of habitats, from acid bog to mountain cliffs and limestone pavement. Frequent to abundant, except in the extreme south.

1. Abundant in the bog by L. Goller. Grassland near Lickeen L. By Drumcullaun L. **2.** Very frequent. **3–6.** Frequent. **7–8.** Abundant.

First record: Wade (1804). 'Lettery Hill, Ballynahinch.'

C. disticha Huds.

1 . 3 4 5 . . . . .

Damp meadows, fen-margins, marshes and river-banks. Local.

1. Frequent in damp grassland W. of Kilnamona. Near Kilfenora. **3.** Frequent in the extreme south, occasional in the centre, and in abundance round Coole L. **4.** Edge of Coolagh fen. By the R. Corrib in Galway city, and by the west bank 2½ km upstream. Shore of L. Corrib near Burnthouse. **5.** Marsh 4 km W. of Spiddal.

First record: ITB (1901). H16.

C. arenaria L.

1 2 . 4 5 6 7 . A B

Sand-dunes, mainly on the mobile parts, and occasionals spreading to stony ground nearby. Common in W. Connemara; local elsewhere.

1. Abundant on the dunes N. of Lehinch. **2.** Fanore dunes. Aran. *Doolin* (Atlas). **4.** Limestone debris near the sea at Rinville. **5.** Dunes W. of Spiddal, and between Spiddal and Barna. **6.** Moyrus. N. end of Furnace I. **7.** On all the principal dunes from Roundstone to Salruck, though not always in great abundance. Inishbofin.

First record: More (1872). 'Great Isle of Aran – *H.C.H.*'

C. diandra Schrank

1 . 3 4 (5) . 7 . . .

Marshes, bogs, ditches and fens; local.

1. Frequent in the marsh at the E. end of Inchiquin L. Ditch 5½ km E. of Lisdoonvarna. **3.** Occasional by the Ballyogan Loughs. Very frequent on Rinroe bog. Templebannagh L. **4.** Abundant in the fen at Coolagh. In a fen in the N.E. part of Galway city, 1968; probably now gone. Fen at Gortachalla L. Bog S. of Carrowmoreknock. (5). *L. Bofin*, 1899 (**DBN**). **7.** By the lake at Bunowen. Near Letterfrack.

First record: Praeger (1906). 'Bog 2 miles W. of Ballyconneely.'

C. paniculata L.

1 . 3 4 5 . 7 8 . .

Marshes, ditches and cut-away bogs; frequent but local.

1. Abundant in marshes by L. Goller and L. Raha, and occasional by Inchiquin L. In a ditch near O'Dea's Castle. **3.** Marsh 2½ km N. of Ruan. A few plants by a turlough at Carrowgarriff. **4.** Near L. Corrib, E. of Moycullen. **5.** In a cut-away bog near the old railway-track N.W. of Rosscahill. **7.** Sparingly at the S.E. end of Garraunbaun L.; also a fair-sized colony in a flush near the Clifden road 2 km to the south-west. **8.** Ditch near Maam Br. By the N.W. corner of Dirkbeg L., and in a coomb 3 km to the south.

First record: Corry (1880). By the margin of Inchiquin L.

C. otrubae Podp.

1 2 3 4 5 6 7 . A B

Marshes and ditches, mainly near the sea. Frequent in W. Connemara; occasional elsewhere.

1. Ditches near O'Brien's Br. *S. shore of Inchiquin L.* (Atlas). **2.** By the turlough E. of Ballyvaughan. Aran. *Poulsallagh* (Atlas). **3.** By the lake at Dromore. One plant by the Kiltartan R. By the Kilcolgan estuary. **4**. N. bank of the Kilcolgan estuary. Saline meadows at Oranmore. **5**. Marshy field near Gentian Hill. Marsh near the sea 5 km W. of Spiddal. **6**. Furnace I. S.W. end of Lettermullan I. Marsh by the 'coral strand' near Carraroe. **7**. Frequent near the W. coast. Inishlackan. Ballynakill harbour.

First record: More (1872). 'On all the isles of Aran; *H.C.H.*'

C. spicata Huds.

. . **3** . . . .

Rough grassland. Occasional in the E. part of the Burren; unknown elsewhere.

3. A few plants by the W. end of Dromore L., 1965 (DW). *By a wall S. of the ruins at Kilmacduagh*, 1954 (**TCD**). By the N.W. shore of L. Atedaun, at the S.E. end of L. George, and by the S.E. shore of Inchiquin L., 1981 (T. Curtis).

First record: Webb (1982).

Very rare in Ireland; some authors, indeed, have expressed doubts as to its occurrence anywhere in Ireland. Specimens in **TCD** from both the first two stations have, however, been verified by A. O. Chater, and the more recent ones match them closely.

It is possible that *C. muricata* L. may occur in the Flora region; there are old records of Ballyvaughan (CH2) and Oughterard (Babington, 1836), but both require confirmation. The *Atlas* record for *C. muricata* in our region is based on an earlier misidentification of the Kilmacduagh specimen of *C. spicata*.

(*C. divulsa* Stokes)

(*1*) . (*3*) . . . .

Very rare; no information as to habitat is available.

(*1*). *Near the S. shore of Inchiquin L.*, 1955 (**TCD**). (*3*). *Garryland* (ITB).

First record: ITB (1901). Garryland.

C. echinata Murr.

1 2 3 4 5 6 7 8 . B

Marshes and bogs. Local on the limestone; abundant elsewhere.

1. Very frequent. **2**. Moorland on Moneen Mt. **3**. Marsh N.E. of Killinaboy. Abundant N.W. of Crusheen. Rinroe and Templebannagh bogs. By L. Briskeen. **4**. Drains by L. Corrib, E. of Srue. E. of Moycullen. 3 km S. of Carrowmoreknock. Acid fen near Ballindooly L. **5–8**. Abundant throughout.

First record: Babington (1836). Maam.

C. remota L.

1 (*2*) . **4** . **6 7 8** . .

Shady or damp places. Occasional to frequent on the N. and S. margins of the region; very rare elsewhere.

1. Frequent. (*2*). *By the turlough at Turlough village* (O'Kelly, 1903). **4**. Edge of scrub at Aughnanure Castle. In scrub 1½ km N.W. of Moycullen. In a ditch by the R. Corrib 3 km above Galway. **6**. On an island in Athry L. Woods at Cashel Ho. **7**. Woods at Ballynahinch. Abundant in a shady marsh by Kylemore L., and by the entrance to the Abbey. Roadsides W. of Clifden. Near Letterfrack. **8**. Frequent by L. Corrib and L. Mask. Island in L. Shindilla.

First record: Corry (1880). Near the Spectacle Bridge, Lisdoonvarna.

C. curta Good.

1 . . . . . . (*8*)   . .

Damp places; very rare.

1. By Doonagore L., 1967 (MS). (*8*). *On tussocks of* C. elata *in the bog between Maam and the head of L. Corrib*, 1895 (Marshall & Shoolbred, 1896).

First record: Marshall & Shoolbred (1896). Between Maam and L. Corrib.

Recent search has failed to reveal the tussocks of *Carex elata*, much less any plants of *C. curta*, in the Maam station.

C. ovalis Good.

1 . **3 4 5 6 7 8**   . **B**

Marshes, meadows and roadsides. Rare on the limestone; very frequent elsewhere.

1. Very frequent. **3.** Marsh at the N.E. corner of L. Atedaun. Marsh by Dromore L. **4.** Shore of L. Corrib E. of Moycullen. Marsh by Ross L. Boggy pasture N.W. of Gortachalla L. **5–8.** Very frequent.

First record: Babington (1836). Maam.

C. nigra (L.) Reich.

1 2 3 4 5 6 7 8   A B

Marshes, turloughs, river-banks and wet grassland. Very frequent, except in the drier parts of the Burren.

1. Frequent. **2.** Abundant in the Carran turlough, and occasional near Turlough village. Occasional on Aran. **3–8.** Very frequent to abundant.

First record: Mòre (1876). Common on Inishbofin.

A plant with the female glumes chestnut-brown instead of black has been collected from beside a small lake E. of Killinaboy (**DBN**).

C. acuta L.

1 . **3** . . . **7** .   . .

Marshes; occasional.

1. Frequent around Garvillaun L. **3.** Marsh near the road S.E. of Ruan. Marsh N.W. of Ballinderreen. Occasional in the saline marsh on the S. side of the Kilcolgan estuary. **7.** Doonloughan (rare).

First record: Webb (1982).

All the records here cited were made by DW between 1965 and 1969. The identity of the plants from the second, third and fifth localities was verified by A. O. Chater. The species had not previously been recorded from the Flora region, and indeed from only two stations west of the Shannon.

Most Floras say that in *C. acuta* the lowest bract always exceeds the inflorescence, but in the west of Ireland this character is most inconstant, the bract varying from barely half as long as the inflorescence to nearly twice as long.

C. elata All. (*C. hudsonii*)

1 2 3 4 5 . (*7*) **8**   . .

Fens, marshes, slow streams and lake-margins. Frequent and locally abundant on the limestone; rare elsewhere.

1. Drumcullaun L. 2. Abundant in a fen near Turlough village. 3. Frequent, and locally abundant. 4. By the R. Corrib in Galway, and 3 km above the city. Coolagh fen. In the stream by Aughnanure Castle. 5. Frequent around a lakelet by the old railway-track N.W. of Rosscahill. (7). *Derryclare L.*, 1894 (**DBN**). 8. Abundant by the shore of L. Mask N. of Clonbur, and also near its S.W. corner. Two clumps in the lake by the old road 3 km E. of Recess. *Maam Br.*, 1925 (**DBN**).

First record: Babington (1836), as *C. cespitosa*. Maam.

Although most plants show the characteristic densely tussocky habit, in some places, as at the S. end of L. Bunny, plants in other respects resembling this species form an almost continuous sward. It is known that partly fertile hybrids can be formed between *C. acuta* and *C. elata*, and it may be that the aberrant forms of both species owe their origin to hybridization.

The two clumps near Recess, mentioned above, have persisted for at least 25 years without any obvious change in size or vigour.

C. aquatilis Wahl.

. . . . **8** . .

Boggy river-banks; very rare.

8. Forming a patch about 10 m long on the bank of an ox-bow *c*. 400 m W. of Maam Br., 1965, 1973.

First record: Marshall & Shoolbred (1896). By the river above Maam.

This species stands out conspicuously by its yellowish-green colour from the *C. rostrata* which surrounds it. It flowers very sparingly. It is rare in Ireland, the nearest known stations being some 100 km distant, on the Shannon by L. Allen and L. Derg.

C. bigelowii Schwein. (*C. rigida*)

. . . . . . **7** (*8*) . .

Mountain-summits and cliff-ledges; rather rare.

7. Widely but rather sparingly distributed on the summits and higher ledges of the Twelve Pins: recorded from Muckanaght, Bengower, Benbreen, the peak N.E. of Bencorr, and the higher of the two Benbauns, 1967–80 (H. J. B. Birks, T. Curtis). (*8*). *Near the summit of Benwee* (Colgan, 1900).

First record: Colgan (1900). Benwee.

It is very curious that this species should have escaped the notice of Hart; he recorded it (1883) repeatedly in Co. Mayo but not in Co. Galway. Nor did Colgan record it for the Twelve Pins. It is one of the most widespread of Irish alpines, though rather less frequent on quartzite than on other rocks.

C. limosa L.

1 . 3 . 5 6 7 8 . B

Lake-margins and very wet bogs. Very rare on the limestone; frequent elsewhere.

1. Summit of Slieve Elva. Occasional on the high ground N.E. of the Cliffs of Moher. *Bog near Kilfenora* (**K**). **3**. Acid fen by the Ballyogan lakes. **5–6**. Frequent. **7**. Frequent in the south and west; rarer elsewhere. **8**. Very frequent, especially in the bogs W. of the Maumturk Mts., and between Maam and the head of L. Corrib.

First record: Graham (1840). In the swamps W. of L. Corrib.

C. pilulifera L.

(*1*) 2 . . 5 6 7 8   . B

Heathy grassland. Frequent in Connemara; rare elsewhere.

(*1*). *Slieve Elva* (Atlas). **2**. Carnsefin. **5–6**. Frequent. **7–8**. Very frequent.

First record: Hart (1883). Common on the mountains of Connemara above 1200 ft.

   Doubtless overlooked in other stations in district 1; it is not a very conspicuous plant.

C. caryophyllea Lat. (*C. praecox*)

1 2 3 4 5 (*6*) 7 8   A B

Dry grassland and limestone pavement. Locally frequent on the limestone; occasional elsewhere.

1. Rocky bluffs near Elmvale Ho., and S. of Liscannor. **2–3**. Frequent. **4**. Frequent in the north; rather rare in the south. **5**. Occasional, mainly near the sea. (*6*). *Near Cashel* (Atlas). **7**. Bunowen, and near the W. end of the Slyne Head peninsula. At two places on the N. shore of Clifden Bay. Inishbofin. **8**. Pasture W. of Teernakill Br. Ferry Br., and 2½ km N. of it. Near the middle of the Dooros peninsula.

First record: More (1876). Inishbofin.

C. hirta L.

1 2 3 4 5 . . 8 . .

Marshes, turloughs and damp pastures. Occasional in the Burren and adjacent areas; very rare in Connemara, and absent from the islands.

1. In a marsh near the coast S. of Lehinch. By the E. shore of Inchiquin L. Near L. Goller. **2**. By L. Aleenaun. In the turlough at Turlough village. Near Carran. Marsh S.E. of Formoyle. **3**. Frequent in the south; occasional elsewhere. **4**. Shore of L. Corrib at Burnthouse. **5**. Roadside near Gentian Hill. **8**. Shore of L. Corrib N. of Oughterard.

First record: ITB (1901). H16.

C. lasiocarpa Ehrh.

(*1*) . 3 4 5 6 7 8   . .

Bogs, lake-margins and fens. Frequent except in the extreme south and west.

(*1*). *By L. Goller* (Atlas). **3**. Frequent, and abundant in some of the fens. **4**. Coolagh fen. By the R. Corrib, 3 km N. of Galway. Shores of L. Corrib, Ross L. and Ballycuirke L. **5**. Frequent. **6**. Occasional. **7**. Marsh N. of Letterdife Ho., Roundstone. L. Nacarrigeen. **8**. Frequent; especially abundant in the bog between Maam and the head of L. Corrib.

First record: Graham (1840). 'In the swamps W. of L. Corrib.'

C. sylvatica Huds.

1 2 3 4 (*5*) . 7 8   . .

Woods, scrub, and crevices in limestone pavement. Frequent in the more sheltered areas, but absent from wide stretches of exposed country.

1. Woods at Moy Ho. and Ennistymon. In scrub by the stream at Derrymore Br. Wood S.W. of L. Raha. **2**. Frequent in hazel-scrub at Carran and elsewhere in the eastern part of the district; also in small quantity in crevices of bare pavement on Carnsefin and Cappanawalla. **3**. Frequent. **4**. On the E. side of the R. Corrib above Galway.

$1\frac{1}{2}$ km N. of Moycullen. By the E. side of Ross L. On an island in the S.W. part of L. Corrib. (*5*). *Spiddal* (Atlas). **7**. Woods at Ballynahinch, Kylemore and Salruck. **8**. Frequent by L. Corrib and L. Mask, but not seen on the islands.

First record: ITB (1901). H16.

C. strigosa Huds.

. **2** . . . .

On limestone rocks at the edge of a turlough; very rare.

2. Filling a small gully in the limestone bluff on the N. side of L. Aleenaun, 1968 (F. H. Perring).

First record: Webb (1982).

It remains to be seen whether this will survive the draining and bulldozing of the lake which began in 1980, but it should have a fair chance.

*C. pendula Huds.

1 . . . **5** . **7** . . .

Damp woods and streamsides; rare.

1. Wooded bank of the Aille R., above Roadford. *Near the Spectacle Br., Lisdoonvarna* (Corry, 1880). **5**. By the stream outside the gate of Killaguile Ho. **7**. In two places near Garraunbaun L.

First record: Corry (1880). Lisdoonvarna.

Although this species may perhaps be native in some parts of Ireland, there is little doubt that in our region it is derived from plants introduced to demesne woods. The station at Roadford is a wild one, but it is downstream from Lisdoonvarna.

C. pallescens L.

. . (*3*) . **5 6 7 8** . .

Scrub and rough, heathy grassland. Frequent in the extreme north-east; rare elsewhere.

(*3*). *Dromore woods* (Corry, 1880). **5**. In scrub on the hill S. of Oughterard. **6**. By the outflow from Invei L. **7**. Kylemore. Inishdawros. E. side of Derryclare L. *Letterfrack* (Atlas). **8**. Frequent around the N.W. part of L. Corrib and the S.W. part of L. Mask, and on some of the islands.

First record: Corry (1880). Dromore woods.

C. panicea L.

1 2 3 4 5 6 7 8 A B

Bogs, moors, damp or heathy grassland and lake-shores. Frequent, though rather local, on limestone; abundant elsewhere.

1. Very frequent. **2**. Occasional, mainly in damp or peaty places. **3–4**. Frequent. **5–8**. Abundant.

First record: Babington (1836). Oughterard and Maam.

Easily distinguished from *C. flacca* in flower or fruit, but in the vegetative condition they are very much alike, both with strikingly glaucous leaves like those of a carnation. In *C. panicea*, however, the tip of the leaf has the form of a 3-sided prism; in *C. flacca* it is gutter-shaped.

C. flacca Schreb.

1 2 3 4 5 6 7 8   A B

Pastures, rough grassland, sand-dunes, limestone pavement and stony lake-shores. Very abundant on limestone; rare on other rocks except near the coast.

1. Frequent by the coast. Roadsides W. of Kilnamona and near Lickeen L. O'Dea's Castle. **2–4.** Abundant. **5.** Occasional. **6.** Frequent near the coast; rare inland. **7.** Frequent near the coast. N. face of Muckanaght. **8.** Frequent near the shores of L. Corrib and L. Mask, and on several of the islands in L. Corrib; not seen elsewhere.

First record: Babington (1836), as *C. recurva*. Oughterard.

On the limestone pavements of the Burren and Aran this species is often very tall (up to 1 m).

C. laevigata Sm. (*C. helodes*)

1 . . . 5 6 7 8   . .

Damp woods and streamsides. Occasional on acid soils; absent from the limestone.

1. Wet grassland near Ennistymon. *Near Lisdoonvarna* (Atlas). **5.** Moycullen woods. By a stream near the Costelloe road *c.* 7 km S. of Oughterard. **6.** In the plantation N.E. of Toombeola. Glendollagh woods. **7.** Kylemore wood. *Near Letterfrack* (Atlas). **8.** On the Hill of Doon; also in relict woodland on the opposite shore, N. of Keeraunnageeragh. Woods by L. Corrib at Gortdrishagh. By the waterfall W. of L. Nafooey.

First record: Corry (1880). By a stream near the Spectacle Br., Lisdoonvarna.

C. binervis Sm.

1 2 3 . 5 6 7 8   . B

Moors, drier parts of bogs, heathy grassland and cliff-ledges. Abundant in suitable habitats, but rare on the limestone.

1. Very frequent. **2.** On all the principal summits from Black Head to Bealaclugga, wherever the peat is fairly deep. **3.** Rinroe bog. **5–8.** Abundant.

First record: Babington (1836). Oughterard and Maam.

C. distans L.

1 2 3 4 5 6 7 .   A B

Rocky or grassy places by the sea, and in the drier parts of salt-marshes. Widespread and locally frequent.

1. Near Lehinch. **2.** Frequent. **3.** Aughinish. S. side of Kilcolgan estuary. **4.** N. side of Kilcolgan estuary. L. Atalia, and shore to the east of it. **5.** Gentian Hill. E. shore of Cashla Bay. **6.** Frequent. **7.** Very frequent.

First record: More (1872). 'Great Aran Island; *H.C.H.*'

Ivimey-Cook & Proctor (1966) reported this species as frequent near Carran, in a community dominated by *C. nigra*. There are three early records by Praeger for its occurrence inland, so that its presence at Carran is by no means impossible, but confirmation is desirable.

C. punctata Gaud.

. 2 . . . . 7 . . .

Rocky places near the sea. Occasional in W. Connemara; very rare elsewhere.

2. Poulsallagh. 7. Rocks by the sea at Belleek. Roadside, very close to the sea, 1 km S.W. of Clifden. Very sparingly at Dog's Bay, among rocks at the edge of the beach. Ballyconneely Bay. Very frequent between the N. shore of Aughrusbeg L. and the road, several hundred metres from the sea, growing on grassy ledges on the landward (south) side of boulders or bluffs.

First record: Rob (1948). Dog's Bay.

Easily confused with *C. distans*, but specimens from all localities cited above have been confirmed by A. O. Chater or R. W. David. It seems probable that the population at Aughrusbeg L. is one of the largest in N.W. Europe.

C. hostiana DC.

1 2 3 4 5 6 7 8   A .

Marshes, bogs, fens and other wet, peaty places; frequent.

1. Bog by Drumcullaun L. Marsh by Lickeen L. By Garvillaun L. On Slievebeg. 2–3. Frequent. 4. Frequent in the north; not seen S. of Ballycuirke L. 5–6. Occasional. 7. Very frequent. 8. Frequent.

First record: Hart (1883). Ben Lettery.

Especially characteristic of nearly neutral peaty soils, as in flushes on blanket-bog or in fens subject to acidification.

Hybrids of this species with members of the *C. flava* group are among the commonest in Ireland. *C. hostiana × lepidocarpa* (*C. × fulva* Good.) has been identified from Gortachalla L., and *C. hostiana × serotina* (*C. × appeliana* Zahn) in great abundance from the most southerly of the turloughs S.W. of Mullaghmore, both by A. O. Chater. *C. demissa × hostiana* almost certainly occurs in the region, but we have no certain record.

C. extensa Good.

1 2 3 4 5 6 7 .   **A B**

Salt-marshes and margins of brackish lakes. Frequent in W. and S.W. Connemara; widespread but local elsewhere.

1. Near Lehinch. 2. Poulsallagh. By the brackish lake N.W. of Bealaclugga. Aran. 3. New Quay. 5 km W. of Kinvara. 4. Oranmore. 5. Gentian Hill. E. shore of Cashla Bay, W. of Keeraunnagark. 6. Very frequent. 7. Frequent.

First record: Mackay (1806). 'In salt-marshes near Cahill.'

Praeger (1900) recorded this species from an inland station (near Oughterard), but later (1934c) withdrew the record. His specimen is, in fact, *C. serotina* with an exceptionally long lowest bract.

C. flava group

This group, consisting of *C. demissa*, *C. lepidocarpa*, *C. flava* and *C. serotina*, presents great difficulties in our region. Although they are closely related, the four species can be distinguished without great difficulty over most of Europe by experienced observers, but in the W. of Ireland many of the distinctions become blurred. It is unlikely that this is because of hybridization, as although hybrids are fairly frequent their fertility seems to be low.

C. demissa Hornem.

1 2 3 4 5 6 7 8   A B

Moorland, bog-margins, stony lake-shores and wet, acid grassland. Very rare on the limestone; abundant elsewhere.

1. Frequent. **2**. Heathy pasture by the Caher R., below Formoyle. Occasional on Aran in peaty pockets on the limestone. **3**. Templebannagh bog. On a mossy tussock near the turloughs S.W. of Mullaghmore. **4**. Frequent on and around the bogs N.E. of Moycullen. **5–8**. Abundant.

First record: Praeger (1934c). H16. Recorded as *C. flava*, but it is clear from the context that this species is intended.

We have, after careful examination, and on the advice of A. O. Chater, rejected all records for this species from districts 2 and 3, except for those cited above. A plant which grows abundantly in the fens and turloughs of the Burren has often been recorded as *C. demissa*, and Ivimey-Cook & Proctor (1966) made it the indicator-species of a new 'nodum'. *C. demissa* is normally calcifuge, and the habitats mentioned are highly calcareous, but the abundance of *Eleocharis multicaulis* and *Juncus bulbosus* in the same habitats shows that this is not in itself a sufficient ground for rejecting the identification. But, after careful study in the field and cultivation in the garden, Chater and DW agreed that, although the plants are unusually robust, the subsessile male spikelet, the very compact inflorescence and the shape and size of the fruit bring them within the range of *C. serotina*.

C. lepidocarpa Tausch

1 2 3 4 5 . . .   (A) .

Fens and other wet, base-rich habitats. Abundant on the limestone; very rare elsewhere.

1. Frequent in a marsh at the E. end of Inchiquin L. **2**. Frequent. **3–4**. Abundant. **5**. A few plants in a marsh by the Drimneen R. Abundant in a fen by L. Naneevin. One plant in wet bog near Lettermore, E. of Screeb (det. Jermy).

First record: Nelmes (1948). Black Head.

Not always as easy to distinguish from *C. demissa* as one would wish, especially in district 3. The last station cited above is very remarkable as being apparently on typical acid blanket-bog without any sign of flushing.

C. flava L.

. . . 4 . . . .   . .

Fens; very rare.

4. One or two plants in Coolagh fen, accompanied by hybrids with *C. lepidocarpa*, 1968 (F. H. Perring and DW; conf. C. A. Jermy).

First record: Webb (1982).

Subsequent visits have revealed only the hybrid, but it is difficult country to search thoroughly. One specimen of the original collection seemed quite typical, with fruits fully 6 mm long. The species is not known elsewhere in Ireland, and is very rare in Britain.

C. serotina Mérat (*C. oederi*)

. **2 3 4 5** . **7 8** **A** .

Fens, turloughs, stony lake-shores and marshes. Very frequent in the Burren lowlands, the extreme west of Connemara, and by L. Corrib; rare elsewhere.

2. Abundant and luxuriant in the fen near Turlough village. Frequent by the brackish lake N.W. of Bealaclugga. Frequent on Aran. **3–4**. Very frequent and locally abundant. **5**. By L. Naneevin, and in the marsh N. of Killaguile. **7**. Frequent by the W. coast, from Cregduff L. to Aughrusbeg L. **8**. Frequent along the shores of L. Corrib, and on many of its islands; also on the shore of L. Mask, N. of Clonbur.

First record: More (1860). 'Upon a islet called Canova in L. Corrib.'

Characteristic especially of marly or stony places subject to flooding by calcareous waters; usually in more or less open vegetation. In W. Connemara, however, it is found in places where the general aspect of the vegetation is calcifuge. It varies greatly in height, and robust forms from fens and turloughs have been widely mistaken for *C. demissa* (see under that species). On the other hand dwarf forms of *C. demissa* from W. Connemara and Aran tend to have unusually small fruits, and are liable, therefore, to be confused with *C. serotina*. The situation is not made easier by the fact that the two species often grow intermingled, as is attested by several mixed sheets in herbaria.

Subsp. *pulchella* (Lönnr.) van Oost. (*C. scandinavica*) has been recorded from districts 3, 7 and 8, and is probably fairly widespread.

C. pseudocyperus L.

. **[2] 3 4** . . **7** . **[A]** .

Ditches and reed-swamps; rare.

[2]. Recorded for Aran in the *Atlas* on the strength of a specimen in **DBN**, but this is almost certainly mislabelled (Webb, 1980*a*). **3**. By Ballyogan Loughs. Among very tall *Phragmites* in a ditch by Templebannagh L. *Dromore* (Atlas). **4**. Fen in N.E. part of Galway city, 1964; probably now drained. *By the old railway-line between Galway and Oughterard* (Marshall & Shoolbred, 1896). *Menlough* (ITB). **7**. Reed-swamp by the most easterly of the lakelets at Bunowen.

First record: Marshall & Shoolbred (1896). By the railway between Galway and Moycullen.

C. acutiformis Ehrh.

. **(2)** . **4 5** . . . . . .

Marshes; rare.

(2). *By the turlough at Turlough village* (O'Kelly, 1903). **4**. Locally abundant around a pond at Rinville. **5**. Locally abundant in the marsh by the Drimneen R., N. of Killaguile.

First record: O'Kelly (1903). Turlough village.

Reported also for the Kilcolgan estuary by Praeger, but probably just outside the limits of our region.

C. riparia Curtis

. . . **4** . . . .

Lake-margins; very rare.

4. At the N. end of Ballycuirke L., 1980 (MS, **DBN**).

First record: Webb (1982).

C. rostrata Stokes

1 2 3 4 5 *(6)* **7 8** . **B**

Lake-margins, ditches and marshes. Rare in districts 2 and 6; frequent elsewhere.

1. Frequent in the north; rather rare in the south. **2.** Carran turlough. **3–5.** Frequent. *(6). Cashel* (Atlas). **7–8.** Frequent.

First record: More (1876). Inishbofin: Church Lake.

C. vesicaria L.

1 2 3 *(4)* . . . **8** . .

Lake-shores, turloughs and marshes; local.

1. N. and E. sides of Inchiquin L. N.W. side of Ballycullinan L. **2.** In several places around the Carran turlough. Fen near Turlough village. *Near Ballyvaughan* (Druce, 1931*b*). **3.** Locally abundant in the south, northwards to L. Atedaun; also in Coole L. *(4). E. of Rosscahill* (Atlas). **8.** Abundant on the limestone shore of L. Mask, N. of Clonbur, especially in bays and small pools. *W. end of L. Nafooey* (Atlas).

First record: Marshall & Shoolbred (1896). Damp places S.'of L. Mask.

GRAMINEAE

Two alien grasses reported from the region deserve mention here, though neither is effectively naturalized. *Spartina pecinata* Link, native to temperate North America, was planted, probably over 50 years ago, in an outlying part of the ground of Costelloe Lodge (district 6), along with many other ornamentals. The area is no longer cultivated, but this grass survives among mainly native vegetation as a large clump, slowly spreading by rhizomatous growth. For details see Boyle (1968). *Gaudinia fragilis* (L.) Beauv., an annual from S. Europe, has once been found below the cliffs of Slieve Carran (district 2). It is an occasional contaminant of imported grass-seed, and has been reported several times as a casual from other parts of S.W. Ireland.

Phalaris

Phalaris arundinacea L.

1 2 3 4 . . **7 8** . .

Riversides, marshes, turloughs and lake-shores. Frequent and locally abundant on the limestone and shale; very rare in Connemara.

1. Frequent and locally abundant. **2.** In the Caher R. near Formoyle. Carran turlough at Castletown. L. Aleenaun. Near Turlough village. **3.** Very frequent throughout. **4.** By the Kilcolgan estuary. Ditch at Rinville. Occasional in marshes E. of Moycullen. **7.** *E. shore of Ballyconneely Bay* (Atlas). N.E. shore of the Slyne Head peninsula. **8.** Limestone shore of L. Mask. On some of the islands in L. Corrib.

First record: Wade (1802), as *Calamagrostis variegata*. Near Ballynahinch.

P. canariensis L., presumably derived from bird-seed, has been seen once or twice on rubbish-dumps, but is not naturalized.

Milium

(*Milium effusum* L.)

Collected by Praeger in 1899, 'in a wood by L. Corrib, N.W. of Oughterard', but not seen since. There is a specimen in **DBN**, but although this record represents an interesting extension of the range of a relatively rare Irish species, it is not mentioned in his article (1900) which describes his expedition to Co. Galway in 1899. There is also a slight discrepancy in dates: the specimen is dated 17 July, but the narrative makes it clear that it was on the 16 July that he was N.W. of Oughterard. Confirmation is therefore desirable.

Alopecurus

Alopecurus geniculatus L.

1 2 3 4 5 6 7 8   A .

Marshes, ditches and wet grassland. Frequent, but unevenly distributed.

1. Frequent. **2.** L. Aleenaun. Margin of Carran turlough. Inishmore. **3.** Frequent in marshes around Corofin, Crusheen and Ennis, and on both sides of Kinvara Bay. **4.** By Ballycuirke L. Brackish marsh E. of Galway. **5.** Marsh on the W. edge of Galway city, 1965; probably now built over. **6–7.** Frequent. **8.** Ditches at Ferry Br., Teernakill Br. and N. of Ballydoo L.; not seen elsewhere.

First record: Hart (1875). Inishmore.

The absence of this species from large areas of suitable ground in districts 4 and 8 is hard to explain, as is also its relative scarcity in the turloughs. On both Inishmore and Inishbofin it has been seen growing on the top of a sodded wall.

‡**A. pratensis** L.   Foxtail

1 2 3 4 . (*6*) **7 8   A** .

Meadows, pastures and roadsides; local.

1. Frequent. **2.** One large patch on Inishmore. **3.** Meadows 4 km N. of Ennis. Pasture by L. Briskeen. Near the shore on the W. side of Kinvara Bay. **4.** Meadow at the S.E. end of Mweenish I., at the head of Galway Bay. (*6*). *Lettermullan* (Atlas). **7.** Frequent and locally abundant near the W. coast, from Dog's Bay to near Claddaghduff. **8.** Roadside near Clonbur. Cong. *Leenane* (Atlas).

First record: ITB (1901). H16.

The status of this species is hard to assess; it is probably introduced in many of its stations, and possibly in all. It seems to have increased considerably in W. Connemara between 1965 and 1980.

Phleum

‡**Phleum pratense** L.   Timothy

1 2 3 4 5 . **7 8   A** .

Meadows, pastures and roadsides. Frequent on the limestone; occasional elsewhere.

1. E. of Inchiquin L. Roadside near Lickeen L. **2.** Frequent. **3.** Very frequent. **4.** Occasional. **5.** Pasture 6 km N.W. of Galway. Meadow 2½ km S. of Oughterard. **7.**

By Bunowen Hill. Roadside N.W. of Clifden. **8**. Meadows by the Hill of Doon and N.W. of Cornamona.

First record: Hart (1875). Inishmore: 'In the field adjoining Mr. O'Flaherty's house, Kilmurvy.'

Most records are referable to subsp. *pratense*; that from Cornamona, and perhaps one or two others, refer to subsp. *bertolonii* (DC.) Bornm. (*P. bertolonii* DC).

P. arenarium L.

1 2 . . . . . .   **A** .

Sand-dunes; rare and local.

1. In small quantity on the dunes N. of Lehinch. **2**. Frequent on the dunes at the E. end of Inishmore.

First record: More (1872). 'Great Island of Aran; *H.C.H.*'

Rare on the west coast, except in Co. Donegal.

The record by Wade (1802) for *P. alpinum* from Benlettery can be confidently dismissed as an error.

Agrostis

Agrostis canina L.

1 2 3 4 5 6 7 8   [A] **B**

Bogs, moors and heathy grassland. Abundant throughout Connemara; local elsewhere.

1. Abundant by L. Goller and Lickeen L., and occasional elsewhere. **2–3**. Occasional in the heathy communities growing on peat in pockets of the limestone pavement. **4**. Frequent on or near bogs in the area N. of Moycullen. **5–8**. Abundant.

First record: Hart (1875). Inishmore. The plant cannot be found there now, and Hart's record, if correct, was probably based on a casual introduction with turf. First confirmed record: More (1876). Inishbofin.

A. stolonifera L.   Fiorin-grass

1 2 3 4 5 6 7 8   **A B**

Marshes, wet meadows, roadsides and waste places, lake-shores and sandy or stony beaches. Abundant throughout.

First record: Mackay (1806). 'Very abundant on the sandy shores...of Cunnamara.'

A. gigantea Roth

1 . **3** . . **6** . .   . .

Cultivated fields; rare.

1. Near Derrymore Br. **3**. Carrowgarriff. **6**. In two places near Carraroe.

First record: Webb (1982).

Possibly overlooked in some other stations, but certainly not a common weed.

A. capillaris L. (*A. tenuis*)

1 2 3 4 5 6 7 8   A B

Heaths, grassland and open woods. Frequent on the limestone; abundant elsewhere.

First record: Hart (1875). Aran Islands.

Plants infested with the smut-fungus *Tilletia decipiens*, and usually of dwarf habit, are frequent in districts 7 and 8. Such plants were described by Linnaeus as *A. pumila*, and some later authors have retained the name at varietal rank.

Ammophila

Ammophila arenaria (L.) Link   Marram, Bent

1 2 . . . 6 7 .   A B

Sandy shores and dunes; local.

1. Between Lehinch and Liscannor. **2.** Fanore. Aran. **6.** Near the graveyard on Mweenish I. **7.** Frequent on the Slyne Head peninsula; also on the coast N.W. of Clifden, from S. of Kingstown Bay to N. of Aughrusbeg L. Dunes N.E. of Gowlaun. Inishbofin. *Roundstone* (P. cat.).

First record: Babington (1836). Near Tully.

Calamagrostis

Calamagrostis epigejos (L.) Roth

. 2 . . . . 7 .   A B

Rocky ground and rough grassland; very rare and confined to the islands.

2. Occasional on Inishmore, and in one area on Inishmaan. **7.** In small quantity in two places on Inishbofin.

First record: More (1872). 'Between the road and the sea, near Killeany, Great Aran Island; H. C. Hart, 1869.'

A rare plant in Ireland; there are only four other known stations, all in the west or north. The two previously recorded stations on Inishbofin were confirmed by DW in 1965–7; it was discovered on Inishmaan by White in 1976, and was seen in several places on the north side of Inishmore (though not in Hart's station) in 1976–8. Praeger cites in ITB a record for Ballyvaughan, ascribed to O'Kelly in 1899. He does not mention the station in later writings, and for that reason, and because of the ambiguity of several of O'Kelly's records in 1895–9, we think that it should be held over until confirmed.

Anthoxanthum

Anthoxanthum odoratum L.

1 2 3 4 5 6 7 8   A B

Meadows, pastures, roadsides, scrub, heaths, bogs and a wide variety of other habitats. Abundant throughout.

First record: Hart (1875). Aran Islands.

Somewhat more abundant on acid than on calcareous soils, but generally speaking one of the most tolerant species in the Irish flora in respect of habitat.

Holcus

Holcus lanatus L.　　Yorkshire fog

1 2 3 4 5 6 7 8　A B

Roadsides, meadows and pastures; abundant throughout.

First record: Hart (1875). Aran Islands.

H. mollis L.

1 2 . . (*5*) . . .　[A] .

Scrub and wood-margins; very rare.

1. Wood-margin on Clifden Hill. *Near the Cliffs of Moher* (Atlas). **2**. Half-way between Castletown and Clooncoose, E. of the road, 1976 (D. Kelly, **TCD**). (*5*). *Near Rosscahill* (Atlas).

First record: Atlas (1962).

Several other records of this species have been communicated to us, but in nearly all cases they turn out to be based on slightly unusual specimens of *H. lanatus*; in one case an authentic specimen was produced, but the finder had forgotten the locality. It is difficult for the botanist from England, or even from E. Ireland, to realize how very rare is *H. mollis* in most of W. Ireland. Least of all is it likely to occur on limestone without any woodland cover; we therefore reject the record of Nowers & Wells (1892) for Aran as an error.

A specimen from Clifden Hill in **TCD** is strongly suggestive of *H. lanatus* × *mollis*, a hybrid known from the Continent, but not so far from the British Isles. It is still under investigation.

Koeleria

Koeleria macrantha (Ledeb.) Schultes (*K. gracilis, K. cristata*)

1 2 3 4 5 6 7 8　A B

Dry or rocky grassland, cliff-tops and sand-dunes. Frequent on limestone and near the sea; very rare elsewhere.

1–2. Occasional. **3**. Frequent. **4**. Occasional. **5**. Coast W. of Spiddal. Abundant at Gentian Hill. **6**. Rocks near the sea at the S.E. corner of Gorumna I., and on the mainland opposite; also on Furnace I. and at Ardmore Point. **7**. Very frequent by the coast. **8**. Cliffs S. of L. Nafooey. Rocks on the S. shore of the N.W. arm of L. Corrib, opposite the Hill of Doon.

First record: More (1876). 'Inishbofin: banks opposite Inish-Lyon, &c.'

Avenula

Avenula pubescens (Huds.) Dum. (*Helictotrichon pubescens*)

1 2 3 4 5 . 7 8　A .

Pastures, roadsides and rocky ground. Very frequent on the limestone; very rare elsewhere.

1. Rock-outcrops near Elmvale Ho. Very frequent on the dunes at Lehinch. **2–3**. Abundant. **4**. Frequent around Ross L., and occasional near L. Corrib. **5**. Abundant on Gentian Hill. **7**. Locally frequent on Omey I. *Near Ballyconneely* (Atlas). **8**. Limestone pavement N.W. of Cong, and on rocks immediately W. of the village.

First record: ITB (1901). H16.

Arrhenatherum

Arrhenatherum elatius (L.) J. & C. Presl   False oat

1 2 3 4 5 6 7 8   A B

Hedges and roadsides; less often in meadows or pastures. Very frequent to abundant.

Most abundant in districts 3 and 4; least so in districts 1, 5 and 7, but fairly common even in these.

First record: Hart (1875). Aran Islands.

Trisetum

Trisetum flavescens (L.) Beauv.

1 2 3 4 5 . 7 8   (*A***) B**

Dry grassland, rocky ground, sand-dunes and roadsides. Frequent on the limestone; rare elsewhere.

1. Occasional near Lehinch. **2**. Fairly frequent, but local. Not seen recently on Aran. **3–4**. Frequent. **5**. Gentian Hill. **7**. Inishbofin (rare). *Near Letterfrack*, 1955 (DW). **8**. Occasional near Cong.

First record: Nowers & Wells (1892). Inishmore, near the Black Fort.

Deschampsia

Deschampsia caespitosa (L.) Beauv.

1 2 3 4 5 6 7 8   (*A***) .**

Marshes, drains, riversides, damp woods and damp meadows. Frequent in suitable habitats S. of Galway Bay, rare elsewhere.

1. Very frequent. **2**. Occasional in woodland or scrub; frequent along the Caher R., and in the turlough S.W. of Ballyvaughan. Not seen recently on Aran. **3**. Frequent. **4**. By Ballycuirke L. **5**. Bog-margin N. of Rosscahill. By the river at Oughterard. **6**. Island in Athry L. By a lake E. of Carna. **7**. Woods at Ballynahinch and Letterfrack. Island in L. Inagh. **8**. Ditch near the tip of the Dooros peninsula. Marsh by the Bealanabrack R., 4 km above Maam.

First record: Hart (1875). Inishmore.

D. flexuosa (L.) Trin.

1 (*2***) . . (***5***) 6 7 8   . B**

Moorland, heathy ground and mountain cliffs. Occasional and locally frequent in the mountains of Connemara; rare elsewhere.

1. Moorland N.E. of the Cliffs of Moher. *Slieve Elva* (Atlas). **(2)**. *Above Black Head* (Atlas). **(5)**. *S.W. of Rosscahill* (Atlas). **6**. Heathy grassland by the sea at Flannery Br. Roadside 4 km N. of Costelloe. Heathland N.E. of Screeb. Island in L. Nahasleam. **7**. Fairly frequent in the mountains and occasional elsewhere. **8**. Frequent in the mountains; rare elsewhere.

First record: More (1876). Inishbofin.

The relative scarcity of this grass in much of S. and W. Ireland, despite the abundance of suitable terrain, is rather remarkable.

D. setacea (Huds.) Hackel

. . . . **5 6 7 8** . .

Lake-margins and very wet bogs; rarely on somewhat drier peaty ground. Frequent over a limited area of S. and C. Connemara; unknown elsewhere.

5. Only on the W. margin of the district, but frequent by L. Nafurnace and L. Nacreeva; also on peat-hags by the Cashla R., near its mouth. **6**. Very frequent in the N. half, and occasional in the south. **7**. Frequent in the Roundstone area; not seen N. of L. Fadda. **8**. Margin of the lake immediately N.W. of Maam Cross. Drains and pools at the mouth of the Owenree R.

First record: More (1869), as *Aira uliginosa*. 'On the swampy borders of Lough Cregg-duff.'

Not known in Ireland outside Connemara, and very rare, though widespread, in Britain. It looks very like *D. flexuosa*, but the morphological differences, though small, are constant, and the habitat is very different. It has been seen since 1959 in 16 distinct stations, mostly lying within a radius of 15 km from the head of Kilkieran Bay, but extending westwards to L. Fadda. The *Atlas* record for 94/11 is presumably based on a specimen in **DBN** labelled by Praeger 'Oughterard', but on the day of its collection he travelled far to the N.W. of the village, and it may well have come from the Owenree R. station given above.

Aira

Aira caryophyllea L.

1 2 3 4 5 6 7 8   A B

Banks, walls, rocks, and short or semi-open grassland. Common in Connemara; occasional elsewhere.

1. Roadside banks S. of Lehinch. Near the Cliffs of Moher. By the stream S. of Clifden Hill. Abundant N.E. of Kilnamona. **2**. Limestone pavement at Black Head and at Turlough village. Grassland by the ruined castle near Glencolumbkille Ho. Aran (rather rare). **3**. Limestone pavement at the E. end of L. Atedaun, and E. of Funshinmore. **4**. Frequent near Aughnanure Castle. On sandy drift by Gortachalla L. Gravelly pasture on Tawin I. On a dyke of igneous rock near Renmore barracks. **5**. Frequent. **6–7**. Abundant. **8**. Abundant in N.E. part; rarer elsewhere.

First record: Hart (1875). Aran Islands.

A. praecox L.

1 2 . 4 5 6 7 8   A B

Banks, walls, rocks and roadsides. Very rare on the limestone; frequent to abundant elsewhere.

1. Frequent in W. half; rare in the east. **2**. Aran (rather rare). *Black Head* (Atlas). **4**. Walls at Aughnanure Castle. Bare peat on a bog 6 km N. of Moycullen. **5–8**. Abundant.

First record: Hart (1875). Aran Islands.

Often seen growing with *A. caryophyllea*, but much more decidedly calcifuge.

Sesleria

Sesleria albicans Schultes (*S. caerulea*)

. 2 3 4 5 . 7 8 A .

Dry grassland, fixed dunes, rocky ground and limestone pavement. Abundant on limestone, calcareous drift and stabilized calcareous sands; unknown elsewhere.

2–4. Abundant throughout on well-drained ground. **5.** Gentian Hill. **7.** Frequent on wind-blown shell-sand on the Slyne Head peninsula. **8.** Limestone rocks immediately W. of Cong.

First record: Graham (1840). On limestone near Galway.

The dominant grass over most of the Burren, and common in calcareous habitats elsewhere. Easily recognized in late April and early May by its steel-blue flower-heads, though occasional albino plants occur, with the flower-heads pale straw-yellow. The recent change of name is unfortunate from this point of view, as the blue colour is far commoner; it is generally agreed, however, that the name *S. caerulea* properly belongs to an allied species of wet ground in C. Europe.

Danthonia

Danthonia decumbens (L.) DC. (*Sieglingia decumbens*, *Triodia decumbens*)

1 2 3 4 5 6 7 8 A B

Grassland of all types, roadsides, heaths and bog-margins. Abundant throughout.

First record: Wade (1802), as *Festuca decumbens*. Ballynahinch.

A species of remarkable ecological tolerance, ranging from dry to very wet, and from calcareous to acidic habitats. The only soils which it seems to avoid are heavy clays.

Molinia

Molinia caerulea (L.) Moench

1 2 3 4 5 6 7 8 A B

Bogs, fens, moorland, limestone pavement and heathy grasslands. Abundant in Connemara; frequent elsewhere.

1. Frequent in the south; rarer in the north. **2–4.** Very frequent. **5–8.** Generally abundant, though relatively rare in some areas near the sea.

First record: Babington (1836). Bogs between Clifden and Roundstone.

The luxuriant *Molinia*-meadows, which develop on any part of the lower slopes of the mountains which is protected from grazing, are a notable feature of the landscape.

Phragmites

Phragmites australis (Cav.) Steudel (*P. communis*)   Common reed   *Giolcach*

1 2 3 4 5 6 7 8 A B

Fens, riversides, lake-margins and in the wetter parts of bogs; occasionally on sea-cliffs where there is seepage. Abundant in S. and W. Connemara; more local elsewhere, though in great abundance in some places.

1. Frequent. **2**. L. Luick, and in the smaller brackish lake N.W. of it. Carran turlough. Aran. **3**. Very frequent. **4**. Locally abundant in L. Corrib and Ross L.; not seen E. or S. of Galway. **5**. Very frequent. **6–7**. Abundant. **8**. Sparingly in L. Corrib, and in a few small lakes, but very local.

First record: Babington (1836). Lakes between Clifden and Roundstone.

Rather unpredictable in its appearance. It varies greatly in vigour: in the Connemara bogs (where it probably indicates the remains of small lakes overgrown by the peat) and in the brackish lakes of the Burren and Aran it is weak and seldom flowers, but in the fens of the eastern part of the Burren it has been seen up to 4 m (13 ft) in height.

Visitors from drier climates are apt to be startled to see it growing not infrequently on the top of a bank or a sodded wall.

Melica

Melica uniflora Retz.

. **2 3 4** . . **7 8** . .

Woods, scrub and limestone pavement, mainly on the limestone. Occasional, but very local.

2. Occasional in the E. half; rare in the west. **3**. Limestone pavement N.E. of Slieve Carran. Garryland wood. **4**. On the hill N.E. of Menlough. Frequent in scrub near Burnthouse. **7**. Oak-woods S. of the river, E. of Clifden, and also at Streamstown. **8**. By the Pigeon-hole near Cong, and in the grounds of Ashford Castle. Woods at Gortdrisha.

First record: ITB (1901). H16.

Nardus

Nardus stricta L.

1 . . **4 5 6 7 8** . **B**.

Grassland, heaths, and on the drier parts of bogs. Abundant throughout Connemara; frequent on the Clare shales; very rare elsewhere.

1. Frequent. **4**. On small hillocks of glacial drift near the E. shore of Ballycuirke L. **5–8**. Abundant.

First record: Babington (1836). On the mountains behind Maam.

The station at Ballycuirke L. is peculiar. The lake lies on the boundary between limestone and metamorphic rocks, but a crisp separation of the two floras is blurred by hillocks of mixed drift. The same hillocks on which *Nardus* grows also support *Carlina vulgaris*.

Parapholis

Parapholis strigosa (Dum.) Hubb. (*Lepturus filiformis*)

1 2 . . **5** . **7** . . .

Drier parts of salt-marshes, and other open habitats just above high-water mark. Rather rare, but perhaps overlooked.

1. Salt-marsh near O'Brien's Br. **2**. Rine Point. **5**. Salt-marsh W. of Keeraunnagark. **7**. On the mainland shore opposite Omey I., where the sand and the rocks meet.

First record: ICP (1966). Near O'Brien's Br.

It is remarkable that, although this species had never before been recorded from the W. coast, it was noted independently by Perring in 1968 and Wadsworth in 1969, neither of them aware of the recent record by Ivimey-Cook & Proctor (1966). It is, however, a very inconspicuous plant, and there is no real reason to suppose that it is a recent immigrant.

It seems probable that the record for Aran by Mackay (1806) of *Rotboellia incurva* refers to this species, but confirmation is required. *Parapholis incurva* (L.) Hubb. has recently been discovered on the E. coast of Ireland, but is unlikely to occur in the west.

Catabrosa

Catabrosa aquatica (L.) Beauv.

1 2 . 4 (5) 6 7 .   . (B)

Wet places; apparently rare.

1. By a lane W. of Ballynalackan Ho. **2.** Near the top of the Corkscrew Hill. **4.** Luxuriant and locally frequent in drains by the R. Corrib *c.* 3 km N. of Galway. (**5**). *Salthill* (Praeger, 1903*a*). **6.** Mweenish I. **7.** Near Aillebrack L. Near Dog's Bay. *Doonloughan*, 1906 (**DBN**). *Inishbofin* (Praeger, 1911). By the Culfin R., W. of Salruck.

First record: Praeger (1903*a*). Near Salthill.

Easily overlooked unless in full flower, and perhaps still to be found in some of its old stations. On the other hand it favours habitats which are easily altered by small changes in drainage, and it is apt to be heavily grazed. The *Atlas* record for 93/06 (Doolin area) requires confirmation; that for 93/17 (Kilfenora region) is probably an error.

Briza

Briza media L.   Quaking-grass

1 2 3 4 5 . 7 8   A .

Dry (or rarely damp) grassland. Abundant on the limestone; rather rare elsewhere.

1. Locally frequent. **2–4.** Abundant, except on Aran, where it is very local and perhaps not native. **5.** Old railway-track at Rosscahill and Oughterard. Heathy pasture 3 km S.S.W. of Oughterard. Frequent on Gentian Hill. **7.** Sandy grassland on the Slyne Head peninsula. Near the head of Ardbear Bay. **8.** Roadside N. of Maam Cross. Meadow near the Hill of Doon.

First record: ITB (1901). H16.

Dactylis

Dactylis glomerata L.   Cock's-foot

1 2 3 4 5 6 7 8   A B

Meadows, pastures, roadsides and wood-margins. Abundant throughout, but especially on the limestone.

First record: Hart (1875). Aran Islands.

Cynosurus

Cynosurus cristatus L.

1 2 3 4 5 6 7 8   A B

Pastures, meadows and roadsides. Abundant throughout, with no clear soil preference.

First record: Hart (1875). Aran Islands.

Lolium

Lolium perenne L.   Perennial rye-grass

1 2 3 4 5 6 7 8   A B

Meadows, pastures, roadsides and cultivated ground. Abundant throughout.

First record: Hart (1875). Aran Islands.

[*L. multiflorum Lam.]   Italian rye-grass

Apart from its occurrence as a sown crop, this species can often be seen self-sown in gateways, on field-margins, etc. It has been observed in all districts except 1 and 5, but it does not appear to persist for more than a year or two. On Aran hybrids with *L. perenne* have been noted.

× Festulolium

× **Festulolium loliaceum** (Huds.) Fourn. (*Festuca pratensis ×Lolium perenne*)

1 . 3 . . . .

Meadows; rare.

1. Below the house at Ballynalackan Castle, 1974 (D. Kelly). **3.** Marshy meadow on the W. side of Dromore L., 1965 (J. Ironside-Wood). By the R. Fergus *c.* 2 km above Ennis, 1980 (MS).

First record: Crit. Suppl. (1968).

Festuca

Festuca ovina L.   Sheep's fescue

1 2 3 4 5 6 7 8   A B

Dry grassland and limestone pavement; frequent.

1. Occasional to frequent. **2–4.** Very frequent to abundant. **5.** Occasional. **6–7.** Frequent. **8.** Occasional.

First record: Hart (1875). Aran Islands.

F. indigesta Boiss., a critical species with a Mediterranean–Iberian distribution, closely related to *F. ovina*, but with stiffer, sharply pointed leaves and a longer awn to the lemma, was recorded from 2 stations in the Burren in 1949 (Markgraf-Dannenberg, 1952). The topographical data as published are confused, but it seems probable that it was collected among hazel-scrub inland from Poulsallagh, and on limestone pavement near the turloughs S.W. of Mullaghmore. It has not been seen since.

F. rubra L.   Creeping red fescue

1 2 3 4 5 6 7 8   A B

Rocks, walls, dry banks, sandy and gravelly places and sea-cliffs; occasionally in damper habitats. Abundant throughout.

First record: Hart (1875). Aran Islands.

Very glaucous variants, perhaps referable to subsp. *pruinosa* (Hackel) Piper, have been noted in a few places by the sea in Connemara and are very conspicuous around small brackish pools on Inishmore.

F. vivipara (L.) Sm.

. 2 . . 5 6 7 8   . B

Mountain grassland, rocky places and streamsides. Common in the mountains of Connemara; rare elsewhere.

2. Rough grassland W. of Turlough village. **5.** Streamside near L. Formoyle. **6.** Rocks by the S. shore of Loughanillaun. **7.** Frequent from Errisbeg and Letterfrack eastwards; not seen in the extreme west except on Inishbofin. **8.** Abundant.

First record: Wade (1802). 'Letter mountain.'

Predominantly a mountain plant, but it descends to near sea-level in several places near the W. coast. Its low-level occurrence on limestone in district 2 is, however, most unusual.

F. altissima All. (*F. sylvatica*)

. . . . . . 7 8   . .

Woods; rare.

7. Occasional in oak-wood at the W. end of Ballynahinch L. **8.** Occasional in the wood on the Hill of Doon.

First record: Webb (1982).

Very rare in W. Ireland.

†**F. pratensis** Huds.

. 2 3 4 . . (7) 8   (A) (B)

Meadows and lake-shores; rather rare.

2. Beside L. Luick. *Aran* (Hart, 1875). **3.** Frequent in the extreme south. *Garryland* (P. cat.). **4.** Coolagh. E. of Moycullen. **(7).** *Near Letterfrack* (Atlas). *Inishbofin* (More, 1876). **8.** By the shore of L. Mask, N. of Ferry Br. *N.W. of Oughterard* (P. cat.).

First record: Hart (1875). Aran Islands.

Doubtfully native, except perhaps in district 3, and seldom persisting for long after sowing.

F. arundinacea Schreb. (*F. elatior*)

1 2 3 4 5 (6) (7) 8   A (B)

Damp grassland. Frequent on the limestone; rare elsewhere.

1. Frequent in a marsh by the E. side of Inchiquin L. In a marsh by L. Raha. **2–4.** Frequent. **5.** Roadsides near Gentian Hill and S. of Oughterard. **(6).** *Lettermullan* (Atlas). **(7).** *Near Letterfrack* (Atlas). *Inishbofin* (More, 1876). **8.** Shore of L. Corrib near Ashford Castle, and on several of its islands N.E. of Oughterard. Wood on S. shore of L. Mask. Between Maamtrasna and the S.W. corner of L. Mask.

First record: More (1876). Inishbofin.

F. gigantea (L.) Vill. (*Bromus giganteus*)

1 2 3 (*4*) 5 . . 8 . .

Woods and other shady places; widespread but rare.

1. Wood at Moy Ho. Shady roadside near Elmvale Ho. **2**. In scrub on Glasgeivnagh Hill. **3**. Wood-margin at Dromore. Edge of Garryland wood. (*4*). *Near Ross L.* (P. cat.). **5**. Moycullen wood. **8**. Wood on S. shore of L. Mask. Wood-margin immediately E. of Clonbur.

First record: ITB (1901). H16.

Vulpia

Vulpia bromoides (L.) Gray

1 (*2*) 3 4 5 6 7 8 (*A*) B

Roadsides, walls, rocky ground, disused quarries and heaths. Very frequent and locally abundant in Connemara; very rare elsewhere.

1. Near Roadford. *Cliffs of Moher* (Atlas). *Slieve Elva* (Atlas). (*2*). *Inishmore* (Nowers & Wells, 1892); probably a casual introduced with turf. **3**. On the railway at Crusheen. **4**. On a dyke of igneous rock near Renmore barracks. **5–8**. Very frequent.

First record: More (1876). Inishbofin.

Rare on limestone throughout Ireland, and especially in the west. The Crusheen plants were on non-calcareous ballast.

(*V. myuros* (L.) Gmel.)

(*1*) (*2*) (*3*) (*4*) . . . . . .

Formerly occasional on the limestone; not seen in the district since 1933.

(*1*). *O'Brien's Br.* (FWI). (*2*). *Ballyvaughan* (More, 1872). *Fanore*, 1892 (**DBN**). (*3*). *Castle L.*, 1905 (**DBN**). (*4*). *Oughterard*, 1933 (**DBN**).

First record: Allin (1871). Burren.

A mainly S.E. plant in Ireland. The *Atlas* record for 94/02 (Maam area) is probably based on a misidentification.

Desmazeria

Desmazeria rigida (L.) Tutin (*Catapodium rigidum*)

1 2 3 4 (*5*) . . 8 A .

Walls, limestone pavement, sand-dunes and roadsides. Frequent on the limestone; unknown elsewhere.

1. Roadside by the dunes N. of Lehinch. **2–4**. Frequent. (*5*). *Gentian Hill* (P. cat.). **8**. Limestone pavement N.W. of Cong.

First record: Hart (1875). Inishmore.

A few specimens from districts 3 and 4 are probably referable to var. *major* (Presl) Lousley.

D. marina (L.) Druce (*Catapodium marinum, Festuca rotboellioides*)

1 2 3 4 5 6 7 . A B

Rocky and sandy habitats by the sea. Frequent in W. Connemara; local elsewhere.

1. Dunes N. of Lehinch. **2**. Abundant on the dunes at Fanore and E. of Ballyvaughan. Occasional elsewhere on limestone pavement near the sea. On the road at Poulsallagh. Frequent on Aran. **3**. Gravel beach at Aughinish, and waste ground by the sea 5 km to the E. **4**. By the railway E. of Galway. **5**. On the shore between Galway and Salthill. Dunes 5 km W. of Spiddal. **6**. Dunes at Moyrus. On the quay on the mainland opposite the N.W. corner of Mweenish I. **7**. Very frequent, especially on the Slyne Head peninsula.

First record: Graham (1840), as *Triticum loliaceum*. Abundant on the shore near Galway.

Poa

‡**Poa annua** L.

1 2 3 4 5 6 7 8 A B

Roadsides and other open habitats. Abundant throughout, though rather less so in blanket-bog areas.

First record: Hart (1875). Aran Islands.

*****P. nemoralis** L.

. . . . . . (*7*) **8** . .

Estate woods; very rare.

(*7*). *Ballynahinch woods* (Praeger, 1906). **8**. Grounds of Ashford Castle, 1974 (DW).

First record: Praeger (1906).

*****P. compressa** was collected at Recess railway station in 1934 (Praeger, 1939), but must be regarded as a casual which has not survived the closing-down of the railway. It is only in a few places in the S.E. of Ireland that it is naturalized.

P. pratensis L.

1 2 3 4 5 6 7 8 A B

Walls, roadsides, sand-dunes and other dry places. Abundant on the limestone and in most of the coastal areas; much scarcer in blanket-bog areas.

First record: Hart (1875).

The above refers to the species in its wide sense. Our records do not allow for a systematic account of the distribution of *Poa subcaerulea* Sm. and *P. pratensis, sensu stricto*, especially as some of the specimens examined show poor correlation of the differential characters given in *Flora Europaea*. There is no doubt, however, that all plants from sand-dunes, and most of those from other habitats near the sea, may be referred to *P. subcaerulea*. In the inland areas of the Burren, both segregates are present, and it may be that *P. pratensis* is the commoner of the two.

P. trivialis L.

1 2 3 4 5 6 7 8 A B

Grassland, ditches, damp roadsides and marshes; occasionally in open woods or as a weed of open ground.

Very frequent to abundant, except in district 2, where it is rather local on account of the scarcity of suitable habitats.

First record: More (1876). Inishbofin, 'possibly introduced'.

Puccinellia

Puccinellia maritima (Huds.) Parl.

1 2 3 4 5 6 7 . A B

Salt-marshes and maritime rocks. Widespread, and abundant in places, especially S. Connemara and the S.E. corner of Galway Bay; more local elsewhere, and apparently rare in N. Connemara.

First record: Hart (1875). Aran Islands.

P. distans (L.) Parl.

. . . 4 [5] . . . . . .

Disturbed ground by the sea; very rare.

4. By a track close to the shore, *c.* 2½ km E. of Galway, 1965 (DW, **TCD**).

First record: Webb (1982).

A specimen in **DBN** collected by Barrington at Salthill forms the basis for the *Atlas* record for the Galway area, and hence for the ascription of the species to H16 in the *Census catalogue*. But as the site is given as a drain on the railway, and there never was a railway at Salthill, Co. Galway, it is clear that the plant was collected at Salthill, Co. Dublin. There is, therefore, no record for H16.

Glyceria

Glyceria fluitans (L.) R. Br.

1 2 3 4 5 6 7 8 A B

Ditches, marshes, river-banks and lake-shores; frequent. Very evenly distributed through the region, being rare only in district 2, where there is little fresh water.

First record: Hart (1875). Aran Islands.

G. plicata Fr.

. 2 (3) . 5 . 7 8 . B

Marshes; rare.

2. S. of Ballyvaughan. (*3*). *Dromore* (Atlas). **5.** Spiddal. **7.** Ditch W. of Renvyle Hotel. Inishbofin. *N. of Roundstone* (Balfour, 1875). **8.** Millpond at Cong.

First record: Balfour (1875). N. of Roundstone.

Possibly overlooked elsewhere, but obviously a rare plant in western Ireland. The *Atlas* record for the Clifden area requires confirmation.

The hybrid *G.* × *pedicellata* Towns. (*G. fluitans* × *plicata*) has been recorded by the

R. Fergus, between Killinaboy and Inchiquin L. (ICP) and in abundance by the turlough at Turlough village (DW, **TCD**).

G. declinata Bréb.

1 2 . 4 5 6 7 8   A .

Ditches and pools. Occasional in the west; rather rare in the east.

1. By L. Raha, and near O'Dea's Castle. Near the Cliffs of Moher. **2**. Poulsallagh. Aran. **4**. By L. Corrib, E. of Moycullen. **5**. E.S.E. of Costelloe. **6**. N.E. of Carraroe. **7**. Occasional, mainly near the coast. **8**. On the football-ground, Leenane.

First record: Atlas (1962).

Bromus

Bromus ramosus Huds.

1 2 3 4 . 6 7 8   . .

Woods and scrub; occasional.

1. Wood on Clifden Hill. **2**. Sparingly in hazel-scrub near the Caher R. Scrub N. of Killinaboy, and on Gleninagh Mt. **3**. Roadside 1½ km S. of L. Bunny. **4**. Scrub and hedges near L. Corrib, E. of Oughterard, and also N.E. of Moycullen. **6**. Woods at Cashel. **7**. Occasional in Ballynahinch wood, and in the wooded ravine W. of Letterfrack. **8**. At the Pigeon-hole, Cong.

First record: ITB (1901). H16.

*B. sterilis L.

. 2 (*3*) . (*5*) . . 8   A .

Rocky or stony places; rare.

2. Widespread on Aran, but not observed there before 1966. *Near Ballyvaughan*, 1895 (ITB). (*3*). *N.E. of Killinaboy* (Corry, 1880). (*5*). *Salthill* (Praeger, 1903a). **8**. Near Cong, 1973 (Curtis).

First record: Corry (1880). 'Between Killinaboy and Glanquin '

The fact that this plant has disappeared from all its old stations and is clearly a recent introduction to Aran enables us to decide with some confidence that it is not native to the region.

B. hordeaceus L. (*B. mollis*)

1 2 3 4 5 6 7 8   A B

Meadows, pastures, roadsides and cultivated ground.

Frequent throughout, except in the inland parts of districts 1 and 2, where it is only occasional.

First record: Hart (1875). Aran Islands.

Mostly subsp. *hordeaceus*. One plant from Inishbofin (**TCD**) has the short lemmas and short, slender awns (7–7½ mm) of subsp. *thominii* (Hard.) Maire & Weiller, but has 6–9 florets in most of the spikelets.

*B. lepidus Holomb.

. 2 3 4 . . 7 8   . B

Meadows and roadsides; rather rare.

2. Burren village. *Near Black Head* (Atlas). **3**. 3 km E. of Aughinish. *S.W. of Mullaghmore* (Atlas). **4**. Near Ross L. **7**. Inishbofin. **8**. S. of Clonbur (**TCD**), and on the Dooros peninsula.

First record: Halliday *et al.* (1967). S. of Clonbur.

*B. racemosus L.

. **2** . . . . **7** . **A** .

Meadows and roadsides; very rare.

2. Roadside on Inisheer, 1976 (DW, **TCD**). **7**. Meadow near the sea N.N.W. of Moyard, 1968 (Perring, det. Melderis).

First record: Webb (1982).

Brachypodium

Brachypodium sylvaticum (Huds.) Beauv.

1 2 3 4 5 6 7 8 A B

Woods and scrub, among rocks, and in crevices in limestone pavement. Abundant on the limestone; local elsewhere.

1. Woods by Moy Ho. and Elmvale Ho. **2–4**. Abundant. **5**. Frequent in Moycullen wood. Hedge near Gentian Hill. **6**. Woods at Glendollagh. **7**. Woods at Kylemore, Ballynahinch and Letterfrack. Island in Derryclare L. **8**. Frequent in woods by the shores of L. Corrib and L. Mask, and on many of the islands in L. Corrib; very rare elsewhere.

First record: Hart (1875). Aran Islands.

Elymus

Elymus repens (L.) Gould (*Agropyron repens*)   Scutch-grass

1 2 3 4 5 6 7 (*8*) **A B**

Sand-dunes, shingle-beaches, roadsides and rough grassland; occasionally as a weed in cultivated ground. Frequent, but local.

1. Near Moy Ho. Dunes N. of Lehinch. Banks of the R. Fergus W. of Killinaboy. **2**. Occasional in the north. Aran (rather rare). **3**. Frequent from Kinvara to Aughinish. Corofin. Dromore. **4**. Frequent. **5**. Frequent by the coast near Spiddal. **6**. Frequent in S.E. part. **7**. Occasional near the coast, and locally abundant near Ballynakill harbour. (*8*). L. *Nafooey region* (Atlas).

First record: Hart (1875). Inishmore.

Very variable. Mainly coastal, and far less common as a weed than in the east.

(*E. caninus* (L.) L.) (*Agropyron caninum*)

. [2] (*3*) . . . . . [A] .

Woods; very rare.

(*3*). *Garryland wood*, 1854 (**DBN**).

First record: CH1 (1866). 'Garryland wood, near Gort; A.G.M.'

We have rejected the record of Nowers & Wells (1892) for Aran, as a less probable habitat for the species would be hard to imagine. The species is very rare in the west, and is known only from woods.

We have received several reports of *E. pycnanthus* (Godr.) Meld. (formerly known as *Agropyron pungens* by most authors), but all specimens so named which we have seen have turned out to be *E. repens*.

E. farctus (Viv.) Melderis (*Agropyron junceiforme, A. junceum*)

1 2 3 . 5 6 7 . A .

Sand-dunes and sandy or gravelly shores. Frequent, and locally abundant.

1. Abundant at Lehinch and Liscannor. **2**. Fanore. E. of Ballyvaughan. Aran. **3**. Finavarra. **5**. Between Galway and Salthill. W. of Spiddal. **6**. Dunes on Furnace I. **7**. On most of the dunes, from Dog's Bay to N.E. of Gowlaun.

First record: Hart (1875). Inishmore.

As elsewhere in Ireland, all specimens are referable to subsp. *boreali-atlanticus* (Sim. & Guin.) Meld.

GYMNOSPERMS

CUPRESSACEAE

Juniperus

Juniperus communis L.   Juniper
. **2 3 4 5 6 7 8   A B**

Heaths, rocky ground and limestone pavement. Occasional to frequent, but irregularly distributed.

2. Frequent around Black Head, and for several kilometres to the south and south-east; mainly on the higher ground, but abundant also near the lower course of the Caher R. Aran (rather rare). **3.** Frequent and locally abundant, mainly in the north and west, but also near the 'racecourse' turlough, W. of Gort; not seen in the south. **4.** Frequent. **5.** Sparingly on heathy grassland S.S.W. of Rosscahill. **6.** Gorumna I. (E. of L. Hibbert). Sparingly near Bunnahown. **7.** Widely distributed, but mostly in small quantity, as on Errisbeg and the bogland north of it, on the Twelve Pins, the hills by L. Fee, and Inishbofin. Locally abundant on rocky heaths N. of Letterfrack. **8.** On the hill 4 km N.N.E. of Maam Cross. On the Maamtrasna plateau, S. of L. Nadirkmore.

First record: Wade (1892). 'On the sides of the Cunnamara mountains frequent. Buffin Island.'

It is probably true to say that all plants in districts 2–4 belong to subsp. *communis*, and all from districts 5–8 belong to subsp. *nana* Syme. Diagnosis can be difficult, as in our region both subspecies are invariably prostrate, and in very exposed situations, as on Aran, the leaves of subsp. *communis* tend to be short and somewhat incurved, as in subsp. *nana*.

PINACEAE

Pinus

*Pinus sylvestris L.   Scots pine   *Giuis*

The scots pine, though not widely planted by the Forestry Department (for it prefers a more Continental climate than our region provides), has been planted here and there throughout the region (except in district 2) in estate woods, along field-boundaries, in small copses, and even on lake-islands. Regeneration and some degree of naturalization takes place in some of these stations, but although one can point to many trees that are obviously planted, and to many saplings that are obviously self-sown, there remains an even larger number whose status is not clear. The most effective naturalization is in districts 3 and 8 (including saplings on several islands in L. Corrib which bear no planted trees). A few saplings have also been seen among native vegetation in district 5.

The pine was a major element in the vegetation for a considerable part of the post-glacial epoch, as is testified not only by the pollen-record, but also by the abundance of pine-stumps in the blanket-peat. The most dramatic display of these

Fig. 19. Yew-trees in the Burren. (*a*) On level limestone pavement, kept down by goat-grazing, despite its toxicity. (*b*) On a cliff on the south-facing side of the Glen of Clab. Here the terrain protects it from grazing, but the wind keeps it down to a height of about 2.5 m. (A holly is growing behind the yew and slightly overtopping it.)

is to be seen by L. Mask, S. of Killateeaun, where the bog in which the stumps were embedded was largely washed away by a temporary rise in the level of the lake.

The pollen-record indicates that the pine became extinct, or at least very rare, soon after the beginning of the Christian era. The earliest plantings in our region probably date from the late seventeenth century. Wade (1802) refers to it as 'scattered in a few places in Cunnamara, of a diminutive stunted growth although apparently very old' – a description more suitable for the yew. The trees one sees today, though none is very large, cannot be called 'stunted'.

TAXACEAE

Taxus

Taxus baccata L.   Yew   *Iúr*

. **2 3 4 5 6 7 8** . .

Cliffs and rocky ground on limestone; in Connemara in scrub and old-established woodland, mainly on lake-islands. Occasional.

2. Cliffs on Mullaghmore, and in the Glen of Clab. Steep slopes on the S.W. side of Gleninagh Mt. Level pavement on the S.E. flank of Slieve Carran. **3**. Occasional as isolated bushes here and there on bare stretches of limestone pavement, mainly in the north and west. A few trees in Garryland wood. **4**. One tree near Aughnanure Castle. Several bushes to the S. and W. of Carrowmoreknock. *By L. Corrib, S. of Burnthouse* (P. cat.). On an island nearby, 1974 (Roden). **5**. On limestone rocks overhanging the river in Oughterard, and in scrub 2 km to the S. **6**. Islands in Athry L. **7**. Frequent on a large number of the lake-islands which are large enough to support woodland or well-developed scrub. **8**. Islands in L. Shindilla, and on several in L. Corrib N.E. of Oughterard. In the wood on the Hill of Doon. One very old tree in the grounds of Gortdrishagh.

First record: Foot (1864). 'Generally on steep cliffs....'

Historical evidence, as well as that from place-names and palaeontological data, makes it clear that the yew must have been much commoner formerly, both in the Burren and in Connemara. Today, although numerous stations can be cited, nearly all of them yield only 1 or 2 trees, and we have not observed regeneration or even fruiting. Even allowing for selective felling it is not clear why the species has declined so sharply.

In the Burren most specimens (as of other trees and shrubs) are severely grazed, in spite of the well-known toxicity of their foliage, unless they are on cliffs steep enough to protect them even from goats. One is forced to conclude that the local goats have learnt just how much they can eat with impunity.

PTERIDOPHYTES

FILICES   Ferns

Most modern classifications of ferns assign the 18 genera represented in our region to 13 different families, of which all but 4 contain only 1 genus. We have not thought it necessary in a book of this kind to burden the reader with these family names.

Numerous teratological variants of some of the ferns, especially *Phyllitis scolopendrium*, were recorded from the Burren during the fern-craze of the late nineteenth century, but as the names then given to them are now mostly forgotten we do not think it necessary to record them.

Ophioglossum

Ophioglossum vulgatum L.   Adder's tongue

1 2 3 . . 6 7 8   A .

Grassy and rocky places; occasional, and locally abundant in turloughs.

1. Near Lisdoonvarna and Kilfenora; also 7 km E.N.E. of Lisdoonvarna (Willmot, 1979). 2–3. Occasional to frequent, and locally abundant in some of the turloughs, as at Tirneevin and Turlough village. Aran (rare). 6. Among rocks by the outlet of Inver L. Behind the 'coral strand' near Carraroe. Grassland by the sea near Canower, and also near Glinsk. 7. In several places on the Slyne Head peninsula. Under a boulder by L. Bollard. *Near Dog's Bay*, 1906 (**DBN**). 8. Meadow 1½ km W. of the 'Narrow Lake'. Inchagoill, in L. Corrib.

First record: Newman (1854). 'Mr Thompson, in company with Mr Ball, found the adder's tongue in the South Isles of Arran, off Galway.'

Botrychium

Botrychium lunaria L.   Moonwort

1 2 . 4 5 6 7 .   (*A*) .

Heathy grassland; rare, and apparently often transient.

1. In some quantity on grassy roadside banks near Knocknalarabana, 1968. *Ballynalackan* (Foot, 1864). 2. By the old lead-mines S.W. of Carran, 1979. Slopes of Gleninagh Mt. *Portcowrugh* (*Inishmore*) (Nowers & Wells, 1892). *Ballyvaughan* (Foot, 1864). 4. On a mound of glacial drift S. of Carrowmoreknock, 1967. 5. 1 km S. of Rosscahill, 1965. *Inveran*, 1892 (ITB). *W. of Galway*, 1895 (ITB). 6. Cliff-top near Ardmore, 1891. 7. One plant by a track on the S.W. flank of the Twelve Pins, 1968. *Ballyconneely*, 1907 (**DBN**). *Near Letterfrack* (Seebohm, 1851).

First record: Seebohm (1851). 'On the lawn in front of J. Ellis's residence at Letterfrack.'

Considering that this species is certainly perennial and probably fairly long-lived, it is remarkably capricious in its appearances. We do not know of any station in our region in which it has been seen more than once, and this is true also of most of its stations in the rest of Ireland.

Osmunda

Osmunda regalis L.   Royal fern

1 . 3 4 5 6 7 8 . B

Lake-shores, bog-drains, riversides, fens and peaty banks. Abundant in Connemara; occasional elsewhere.

1. Drumcullaun L. Lickeen L. **3.** Rinroe bog. Fen S.W. of Mullaghmore. Fen by the Ballyogan Loughs. N. of Crusheen. **4.** Frequent between L. Corrib and the Moycullen–Oughterard road. Very luxuriant at Aughnanure Castle. **5–8.** Generally abundant, though somewhat local.

First record: Babington (1836). Oughterard.

Hymenophyllum   Filmy fern

Hymenophyllum wilsonii Hook. (*H. peltatum*)

. 2 . . . . 7 8 . B

Rocky ledges, mountain cliffs and ravines; occasionally in woods or by mountain streams. Frequent in Connemara, especially in the north; very rare elsewhere.

2. Sparingly on tree-trunks or fragments of sandstone near the margin of Poulavallan, on the S. side (Webb, 1963*b*; confirmed in 1967 and 1968). **7–8.** Frequent in the northern and central parts of both districts, both on mountains and in suitable habitats on lower ground; not recorded S. of Benlettery or Carnseefin.

First record: Mackay (1836). 'On the Kerry mountains, Cunnamara, Glencree....'

H. tunbrigense Sm.

. . . . . 6 7 8 . .

Shady vertical rocks and tree-trunks. Occasional in Connemara; not seen elsewhere.

6. Woods at Glendollagh Ho. Islands in Athry L. **7.** Woods at Ballynahinch, Kylemore, E. of Clifden, and on the W. side of Derryclare L. On most of the larger, and on a few of the smaller lake-islands. In the ravine on the S.W. side of L. Fee. On Benlettery at *c.* 530 m (1750 ft). **8.** Frequent in the wood on the Hill of Doon. By a stream below the N.E. side of Carnseefin.

First record: Graham (1840). 'On the shores of a little lake westwards of Halfway House [Maam Cross].' Mackay's record (1806) under this name almost certainly refers to the preceding species, which at that time was not generally recognized as distinct.

All stations lie within an area measuring 40 km from E. to W. and 13 km from N. to S. Nearly all are lowland; the Benlettery station is at an altitude very unusual for this species in Ireland.

It can be distinguished with certainty from *H. wilsonii* by the raggedly toothed margin to the valves of the indusium, but specimens without spores can be distinguished with a little experience by their broader, flatter leaves of a distinctly bluish shade of green.

Pilularia

Pilularia globulifera L.

. . . . . **6 7 8** . .

Shallow water in pools or at the margins of lakes or rivers; rare.

6. Marshy margin of Glendollagh L., near its W. end, 1968–76. **7.** Edge of the Ballynahinch R., near where it leaves the lake, 1980. *In the stream draining Poulnacappul L.* (Seebohm, 1851). **8.** In several places in or near the S.W. corner of L. Mask: in the stream S. of Srahnalong, 1966; near the S.W. corner of the 'Narrow Lake', 1965; in pools near the mouth of the Owenbrin R., 1967; *in a swamp by L. Mask N. of Clonbur* (Marshall & Shoolbred, 1896).

First record: Wade (1804). 'In a sandy, boggy situation near the salmon-leap, Ballynahinch.'

Pteridium

Pteridium aquilinum (L.) Kuhn   Bracken   *Raithneach*

1 2 3 4 5 6 7 8   A B

Heaths, rough grazing, abandoned pastures and limestone pavement. Abundant almost throughout.

Absent from wet peatlands and heavy clays, and from the more remote mountain-slopes in central Connemara. It is characteristic especially of undergrazed or abandoned farmland. Although it shows an unmistakable calcifuge tendency in Britain and Ireland as a whole, it is abundant throughout the Burren. Patches are, however, seen here and there suffering from severe chlorosis.

First record: Andrews (1845). Inishmore.

Cryptogramma

Cryptogramma crispa (L.) Hook.   Parsley fern

. . . . . . **7** . . .

7. A single plant on the N. side of the hill W. of Doughruagh (Scannell & White, 1976).

First record: Wade (1802). Ballynahinch. This has usually been written off as an error (and reasonably enough, in view of Wade's other records for the species), but the recent find suggests that it may perhaps have been correct.

Very rare in Ireland, and in some of its stations little more than casual. The nearest known stations are in Cavan and Donegal.

Adiantum

Adiantum capillus-veneris L.   Maidenhair fern   *Dubh chosach*

. **2 3 4** . . **7** .   **A** .

Crevices in limestone rocks. Locally frequent in the Burren and on Aran; very rare elsewhere (Plate 4).

2. Recorded from most parts of the districts, and fairly frequent in a few localities, as Black Head, Poulsallagh and S. of Carran. Frequent throughout Aran. **3.** Frequent on bare pavement E. and E.N.E. of Killinaboy. Pavement N.N.E. of Slieve Carran.

4. Abundant in a seepage-line in a limestone quarry 5 km E. of Galway. **7**. On a boulder near the N.E. corner of L. Bollard.

First record: Lhwyd (1712). 'In the Isle of Aran near Galloway we found a great plenty.'

The typical habitat is in deep crevices of horizontal limestone pavement, where the tips of the fronds barely reach the surface. Here and there, however, it may be seen in horizontal seepage-lines or other crevices in vertical rock-faces, more or less exposed to the air. Its occurrence on an isolated boulder by L. Bollard is unusual; it was first found here in 1836, and was confirmed in 1967; unfortunately reports as to the nature of the rock vary. It is often described as limestone, but is more probably an ultrabasic lava.

Blechnum

Blechnum spicant (L.) Roth
1 2 3 4 5 6 7 8 (*A*) **B**

Bogs, hedge-banks, woods, heathy ground, streamsides and lake-shores. Abundant almost everywhere on acid soils; very local on the limestone, and only where there has been some development of peat.

1. Very frequent. **2**. Occasional in the wood at Poulavallan. A few plants on the S.W. slope of Gleninagh Mt. *Formerly on Aran*. **3**. Occasional on Rinroe bog. **4**. In small quantity on the bogs between L. Corrib and the Moycullen–Oughterard road. **5–8**. Abundant.

First record: Ogilby (1845). Inishmore: 'not scarce'.

Ogilby's record for Aran was confirmed by Hart (1875), but it has not been noted since then. It is best explained as a temporary importation with peat from Connemara.

The variant in which there is no distinction between fertile and vegetative fronds was seen in Derreen wood in 1977 (MS).

Phyllitis

Phyllitis scolopendrium (L.) Newm.   Hart's-tongue fern
1 2 3 4 5 6 7 8   **A B**

Hedges, walls, woods, scrub and limestone pavement. Frequent and locally abundant.

1. Frequent. **2–4**. Abundant. **5**. Frequent. **6**. In small quantity under trees near the entrance to Glendollagh Ho. **7**. Frequent and locally abundant. **8**. Very frequent.

First record: Andrews (1845). Inishmore.

Asplenium

Asplenium marinum L.   Sea spleenwort
1 2 . . (*5*) **6 7** .   **A B**

Maritime rocks on exposed coasts. Frequent on Aran and the western part of the Burren; rather rare elsewhere.

1. S. of Hag's Head and near Liscannor. **2**. Frequent from Fisherstreet to Black Head, and on Aran. *Ballyvaughan* (ITB). (*5*). *Near Inveran* (Atlas). **6**. Near the 'coral strand'

Fig. 20. Limestone pavement near the sea at Black Head. On account of the salt spray there are few plants in the crevices, but deep down in many of them may be found plants of *Asplenium marinum*.

at Carraroe. *St Macdara's I.* (ITB). **7**. Towards the W. end of the Slyne Head peninsula. Near Ardmore. Inishbofin. *High I.* (Seebohm, 1851).

First record: Wade (1802). 'Abundantly on Buffin Island.'

On Inishmore this plant grows, accompanied by other halophile species, on limestone pavement at the top of a 30 m (100 ft) cliff above Blind Sound. It is also found on walls at some distance from the sea on Inishshark.

A. trichomanes L.   Maidenhair spleenwort

1 2 3 4 5 . 7 8   A .

Limestone rocks and mortared walls; occasionally on the ground in scrub or in hedges. Common in suitable habitats, but absent from considerable areas of S. Connemara.

1. Walls near Lehinch, at Ennistymon, and S.S.W. of Corofin. Rocks W. of Killinaboy and N.E. of Kilnamona. **2–4**. Abundant. **5**. Walls at Spiddal and Gentian Hill, and S.W. of Rosscahill. In a hedge on slightly calcareous drift 4 km W. of Oughterard. **7**. Very frequent, but not on Inishbofin. **8**. Abundant.

First record: Andrews (1845). Inishmore.

Decidedly calcicole, but not as rigorously as *A. ruta-muraria*. All plants examined appear to be referable to subsp. *quadrivalens* Mayer, even though some of them were growing on non-calcareous substrata.

A. viride L.

. . . . . . 7 8   . .

Mountain cliffs and screes; rare.

7. Rather sparingly on the N. face of Muckanaght and the N.W. face of Bengower.
8. Frequent on ledges and screes on the W. side of Lissoughter. '*Glenlosh*' (*probably an error for Glenlusk, on the E. side of the Maumturk Mts.*) (CH2).

First record: CH1 (1866). Benlettery. (Bengower is probably intended.)

A. adiantum-nigrum L.

1 2 3 4 5 . 7 8   A B

Rocks, limestone pavement, walls and banks; widely distributed, but rather local.

1. Wood near Roadford. W. of Kilnamona. W. of Killinaboy. N. of the Cliffs of Moher. **2**. Occasional, mainly in the north and east. Aran (rather rare). **3**. Occasional. **4**. Menlough Hill. N.W. of Moycullen. **5**. Near the harbour at Spiddal. *Inveran* (Atlas). *By L. Bofin* (P. cat.). **7**. Frequent in the extreme west, from Ballyconneely to Cleggan and Inishbofin. On an island in Derryclare L. **8**. On dolerite dykes at Knocknahillion, and at the Maumeen gap. Carrowgarriff. Carnseefin. Ferry Br. Island in L. Shindilla.

First record: Wade (1804). 'I met with it, but sparingly, in Cunnamara.'

Found on various types of rock, but it seems to require a minimum of base, and is not, therefore, to be found on quartzite or poor sandstones.

Plants differing slightly from typical *A. adiantum-nigrum* in the shape of their pinnules have been recorded from the Connemara serpentines as *A. cuneifolium* Viv. (Scannell, 1978), but it appears that the difference in chromosome number which was held to justify specific separation is not constant (Fern Atlas, p. 54), and recent work by Sleep (1980) suggests that all records of *A. cuneifolium* from Britain and Ireland are based on a misunderstanding of its true nature.

The record of *A. billotii* Schultz from Black Head published in the *Fern Atlas* is erroneous. This species is known in Ireland only from the south-east.

A. ruta-muraria L.

1 2 3 4 5 6 7 8   A B

Limestone rocks (rarely on other basic rocks); also on limestone or mortared walls. Common in suitable habitats.

1. Occasional on walls near the coast, and on rocks on the N.E. margin; also on walls N.E. of Kilnamona. **2–4**. Abundant. **5**. Walls at Spiddal. Rocks at Gentian Hill. **6**. Wall at S.E. corner of Gorumna I. Wall S.E. of Screeb. **7**. Occasional throughout. **8**. Very frequent.

First record: Wade (1802). 'Renvi castle and other places, Cunnamara.'

A. septentrionale (L.) Hoffm.

. . . . . . 7 . . .

Exposed igneous rocks; very rare.

7. On outcrops of ultrabasic volcanic rock at *c*. 110 m (350 ft) on the ridge E. of L. Nalawney.

First record: Bannister & McAllister (1966).

Discovered in 1965. In 1966 DW counted 35 plants, mostly very stunted, and with fronds never more than 3 cm long, distributed over 3 separate outcrops within a radius of 300 m, and mostly growing on nearly vertical south-facing rocks. The station is about half-way between the N.E. shoulder of Errisbeg and the spur (417 ft) to the north-west.

Unknown elsewhere in Ireland, except for one record from Co. Down, where the plant was undoubtedly of garden origin.

Fig. 21. *Ceterach officinarum* (rustyback), usually seen on mortared walls, here growing in its natural habitat, the limestone rock of the Burren.

Ceterach

Ceterach officinarum DC.   Rusty-back

1 2 3 4 5 . 7 8   A (*B*)

Limestone rocks and mortared walls; rarely on unmortared walls of siliceous rock. Abundant in the Burren; frequent by L. Corrib and L. Mask; occasional elsewhere.

1. Occasional on walls, mainly around Ennistymon and Lehinch. On rocks N.E. of Kilnamona. **2–3.** Abundant, though surprisingly scarce on Inishmore. **4.** Occasional. **5.** Wall near Killaguile Ho. **7.** Frequent on walls around Clifden. Wall near the W. end of Ballynahinch L. **8.** Very frequent in the east and centre; also at Leenane and at Maam Br.

First record: Wade (1802). 'Many places in Cunnamara.'

Athyrium

Athyrium filix-femina (L.) Roth   Lady-fern

1 2 3 4 5 6 7 8   A B

Woods, hedges, ditches, bog-margins, etc. Rare on calcareous ground; abundant elsewhere.

1. Abundant. **2.** Wood at Poulavallan and in the Glen of Clab. Scrub near Carran and on Cappanawalla. A single plant on pavement on Inishmore. **3.** Wood S.E. of Ruan. By a bog-pool near the Ballyogan Loughs. One plant on pavement S.W. of Mullaghmore. **4.** Near the canal N.E. of Moycullen. **5–8.** Abundant.

First record: Seebohm (1851). 'One of the commonest ferns near Letterfrack.'

Cystopteris

Cystopteris fragilis (L.) Bernh.

1 2 3 . . . 7 8 . .

Limestone pavement, walls and banks. Very frequent in the Burren, especially in the western and central parts; rare elsewhere.

1. Shady bank near Moy Ho. S. of Lisdoonvarna (Willmot, 1979). **2**. Very frequent and locally abundant, but not on Aran. **3**. Walls near Caherglassaun L. and near Rinroe Br. Pavement E. of Funshinmore. **7**. North-facing cliff on Muckanaght, Rock-outcrop on Cregg Mt. On a rock near Glendollagh L. **8**. In hollows in a roadside bank by L. Corrib opposite the Hill of Doon. Frequent at the Maumeen gap.

First record: More (1860). Black Head, and between Ballyvaughan and Kinvara.

Particularly abundant on limestone rocks beside the swallow-holes which mark the junction between the limestone and shale on the border of districts 1 and 2.

Dryopteris

Dryopteris filix-mas (L.) Schott   Male fern

1 2 3 4 5 6 7 8   A B

Woods, hedges, banks, streamsides and limestone pavement. Frequent to abundant in most districts.

Especially abundant in districts 2, 3 and 8 (though rather scarce on Aran); very frequent in 1, 4, 6 and 7. Only in district 5, where it is almost confined to the coastal strip, is it rather scarce.

First record: Seebohm (1851). Abundant near Letterfrack.

D. pseudomas (Wall.) Holub & Pouzar (*D. borreri*, *D. assimilis*)

1 2 3 4 5 6 7 8   A B

Hedges, woods and among rocks. Rather rare on the limestone; widespread and locally frequent elsewhere.

1. Frequent. **2**. Occasional in the east; also on pavement at Formoyle. **3**. Wood at Dromore. Pavement at Funshinmore. E. of Killinaboy, and near the S. end of L. Bunny (Willmot, 1979). **4**. Ardnasillagh. S. of Carrowmoreknock. **5–6**. Frequent. **7–8**. Very frequent, but only one plant seen on Inishbofin.

First record: Pugh (1953). Five stations in the Flora region are given.

Seldom seen in great abundance (Ballynahinch wood providing about the only exception). It is much less tolerant of calcareous soils than is *D. filix-mas*, and also rather less tolerant of exposure to wind.

The hybrid with *D. filix-mas* (*D.* × *tavelii* Rothm.) probably occurs here and there. The *Fern Atlas* records it for the Kylemore area, and Willmot (1979) found a single plant near Ennistymon.

D. dilatata (Hoffm.) Gray (*D. austriaca*)

1 2 3 4 5 6 7 8   (*A*) B

Woods, banks, hedges, streamsides, bog-margins and rocky or heathy ground. Rare on the limestone; abundant elsewhere.

1. Very frequent. **2**. Frequent at Poulavallan and in the Glen of Clab. Occasional in scrub near Carran. **3**. Dromore (Willmot, 1979). **4**. 1 km N.E. of Moycullen, and in meadows by L. Corrib further east. On a limestone island in L. Corrib. **5–8**. Abundant throughout.

First record: Seebohm (1851), as *Lastraea multiflora*. Common near Letterfrack.

D. carthusiana (Vill.) Fuchs (*D. spinulosa*)

. . . . **5** . **7 8** . .

Bogs, streamsides and among rocks. Rather rare, and only in Connemara. **5**. At the base of a wall half-way between Spiddal and Moycullen. Among rocks S.W. of Tonabrocky. **7**. Bog on Inishlackan. Among rocks near L. Conga. By L. Tanny. *Letterfrack* (Seebohm, 1851). **8**. By a lake 1 km N. of Maam Cross. By a stream W. of Glann. On cliffs in the Maumeen gap.

First record: Seebohm (1851). Frequent near Letterfrack.

Our records, which suggest that this fern is rare but widespread (they were made by 4 different observers between 1964 and 1978), are hard to reconcile either with Seebohm's description of it as frequent, or with Praeger's failure to record it anywhere in Connemara.

D. aemula (Ait.) Kunze

1 2 (*3*) . **5 6 7 8** . **B**

Woods and shady banks. Very rare on the limestone; local elsewhere, but abundant in a few places.

1. Woods at Lisdoonvarna and near Roadford. Scrub on Slievebeg. By Doonagore L. **2**. Poulavallan. (*3*). *Dromore wood*, 1905 (**DBN**). **5**. Abundant in the wood N.W. of Moycullen. **6**. Occasional in the woods at Glendollagh, and S. of the main road at Recess. On islands in Athry and Nahasleam Loughs. *Near Carraroe*, 1957 (DW). **7**. Frequent. **8**. Very abundant on the Hill of Doon, and frequent on shady banks and in scrub on the S side of the lake, opposite the hill, and thence more sparingly eastwards to Glann. On several islands in L. Corrib. By a stream at the Maumeen gap, and on shady scree on Lissoughter.

First record: Seebohm (1851). 'Among loose rocks on the banks of a mountain stream [probably N.W. of Kylemore].'

Foot (1864) describes this species as 'pretty abundant among the deep heath on the hills over Gleninagh' and 'very abundant on the coal-measures, both in shady ravines and in exposed boggy places'. This is so much at variance with later observations as to throw some doubt on his identification.

Polystichum

Polystichum lonchitis (L.) Roth   Holly fern

. . . . . . **7 8** . .

Mountain cliffs; very rare.

7. Rather sparingly on the N. face of Muckanaght. **8**. At the base of a slaty cliff near the summit of Benwee, 1981 (Roden).

First record: Hart (1883). Muckanaght.

In Ireland confined to the western mountains, and rare in most of its stations. It has been recorded in Praeger, 1934*d*, from Mweelrea, not far to the north from our region, but we can find no further details relating to this station.

P. setiferum (Forsk.) Woynar

1 2 3 4 . . **7 8 A** .

Hedges, scrub, woods and limestone pavement. Rare in Connemara; occasional and locally frequent elsewhere.

1. Abundant at the base of Clifden Hill, and in hedges near O'Dea's Castle; also in Ennistymon wood. Rare elsewhere. **2**. Frequent in the south and west; occasional on Aran. **3**. Hedges 3 km N.N.W. of Ennis. **4**. Occasional in the north. **7**. Wooded ravine W. of Letterfrack. Moyard. Derreen wood. **8**. Hedge by the Hill of Doon. By the Pigeon-hole at Cong.

First record: Oliver (1851). Inishmore.

It is curious that this fern, which is predominantly western in Britain, should be so much commoner in eastern Ireland than in the west. In much of the south-east it is the dominant hedgerow fern, but it seems to require better shelter than most of Connemara affords.

P. aculeatum (L.) Roth (*P. lobatum*)

1 2 3 4 (*5*) . . **8 A** .

Frequent on limestone pavement; very rare on mountain cliffs.

1. N. of Lisdoonvarna; between Kilfenora and Ennistymon; 5–6 km N. of Kilfenora (all *fide* Willmot, 1979). **2**. Frequent throughout, except on Aran, where it is very rare. **3**. Occasional in the area between Killinaboy and Gort. Ballinderreen. **4**. Rocky scrub N. of Ross L. Pavement at Menlough. Rocks at Aughnanure Castle. (*5*). *W. of Oughterard, S. of the river* (Hurst, 1902). **8**. Cliffs at the Maumeen gap, and S. and W. of the W. end of the 'Narrow Lake'.

First record: O'Mahony (1860). Frequent in the Burren. Seebohm (1851) records it as 'not infrequent near Letterfrack', but in the absence of any subsequent confirmation this record is best set aside.

Although this species is widespread in Ireland, the Burren limestones constitute one of the very few areas where it can be said to be common.

The hybrid *P.* × *bicknellii* (*P. aculeatum* × *setiferum*) has been recorded by Willmot (1979) from a disused quarry N.W. of Killinaboy (district 1).

Thelypteris

Thelypteris limbosperma (All.) Fuchs (*T. oreopteris*)   Mountain fern

(*1*) . . . **5** . **7 8** . .

Streamsides, drains, mountain cliffs and rocky slopes. Frequent, though local, in N.E. Connemara; rare elsewhere, and completely absent from the limestone.

(*1*). *Lisdoonvarna* (ITB). **5**. By a bog-drain 6 km N.W. of Rosscahill. **7**. E. shore of Derryclare L. Frequent in a drain near the S.E. corner of L. Fee. *Near Letterfrack* (Seebohm, 1851). **8**. Frequent and locally abundant from Glann and Maam Cross northwards to the cliffs of the Maamtrasna plateau, but not seen in the W. part of the district.

First record: Seebohm (1851). Very sparingly on the banks of 1 or 2 small streams near Letterfrack.

The record in the *Fern Atlas* for Black Head is an error.

T. palustris Schott (*T. thelypteroides*)   Marsh fern

(*1*) **2 3** (*4*) . . . .

Marshes and margins of turloughs. Rare, and seen recently only in the E. part of the Burren.

(*1*). *W. shore of Ballycullinan L.* (Praeger, 1905). **2**. Common among rocks at the N. end of the turlough at Turlough village. **3**. In the marsh at the S. end of L. Bunny. By. a very small turlough 4 km E.N.E. of Killinaboy. On the W. shore of L. George. *L. Fingall* (P. cat.). *By the R. Fergus above Ennis* (Mackay, 1806). (*4*). *By the lake at Ballindooly* (Phillips, 1924).

First record: Mackay (1806). 'In a marsh by the R. Fergus, a little above the bridge at Ennis.'

T. phegopteris (L.) Slosson (*Phegopteris connectilis, P. polypodioides*)   Beech fern

. . . . . . **7 8** . .

Mountain cliffs. Occasional in N.E. Connemara; very rare elsewhere.

7. N. face of Muckanaght. Corrie on the S.W. side of Benchoona. **8**. On the cliffs south of the W. end of the 'narrow lake'; also on Lugnabrick, further to the west. By the waterfall W. of L. Nafooey. On the N. faces of Benbeg and Bencorragh. Cliffs above L. Nadirkmore. On the E. side of the northerly spur of Benwee. *Rocks over Maamtrasna Bay, descending to 30 m (100 ft)* (Praeger, 1933).

First record: Seebohm (1851). 'A few small plants in a wet chasm of the rocks' on one of the Twelve Pins, most probably Benbrack.

Polypodium   Polypody

The species in this genus were not generally recognized as distinct until about 25 years ago, and for a few years thereafter there were varying opinions on the characters for their discrimination. Most of our early records, therefore, cannot be assigned to species unless they are backed by a herbarium specimen. We have, however, enough records of the segregates to give at least a provisional indication of the pattern of the geographical and ecological distribution within the region.

All three species are found mainly on rocks or walls, and to a smaller extent as epiphytes on trees.

Polypodium vulgare L.

1 2 . **4 5 6 7 8** . **B**

Frequent in most of Connemara; occasional elsewhere.

1. Between Kilfenora and Ennistymon. Scrub at W. end of L. Goller. W. of Kilnamona. Near Elmvale Ho. **2**. Near Glencolumbkille Ho. **4**. W. of Ballycuirke L. **5**. Frequent. **6**. Occasional. **7**. Very frequent and abundant in parts of the west. **8**. Frequent.

First record: Shivas (1962). H16.

P. interjectum Shivas

1 2 3 4 . **6 7 8 A B**

Fairly frequent in W. Connemara; widespread but rather rare elsewhere.

1. Immediately S. of Ballynalackan Ho. Bridge over the R. Aille between Roadford and Lisdoonvarna. **2**. Black Head. E. of Carran. Inishmore (rare). **3**. By L. Atedaun.

S.E. of L. Cleggan. Kinvara. **4**. By Ballycuirke L. **6**. Mweenish I., and to the north of it at the head of Ard Bay. S. of Glendollagh Ho. **7**. Occasional to frequent; epiphytic at Kylemore and Derreen. **8**. Leenane. Woods at Cong, and at Gortdrishagh.

First record: Stirling & Beckett (1966).

P. australe Fée

1 2 3 4 . . 7 8 A .

Frequent in the W. part of the Burren and on Aran; rare to occasional elsewhere.

1. Wood on Clifden Hill. **2**. Frequent and locally abundant. **3**. Dromore. W. side of Caherglassaun L. **4**. Walls by Aughnanure Castle. **7**. Epiphytic in the wood W. of Derryclare L. **8**. Island in L. Corrib N.E. of Oughterard.

First record: Webb (1980*a*). Aran Islands.

EQUISETACEAE

Equisetum   Horsetail   *Gliogán*

Equisetum telmateia Ehrh.

1 2 . 4 . . 7 (*8*) **. B**

Clay banks, damp roadsides and by streams; very local.

1. Very frequent in the north, as at Lisdoonvarna, Kilfenora, Ballynalackan and by the river above Roadford; also 1 km E. of the Cliffs of Moher. **2**. Immediately north-west of Ballynalackan Castle. Frequent by the river at Formoyle. **4**. Abundant in hedges N.E. of Moycullen. **7**. Roadsides 2 km E. of Moyard. *Roundstone* (P. cat.). Inishshark. (*8*). *Below Carnseefin* (P. cat.).

First record: Seebohm (1851). Near Letterfrack.

E. arvense L.

1 2 3 4 5 6 7 8 A B

Roadsides, banks and other dry, grassy or stony places. Occasional to frequent except in blanket-bog areas.

1. Frequent. **2**. Occasional. **3–4**. Frequent. **5**. Spiddal, and 5 km further west. S. of Oughterard. **6**. Near Carraroe. **7**. Occasional in the extreme west, from the Slyne Head peninsula to Cleggan. Inishbofin. **8**. Frequent.

First record: Seebohm (1851). Abundant near Letterfrack.

E. sylvaticum L.

1 . . . . . 7 8 . .

Streamsides and damp, grassy places; rather rare.

1. Occasional. **7**. Frequent by a stream near the school at Gowlaun. Wood W. of Ballynahinch L. **8**. By a moorland flush on the hill W.S.W. of Leenane. *Clonbur* (Marshall & Shoolbred, 1896). *N.W. of Oughterard*, 1899 (**DBN**).

First record: Seebohm (1851). 'Sparingly on the banks of small streams' in the Letterfrack area.

E. palustre L.

1 2 3 4 5 6 7 8 A .

Bogs, fens and marshes; widespread but local.

1. Fairly frequent in a band extending from near Ennistymon to the S. slopes of Slieve Elva (Willmot, 1979). By Doonagore L. **2**. By the Caher R. By the Carran turlough. Inishmore. **3**. Edge of Rinroe bog. Abundant in blown sand at the N. end of L. Bunny. Marsh at Ballinderreen. Marsh by the R. Fergus above Ennis. **4**. Drain by L. Corrib, E. of Oughterard. By the canal N.E. of Moycullen. **5**. Abundant in a fen by L. Naneevin, and in bogs and marshes to the W. and N.W. **6**. In wet scrub by the road W. of Oorid L. **7**. Marsh at the W. end of the Slyne Head peninsula. N. of Clifden. By the lake N.E. of Letterfrack. **8**. Marsh by Teernakill Br.

First record: ITB (1901). H16.

E. fluviatile L. (*E. limosum*)

1 2 3 4 5 6 7 8 . B

Lakes, slow streams, marshes and bogs. Very frequent in most districts.

1. Very frequent. **2**. At the edge of the more or less permanent water in the Carran turlough. **3–4**. Frequent. **5**. Marsh W. of Ross L. Lake 6 km N. of Gentian Hill. **6**. Abundant in lakes by Glinsk and Knockboy, and frequent around Carraroe. **7**. Very frequent in the west; rare elsewhere. **8**. Very frequent.

First record: Seebohm (1851). Abundant in several lakes near Letterfrack.

The *Atlas* record for Aran is almost certainly an error.

E. × litorale Rupr. (*E. arvense ×fluviatile*)

. 2 3 4 5 . 7 8 . .

Streamsides, hedges, wet meadows and stony lake-shores. Locally frequent, often in the absence of one parent, and sometimes of both.

2. Frequent by the Caher R. **3**. At both ends of L. Bunny. **4**. By a stream 3 km N.N.E. of Galway. **5**. Abundant in a wet meadow 3 km W. of Barna. **7**. In a hedge N.W. of Moyard, and in the wooded ravine W. of Letterfrack. N. and S. of Clifden. Roadside bank W. of Ballyconneely. **8**. In two places near Maam. Frequent on the W. and N.W. shores of L. Corrib, and on several of its islands.

First record: Praeger (1932*b*). On an island in the N.W. part of L. Corrib.

This hybrid, which is fairly frequent in Ireland, spreads extensively to form large clones, though it lacks the vigour of either parent. Its habitat is usually intermediate between those of the parents, but varies a good deal.

E. variegatum Weber & Mohr

. 2 3 . . . . . (A) .

Wet, usually stony places. Occasional in the Burren; unknown elsewhere.

2. Abundant along the Caher R. Carran. *Inishmore* (Praeger, 1895*b*). **3**. In a wet field near the N. end of L. Bunny, and here and there along its E. shore. Between boulders on the shore of a small lake N.E. of Killinaboy. Under a boulder at the edge of a marsh near Castle L. By the W. shore of L. George.

First record: Praeger (1895*b*). Inishmore (Turloughmore). .

The map of this species in the *Fern Atlas* is misleading, partly because of the incorporation of some erroneous records, and partly because of the erratic results arising from conventional conversion of the original records to the Irish grid.

E. × *trachyodon* Braun, which is usually (and reasonably) interpreted as a hybrid between *E. hyemale* and *E. variegatum*, occurs sparingly on the E. shore of L. Bunny, along with *E. variegatum*, and also at the S.W. corner of L. Bollard, remote from either parent. The nearest known station for *E. hyemale* lies only some 17 km E. of L. Bunny, but there is none recorded within 100 km of L. Bollard. It was first recorded at L. Bollard by More (1872), but erroneously, as *E. variegatum*; it was first correctly named by Praeger (1939).

LYCOPODIACEAE

Huperzia

Huperzia selago (L.) Schrank & Mart. (*Lycopodium selago*)   *Crúibíní sionnagh*

(*1*) 2 . 4 5 6 7 8   . B

Moorland, rocky mountain-ledges and lowland bogs. Frequent in much of Connemara; rare elsewhere.

(*1*). *Slieve Elva* (Atlas). **2**. Near the summit of Gleninagh Mt. **4**. Bog S. of Carrowmoreknock. **5**. Bog 4 km S.E. of Costelloe. By Slievaneena L. In a cutting of the old railway W. of Oughterard. 4 km W.N.W. of Galway city. **6**. Occasional. **7**. Frequent, but rare on Inishbofin. **8**. Widely distributed, usually in small quantity, but abundant in a few places, as on Leckavrea and the mountains east of it, and on the mountains S.W. of Leenane.

First record: Wade (1802). 'Mountains of Cunnamara, common.'

Lycopodium

[Lycopodium clavatum L.]

There are two records of this species for the region: one, cited in CH2 (undated, but probably *c.* 1890) for Killaguile (district 5), the other in the *Atlas* for the Kylemore region (district 7). The authority for the latter cannot be traced, and it seems probable that it was an error. The Killaguile record may well have been correct, but as the species has been declining rapidly in Ireland for at least the past 50 years it is unlikely that it survives there now.

Hart (1883) recorded it from Leenane, just outside our region. The nearest stations in which it has been seen growing recently are in Co. Donegal and Co. Wicklow, both about 175 km distant.

Diphasium

Diphasium alpinum (L.) Rothm. (*Lycopodium alpinum*, *Diphasiastrum alpinum*)

. . . . . . 7 8   . .

Heathy grassland or bare peat on mountains; rare.

7. S. side of Benchoona, 1966 (GH). **8**. *Maumturk range near Maumeen* (Hart, 1883). Widespread on the Maamtrasna plateau, 1965 (GH); 1978, 1981 (Roden).

First record: Wade (1802). 'Sparingly, on the high part of Lettery mountain, and near the summit of Maam Turc.'

Wade's record for Benlettery has never been confirmed: he probably meant Bengower. There is in **DBN** a specimen collected by Shuttleworth in 1832 labelled 'summit of Benree, Renvyle', but the hill cannot be identified; it may well have been Benchoona. 'Renvyle' means no more than that it was his base of operations.

Lepidotis

Lepidotis inundata (L.) Börner (*Lycopodium inundatum, Lycopodiella inundata*)

. . . . . **6 7 8**   . (*B*)

Wet peaty ground; rare.

6. At the edge of a pool filled with *Phragmites*, on a wet bog 1 km S. of Carraroe. **7**. By the lakes on Altnagaighera. *Inishbofin* (Praeger, 1911). **8**. In fair quantity at the edge of damp, sandy flats E. of the road at Killateeaun. Widely scattered in flushes on moorland at *c*. 120 m (400 ft) 2 km W. of the preceding station. Near the S.W. corner of L. Mask, E. of Maamtrasna. On wet bog by the road immediately S. of Maam Br. By the N.W. arm of L. Corrib at the mouth of the Owenree R., and *c*. 2 km to the N.W. of this, opposite Castlekirk I. *Shore of L. Nadirkmore* (Praeger, 1933).

First record: Wade (1802). 'In the turfy ditches on the public road, between Oughterard and the Recess.'

All the stations in district 8 lie within a circle of radius 8 km, and constitute the area in which this very local species is commonest in Ireland.

SELAGINELLACEAE

Selaginella

Selaginella selaginoides (L.) Link

. **2 3 4 5 6 7 8**   . B

Bogs, lake-shores, fens, heathy grassland, mountain cliffs and sand-dunes. Very frequent in E. Connemara; rather local elsewhere.

2. Roadside at Black Head, and in a *Schoenus*-flush on the hillside above it. **3**. Frequent in the *Schoenus*-zone of the fens by several of the larger lakes. On pavement near Rinroe Br. Damp grassland at Owenbristy. **4**. Fen by Ballindooly L. Very frequent in the bogs N. of Moycullen. By Ballycuirke L. **5**. Frequent in bogs. **6**. Sparingly by a lake immediately S.W. of Maam Cross, and in the hills 5 km to the S.W. **7**. Occasional throughout, and locally frequent in the west. **8**. Very frequent.

First record: Wade (1802). 'Maam Turc mountains, Cunnamara.'

Although it is near the southern limit of its range in Ireland, this species shows, within the Flora region, a tolerance of an astonishingly wide range of soils, in respect of both acidity and water-content.

ISOETACEAE

Isoetes

Isoetes lacustris L.

. . . . . **6 7 8** . .

In lakes with acid or neutral water and clean, rocky beds; usually in fairly deep water. Frequent in W. and C. Connemara; not seen elsewhere.

6. Occasional, mainly in the east. **7**. Frequent. **8**. L. Nadirkmore. L. Nacarrigeen. Plentiful in a lakelet in the Maumturk range 3 km S.E. of Letterbreckaun.

First record: Seebohm (1851). Abundant in several of the lakes near Letterfrack.

I. setacea Lam. (*I. echinospora*)

. . . . . **6 7** . . **B**

In similar habitats to the preceding; rather rare.

6. Stream S. of Maam Cross (**TCD**). In a small lake N.W. of Carna (**DBN**). *Glendollagh L.* (Praeger, 1912). **7**. Inishbofin (L. Nagrooaun) (**TCD**). Cregduff L. (**DBN**). Kylemore L., and in a small lake N. of the road 4 km S.E. of Clifden, 1977 (J. Ryan).

First record: More (1876). Inishbofin.

Possibly rather more frequent than the above records would suggest. We have received verbal reports of its presence in some other places, but we are reluctant to accept any records which have not been verified by microscopic examination of the megaspores. In our experience both species can vary widely in the colour, length, shape and orientation of the leaves.

MOSSES AND LIVERWORTS

by

A. R. Perry

National Museum of Wales

Connemara and the Burren are of exceptional interest to the bryologist. The wide variety of habitats, ranging from the maritime rocks and sand-dunes of the coast through bog, marsh, fen and moorland to the mountains, encourages a wide variety of species, and the topographical diversity is further enhanced by the geological diversity of the region. The further west one travels, in Britain and Ireland alike, the richer becomes the bryophyte flora. The prevailing westerly or south-westerly winds, carrying a warmth derived from the Gulf Stream and associated with a high atmospheric humidity, suit the growth of these plants, which are mostly intolerant of desiccation. Many British and Irish species are confined to the western parts of the islands, or at least are much more abundant there, and these 'Atlantic' species are well represented in Connemara and the Burren and form one of the most interesting elements of the flora.

The Twelve Pins will serve as representative of the montane ground. They cover a roughly circular area in western Connemara, and Benbaun, the highest peak, rises to nearly 730 m (2400 ft). The Twelve Pins have the distinction of providing the habitat for a number of bryophytes that are rare or very local on a national scale. The slopes are steep, and consist largely of quartzite, which seldom bears much vegetation, but in two places there are outcrops of mica-schist, a more hospitable substrate which supports a rich bryophyte flora. Muckanaght is probably the best mountain for bryophytes. On its north side there is weathering schist up to a height of about 600 m (2000 ft), and it is on shaded ledges of this base-rich rock that the uncommon though widely distributed moss, *Orthothecium intricatum*, is found. In size and form it resembles the common *Hypnum cupressiforme*, var. *resupinatum*, but its reddish tinge and habitat distinguish it. On shaded rock-faces there are several notable leafy liverworts. One of these, *Cephaloziella pearsonii*, a small plant which forms flattened, reddish-brown patches closely attached to the rock, is one of a number of species with a distribution restricted to the Atlantic parts of Europe and the highest peaks of the Appalachian Mountains in the United States. On Muckanaght it grows near one of its usual associates, *Plagiochila tridenticulata*, which has a similar world distribution. *Eremonotus myriocarpus*, with a similar western distribution in the British Isles, also grows here; it can be confused with the *Cephaloziella*, but often has perianths, which are unknown in the latter. Another small liverwort, *Anthelia juratzkana*, has its southernmost Irish station on Muckanaght, being otherwise known only from Donegal (though it is also recorded from Britain). It forms small, bluish-grey patches on earthy, sheep-trampled ledges, but can easily be overlooked as its stems do not exceed 5 mm in length.

Several rare mosses grow on the same mountain. One, *Ditrichum zonatum*, var. *scabrifolium*, endemic to the British Isles, was found for the first time in Ireland on Muckanaght in 1968. It grows at a height of 490 m (1600 ft) in dense tufts on dripping schist, and often shows annual growth-zones. Another moss, *Leptodontium recurvifolium*, frequently reaching a handsome 10 cm in height, grows on several of the higher mountains of Connemara. It forms lax, green or yellowish tufts on base-rich rocks and ledges, on banks or in rock-crevices, often near a waterfall. It is known from western Britain as well as western Ireland, but elsewhere only from Spain; it is recognizable by its undulate, coarsely toothed leaves.

A characteristic community of leafy liverworts is widespread in the mountains of western Ireland; they require a very high humidity for their survival, and even where the annual rainfall is of the order of 2000 mm seek shaded and sheltered habitats, for example beneath the canopy of *Calluna*. Typical species of this group to be found above 450 m in the Twelve Pins are *Mastigophora woodsii*, *Herberta adunca* and *Scapania ornithopodioides*. One of the most memorable sights on the mountains is the startling colour contrast provided by reddish-brown or rose-tinged hunks of *Herberta* mixed up with the deep purple leafy liverwort *Pleurozia purpurea*. The local speciality of this liverwort turf is, however, *Adelanthus unciformis*. This beautiful leafy liverwort was first discovered in Ireland in 1903 by H. W. Lett on Achill Island, Co. Mayo, and described as a new species, *A. dugortiensis*; it has since then, however, been equated with *A. unciformis* from the southern hemisphere and Ruwenzori. It is relatively common on the Twelve Pins above 550 m (1800 ft), under the *Calluna* canopy, especially on the north-facing slopes of Benbreen and Bengower. Its only other known station in the British Isles is in Co. Donegal, where it was discovered in 1962.

Connemara, like most other parts of Ireland, has suffered severely from deforestation, and there are few bryophyte-rich woods left, though no doubt once they abounded. There remains, however, an exceptional sessile oak-wood behind Kylemore Abbey. Though it faces south, and might be expected to dry out readily and consequently have a sparse bryophyte cover, this is in fact not the case; drying out is presumably prevented by the high rainfall and the densely tangled shrub layer, formed mainly of *Rhododendron ponticum*. The trunks of this naturalized evergreen form a substrate in highly shaded conditions which is readily colonized by liverworts such as *Telaranea sejuncta*, known elsewhere in the British Isles only from Cornwall, Kerry, Cork and Mayo, and *Cephalozia hibernica*, known otherwise only from Kerry, the Azores and Madeira. Three other rare liverworts grow here, usually on shaded boulders or tree boles; these are *Lejeunea flava*, *L. holtii* and *L. diversiloba*. The wood can, in fact, be regarded as a *Lejeunea* paradise, as all but two of the nine species and varieties known from the British Isles occur here, often in great profusion. They are closely attached to the substrate, and evidently enjoy the local microclimate which gives them the high humidity and freedom from winter cold which they require. Two mosses of note grow on and among the rocks above Kylemore Abbey, as well as in some other stations in Connemara. The first is *Oxystegus hibernicus*, a plant up to 10 cm high with long, flexuous leaves, which forms lax tufts on damp rocks and in crevices. It is known from western Scotland as well as western Ireland, but nowhere else. The second is *Campylopus setifolius*, a robust plant which forms dark or olive-green flattened tufts in damp places on rock-ledges or screes, often where water drips. It grows in Spain as well as in Britain and Ireland.

Connemara is famous for its bogs, the greenness of which is due in large part to the bryophyte cover, which may be more or less continuous over considerable areas. Species of *Sphagnum* abound, with the uncommon *S. pulchrum* forming hummocks in blanket-bog at Roundstone. It may be recognized by its orange-brown colour and five-ranked leaves. Microscopic examination of the Connemara peat just below the surface shows that it consists largely of the remains of *S. imbricatum*, once widespread but now of very restricted distribution; its decline is probably due to a change of climate. Today the major peat-formers are *S. papillosum, S. palustre, S. recurvum, S. cuspidatum, S. subnitens* and *S. auriculatum*, Small liverworts are abundant in the

Sphagnum hummocks, creeping over and among the stems. Among them is a species of *Calypogeia*, aptly named *C. sphagnicola*, which usually grows as solitary strands, and which is probably widespread in Connemara. Several species of *Cephalozia*, another genus of leafy liverworts, grow in similar situations or on bare peat. One of them, *C. loitlesbergeri*, which has been found on a bog near Tullywee Bridge (east of Recess), is one of the least common of the group. Many of the Connemara bogs are covered with a thick, continuous turf of the moss *Campylopus brevipilus*, which is usually conspicuous by virtue of the short hyaline apex to its leaves. It is a native species, but seems to be being ousted by the very similar but alien species, *C. introflexus*, which reached the British Isles fairly recently either from America or the southern hemisphere. It was first recorded in Co. Dublin in 1942, but since then has spread rapidly westwards, mainly on bare peat or burnt moorland, to the detriment of some of the native species. It has a much longer and more conspicuous hyaline leaf-apex than has *C. brevipilus*, and unlike the latter frequently produces capsules.

If we turn to the Burren we find that much of its bare limestone is devoid of bryophytes, for they are unable to colonize the smooth surface exposed to the drying effects of the sun. In spite of this, however, one does not need to go far to see big cushions of the common mosses *Tortella tortuosa* and *Breutelia chrysocoma* in relatively exposed situations. Such cushions, however, are always associated with a fissure or crevice in the rock, often quite shallow but nonetheless deep enough to retain a little moisture, which enables the initial colonization to take place and favours the further growth of the plant. On shaded rock-faces (often with overhangs), in crevices, and in the shade and shelter of higher plants there is a fairly rich bryophyte flora. Many lime-loving species can be found on shaded rocks near Black Head. *Neckera crispa* is one of the most conspicuous mosses here, and its flattened, glossy shoots with undulating leaves form cascading curtains on the rocks. *Marchesinia mackaii*, a dark green, almost black leafy liverwort named after J. T. Mackay, author of the *Flora Hibernica* of 1836, grows firmly attached to the shaded rock-faces, often covering considerable areas. *Cololejeunea calcarea*, a minute leafy liverwort forming very small yellow-green patches, also grows here, and with it may be found two rare species, *Pedinophyllum interruptum*, a leafy liverwort with creeping stems, and *Orthothecium rufescens*, a striking moss with glossy, pinkish leaves, whose only other Irish stations lie in the north-west. On drier, more exposed surfaces of the limestone *Bryum capillare*, var. *rufifolium*, a moss which forms compact tufts, usually deep red or variegated in colour, occurs in its only Irish locality.

Much of the more sheltered part of the Burren is covered by hazel-scrub, in which there are conspicuous bryophyte mats on the hazel-stems, on boulders and on the ground; indeed mosses often cover more of the ground than do herbaceous plants. Even outside the shelter of the hazel one is struck by the way in which shoots of hawthorn or blackthorn, attempting to regenerate after grazing, have to fight their way through a substantial layer of mosses such as *Hylocomium splendens*, *Ctenidium molluscum*, *Hypnum cupressiforme* and *Rhytidiadelphus triquetrus*. The bryophyte cover of boulders in the scrub is often impressive, and in addition to the species just mentioned others such as *Homalothecium sericeum*, *Eurhynchium striatum* and *Hylocomium brevirostre* abound. The epiphytic flora on the boles and branches of the hazel is also worth studying for its rich variety, especially in the small cushion-forming genera *Orthotrichum* and *Ulota*. The widespread *Ulota crispa*, *U. bruchii* and *U. phyllantha* are common species here, often found in close proximity to the much rarer *U. calvescens*, a speciality of the area. *Orthotrichum striatum*, *O. pulchellum* and *O. lyellii* all occur, the last quite distinctive in the field by its dusty appearance, caused by the powdery covering of multicellular gemmae on the surface of its leaves.

Fig. 22. Hazel-scrub in the Burren, with a rich growth of mosses on the stems and on the ground.

The turloughs, so characteristic of the eastern part of the Burren, hold some bryological surprises. Not far from L. Bunny, for example, they yield the moss *Calliergon trifarium* in its only Irish station. Its usual habitat in Britain is in basic montane flushes and fens in the highlands of Scotland at a considerable altitude. Its occurrence in the Burren at less than 30 m (100 ft) above sea level, in an area whose flora includes a definite Mediterranean element (such as the moss *Scorpiurum circinatum*) echoes the behaviour of several vascular plants characteristic of higher altitudes which descend in the Burren to rub shoulders with others of Mediterranean affinities. When the water level sinks in the summer it leaves much of the turlough with a black covering of the aquatic mosses *Cinclidotus fontinaloides* and *Fontinalis antipyretica* left high and dry. In these conditions the *Fontinalis*, normally a very shy fruiter, often produces capsules. Its variety *cymbifolia*, endemic to the British Isles, is found on stones submerged in the river Fergus near Killinaboy, in close company with two leafy liverworts, *Porella cordaeana* and *P. pinnata*.

In the flattish limestone country north-east of Killinaboy there are some calcareous fens which dry out to some extent in summer, but seldom completely. They share with the calcareous marly margins of the larger lakes further east a typical and characteristic bryophyte flora. In these habitats the conspicuous mosses *Drepanocladus sendtneri*, *D. lycopodioides*, *D. vernicosus*, *Cratoneuron commutatum* and *Scorpidium scorpioides* thrive, and are relatively frequent, often forming a squelching turf. *Bryum neodamense*, a very rare moss, forms reddish-green patches in a *Schoenus*-fen on the shores of L. Bunny.

The sand-dune systems of the region are not very extensive, and do not show any prominent development of slacks. Those at Dog's Bay in Connemara and at Fanore in the Burren, however, boast the usual mosses associated with sand-dune turf: *Tortula ruraliformis*, *Homalothecium lutescens* and *Brachythecium albicans*, all of which occur in some quantity. Less common mosses found in both systems are *Entodon concinnus* and *Tortella flavovirens*, while *Thuidium philibertii* and *Brachythecium mildeanum* are recorded so far only from Dog's Bay and *Barbula acuta* only from Fanore.

A man-made habitat worthy of more attention than it usually receives is the type of lay-by which is common on minor roads, especially those of single-track width. These often have a covering of fine gravel embedded in compacted soil, and can for such a small habitat show a remarkable wealth of bryophytes. The specialities to be found here include the leafy liverworts *Haplomitrium hookeri* and *Fossombronia incurva* (the latter in such conditions resembling minute lettuce plants the size of a pin's head); also the thallose liverwort *Riccardia incurvata*, which consists of a simple or sparingly branched thallus no more than 1 mm in width, and the moss *Archidium alternifolium*, which produces globose capsules immersed among the leaves.

Out of a total of about 700 mosses and just over 300 liverworts so far recorded from the British Isles, 382 mosses and 189 liverworts are known to occur in Connemara or the Burren. Though the region has been studied by bryologists more closely than have many parts of Ireland, a great deal still needs to be found out about it, and many species new to the region remain to be discovered, especially in Connemara. As an example of this potentiality, an exciting new discovery was made as recently as 1973, along a rivulet at Roundstone, of the elegant and conspicuous moss *Myurium hochstetteri*, new to Ireland, and in Britain known only from north-west Scotland.

LICHENS

by
M. E. Mitchell
University College, Galway

The mild, wet climate of Connemara and the Burren favours the existence of a varied lichen flora, but the distribution of species is dependent on other factors, of which the most important is the nature of the parent rock. This factor has also influenced the study of the flora, because the Burren, which consists largely of uniform tracts of limestone pavement, has not attracted the attention of lichenologists to at all the same degree as has Connemara, where the lithological variation is reflected in a wide range of natural habitats.

The pavement of the Burren does indeed appear at first sight rather dull lichenologically, owing to the predominance of a small number of endolithic species, but on close investigation the flora is seen to be quite diverse. Colonization by endolithic lichens results in extensive pitting of the surface of the clints by the minute foveolae in which such species produce their ascocarps. As the ascocarps age and are eventually washed away, a gradual coalescence of adjacent foveolae leads to the formation of larger pits, which may eventually develop into solution-cups; the role of these biogeochemical processes in the weathering of the Burren limestone has still to be assessed. On the floor of the shallower cups *Collema fragile*, *C. multipartitum*, *Leptogium diffractum* and *L. schraderi* occur, while in older cups, where some soil has accumulated, such species as *Agonimia tristicula*, *Collema cristatum*, *Lempholemma cladodes* and *Leptogium plicatile* are found. Though much of this flora is inconspicuous, thalli of *Aspicilia calcarea*, *Caloplaca aurantia* and *C. heppiana* occur on the pavement, and these white and orange species also add colour to the many ancient ruins and countless dry-stone walls. The granite erratics which occur sporadically throughout the region, though most frequently in the east, support an entirely different flora from that of the pavement, except where they serve as perches for birds, when a distinctive assemblage of species similar to those occurring under like conditions on limestone blocks becomes established.

Because of the rapid seepage of rainfall through joints and fissures into an underground drainage system, surface water is of restricted occurrence in the Burren. There are, however, a number of shallow lakes in the south-east, including L. Bunny, which on the pavement of its eastern shore has a distinct zone of *Dermatocarpon* spp., corresponding apparently to the winter water-level. Apart from these lakes and the R. Fergus which drains some of them, the only other permanent, overground water is in the north-west, where the Caher R. runs for most of its course of 6 km over drift deposited in a pre-glacial valley. Near its mouth, however, the river has cut through the drift and runs on a limestone floor strewn with glacial boulders. This reach of the river remains the only locality known outside Continental Europe for *Arthopyrenia caesia*, and other interesting micropyrenocarpous species also occur here.

At Fanore, lichens do not become established on the mobile dunes, but the transition-zone to fixed dune-grassland is dominated by *Squamarina cartilaginea* and *Collema tenax*, with *Catapyrenium lachneum*, *Peltigera spuria*, *Polyblastia gelatinosa* and *Toninia coeruleonigricans* also well represented. On the terraced platforms which constitute much of the Burren shore, *Arthopyrenia halodytes*, *Caloplaca marina*, *Lichina confinis*, *L. pygmaea* and *Verrucaria maura* are consistently present.

One of the many striking aspects of the Burren landscape is the extensive occurrence of hazel-scrub, which is particularly well developed in the central area; here the older stands are almost impenetrable. Young twigs of the hazel are colonized first by *Arthopyrenia punctiformis* and *Tomasellia gelatinosa*, which are succeeded by *Arthonia tumidula* and *Pyrenula nitida*. In the final stages of colonization *Nephroma laevigatum*, *Pannaria rubiginosa* and *Parmeliella plumbea* become dominant. The widespread and often luxuriant development of such species clearly demonstrates that the Burren is not subject to any appreciable amount of environmental pollution. The presence of the pollution-sensitive species *Caloplaca cerina*, *Parmelia exasperatula* and *P. perlata* at various sites within the Galway urban area underlines the fact that even here the quality of the air has not deteriorated to any significant extent.

The topographical diversity of Connemara and the variety of its igneous, meta-morphic and sedimentary rocks provide a far more extensive range of habitats than is found in the Burren, and the lichen flora is correspondingly richer. One feature common to both areas is, however, the great scarcity of woodland. The conifer plantations are of scant lichenological interest, and the remnants of native oak-wood are few and scattered. Of the latter, the most interesting for the lichenologist is that which covers a strip of metamorphic limestone about 1 km long on the western shore of Derryclare L. The environment is particularly humid, and the epiphytic flora is similar to that of the Killarney woods. Twelve species of *Parmelia* have been noted, among them *P. endochlora* and *P. laevigata*, but perhaps the most striking feature of the wood is the luxuriant growth of *Sticta canariensis*, together with its blue–green phycotype which is colonized here by the interesting lichenicolous fungus *Hemigrapha astericus*. The wood lies at the foot of the Twelve Pins, but these quartzite mountains have received little attention from lichenologists. Indeed, apart from Lissoughter and the uplands of the Carna peninsula, which have been worked to some extent, the only high ground to have been investigated in any detail is Doughruagh, a basic intrusion north of Kylemore.

Rivers and streams in Connemara are mostly fast flowing, and consequently their beds are composed of eroded rocks and boulders. Where these project and are subject only to periodic submergence, as in the Bunowen R , west of Leenane, they are conspicuously capped by *Pertusaria pseudocorallina*, and this effect is often heightened by a black zone of *Ephebe lanata* immediately below. Other species found in this habitat include *Catillaria chalybeia*, *Hymenelia lacustris* and *Porina chlorotica*.

The Connemara lowland is a heavily glaciated surface now largely covered by blanket-bog, but its history can still clearly be read in the massive blocks, rugged outcrops and wide expanses of ice-scoured granite which interrupt the cover. The bog and heath communities are remarkable for the luxuriant development of *Cladonia arbuscula* and *C. portentosa*. These two species occur with the same monotonous regularity as does *Calluna vulgaris*, the epiphytic flora of which affords welcome relief and merits careful investigation. The practice of burning heathland to improve grazing frequently produces areas of peaty soil dominated by *Lecidea granulosa*, *L. uliginosa* and *Cladonia* spp.

The huge blocks and boulders which accentuate the desolate nature of much of the terrain are colonized largely by foliose species, in particular *Hypogymnia physodes*, *H. tubulosa*, *Parmelia omphalodes*, *P. saxatilis* and *Platismatia glauca*. Horizontal exposures, such as those extensively developed on the north-east shore of Cashla Bay, are characterized by the presence of mainly crustose lichens, notably *Fuscidea cyathoides*, *Opegrapha conferta*, *Rhizocarpon geographicum* and *R. obscuratum*, though some foliose species, among them *Parmelia conspersa*, *P. glabratula* subsp. *fuliginosa* and *P. revoluta*, also commonly occur.

The distinctive zonation of marine lichens frequently seen on steep, rocky shores is seldom well developed in Connemara because of the nature of the coastline. Above high-water mark, the northern shore of Galway Bay is largely a boulder beach, while much of the west coast is represented by a gently sloping land-surface submerged in post-glacial times. Suitable conditions occur, however, in some places, as on the south coast of Killary harbour, for the establishment of the characteristic transition from a black, littoral zone dominated by *Verrucaria* spp. through an orange zone of *Caloplaca* spp. and *Xanthoria parietina* to the greyish-green upper section of the shore dominated by *Ramalina siliquosa*.

Only a few of the more interesting and obvious aspects of the lichen flora, which comprises some 500 species, have been mentioned in this *vue d'ensemble*; it should be understood that our knowledge of the distribution of lichens in Connemara and the Burren is still very incomplete and that much work remains to be done. But it must also be emphasized that many species are of rather limited occurrence, and both resident and visiting botanists should at all times bear in mind the need for stringent conservation measures.

MARINE ALGAE

by

M. D. Guiry
University College, Galway

The seaweeds of Connemara and the Burren contribute significantly to the landscape and to the economy of the region, and are also of considerable botanical interest. The much indented coast of western Connemara is characteristically lined with a golden or greenish-yellow band of *Ascophyllum nodosum*, aptly known as *feamainn bhuí* (yellow seaweed), and, as Praeger has remarked, it is often impossible to tell without the aid of the algal zone whether one is looking at an arm of the sea or a freshwater lake. Many of the beautiful white beaches of Connemara, like those at Mweenish Island and Mannin Bay, are composed largely of the cast-up, dead fragments of the calcareous red algae known as nullipores or maerl. Such beaches are usually known as 'coral strands', but they are of plant, not animal, origin.

The diversity of species in the region is very high; so far 336 species have been reported – about 70 % of the total Irish flora of benthic marine algae. Of these, 54 belong to the Chlorophyta (green algae), 97 to the Phaeophyta (brown algae) and 185 to the Rhodophyta (red algae). This diversity is partly because of the extensive range of habitats available, but the mildness of the climate, the relatively high duration of cloud cover, the narrow range of sea temperatures and the proximity of clear, nutrient-rich, oceanic water also contribute significantly. One is bound to add that the marine algae of this region are better known than those of many other parts of Ireland.

The limestone shores of the Burren are particularly variable in their topography. The strata are for the most part horizontal, giving rise to a terraced structure with a very marked pattern of zonation, and on such shores the pools are shallow and contain few species of seaweed. The communities and zonation pattern of these shores are of considerable interest, but have not yet been studied in detail. On many of them the dark purple sea-urchin, *Paracentrotus lividus*, occurs in vast numbers in shallow pools encrusted with the coralline red alga *Lithophyllum incrustans*. Where the urchins burrow into the relatively soft rock, as at Fanore and Black Head, they give rise to a very irregular topography and probably to a correspondingly greater diversity of algal species.

In much of Connemara, on the other hand, granite dominates the intertidal zone, and frequently there is a storm beach at high-water mark. Because of the harder nature of the rock *Paracentrotus* is very rare in the intertidal zone, and where it occurs is unable to burrow. Boulder-strewn shores are commoner than in the Burren, and some of these granite shores, as, for example, on Mweenish Island, have an extraordinary range of tidal pools of varying dimensions, some of them regular lagoons, and this greatly increases the number of seaweeds to be found.

Rapids, similar to the famous ones at L. Ine, Co. Cork, are found in several places, as at Flannery Br. and by Tawin Ho.; the speed of their current is not as high as at L. Ine, but their algal communities would well repay investigation.

Subtidal habitats are equally diverse, the most interesting and characteristic, perhaps, being the beds of detached coralline algae (maerl), which are found at depths

Fig. 23. Maerl (fragmented calcareous algae) on one of the 'coral strands' near Ballyconneely.

of 0–25 m in the Burren north-east of Ballyvaughan, and in Connemara from Cashla Bay westwards. The most abundant species are *Lithothamnion corallioides* and *Phymatolithon calcareum*, but in this group there are many ecotypes, which may, under similar conditions, resemble those of another species, so that microscopic examination is often necessary for determination. P. Cooke tells me that some other species are also found in maerl, especially in Connemara. Beds of these algae are also found at a depth of 35 m in the clear oceanic water on the leeward side of the Aran Islands. In Muckinish Bay it is possible to examine the living algae in the intertidal zone, but this is very unusual. The maerl species bear a rich and characteristic flora of algal epiphytes, the dominants being *Peysonnelia* spp., *Cruoria* spp., *Gelidium* spp., and, during the warmer months, *Dictyota dichotoma, Nitophyllum punctatum, Halarachnion ligulatum, Dudresnaya verticillata* and *Stenogramme interrupta*.

In areas where sand abounds, as in Mweenish Bay, beds of *Zostera marina* are common in both the subtidal and the lower part of the intertidal zones, and there is a fascinating associated algal flora. Small, pink, encrusting coralline algae belonging to the genera *Fosliella* and *Dermatolithon* cover the leaves, while their margins may be fringed with *Rhodophysema georgii* or *Audouinella virgatula*. Other epiphytes include *Cladosiphon zosterae* and *Champia parvula*.

Clear, deep water abounds in the outer part of Galway Bay and around the Aran Islands, and here algae can be found to a depth of at least 35 m. The inner part of the bay is more turbid, with sand generally dominating the subtidal zone, and the number of algal species is small.

There is no true relict component in the marine algal flora of the west of Ireland. A number of reputedly southern species such as *Bifurcaria bifurcata, Helminthocladia calvadosii, Dilophus spiralis, Lithophyllum dentatum* and *Vaucheria piloboloides* are found in the Burren or Connemara, but none is common except the *Bifurcaria*, and it is found only on the Co. Clare coast. It extends northwards as far as north Co. Donegal, but the other species probably reach the northern limit of their European distribution in Connemara. On the other hand some southern species such as *Padina pavonica*, *Halopitys incurvus* and *Laminaria ochroleuca*, which are common on the south coast of England, are not found in the Burren or Connemara, although the first two occur very rarely further south in Ireland. If, as seems likely, there are no barriers to dispersal there must be ecological reasons for the absence of these species from our region. So-called northern species, such as *Ptilota plumosa* and *Phyllophora truncata*, are fairly common in certain habitats both in Connemara and the Burren, but occur also on the south coast of Ireland; in Britain, however, they seem to reach their southern limit on Anglesey.

There is no 'American element' in the algal flora, but a number of introduced species and subspecies have been recorded. Tetrasporophytes of *Asparagopsis armata*, an Australian species introduced into European waters some time in this century, were reported for the first time in the British Isles by De Valéra from the coast of Connemara in 1939; the gametophyte was found in 1941. She concluded that Atlantic liners, which at that time called at Galway, may have been the vectors. *Bonnemaisonia hamifera*, originally described from Japan and known from Britain since 1870, was surprisingly discovered on Clare Island (Co. Mayo) by Cotton in 1911 and was found on the Aran Islands by de Valéra in 1938. Since then both these species have become widely distributed in Ireland, but are restricted to certain habitats. *Antithamnion spirographidis*, another probable introduction from the southern hemisphere, was first found in Europe at Cherbourg in 1910, and in Ireland west of Spiddal in 1938; it is now widely distributed, though under-recorded, but fertile plants are rare. Two subspecies of *Codium fragile* have been introduced into the region; subsp. *atlanticum* was probably the first to arrive (before 1893), but later (around 1957) subsp. *tomentosoides* began to displace it, and is now much the commoner of the two.

Two fucoids, *Ascophyllum nodosum* ecad *mackaii*, and *Fucus vesiculosus* var. *muscoides*, are well-known botanical curiosities of the region. The first was described as *Fucus mackaii* by Dawson Turner in 1808 on the basis of material sent him from Roundstone by Mackay; the latter was described by Cotton from Clew Bay. The ecad *mackaii* occurs in very sheltered, muddy inlets, both in the Burren and in Connemara, as large, floating clumps, often without the usual air-bladders, which apparently originate as fragments detached from ordinary plants of *Ascophyllum*. *Fucus vesiculosus* var. *muscoides*, which seldom exceeds 3 cm in height, and is often very much less, is most often found growing in salt-marsh turf. It has been reported from Finavarra in the Burren and from numerous localities in Connemara.

To sum up, there is nothing very remarkable about the individual species of marine algae found in Connemara and the Burren; as far as can be ascertained there are only four species which have not been reported from other parts of the British Isles – *Dermocorynus montagnei*, *Lithophyllum dentatum*, *Audouinella rhipidandra*, and a still undescribed foliose red alga, perhaps belonging to the Kallymeniaceae. What are remarkable are the large number of species often present on a single shore and the size-range of many of the plants. I once sent specimens from the region for illustration for a marine algal flora of the British Isles, but they were rejected as being too luxuriant to be typical.

A word, in conclusion, on economic aspects: seaweed is widely gathered, alike in the Burren and in Connemara, for use as an agricultural fertilizer, as fodder for animals, as food for human consumption and for the extraction of substances used in industry. Drift seaweed (consisting chiefly of cast-up plants of *Fucus* species, *Ascophyllum nodosum* and *Laminaria* species) is used to make 'lazy-beds' for potatoes

Fig. 24. A small-holding in southern Connemara. In the foreground, potatoes have been planted in lazy-beds (broad, flat-topped ridges), and down the centre of each has been spread a line of seaweed as fertilizer. The hay-cock by the house is secured with ropes against the wind; to the left of it is the remains of a turf-stack. The rock is granite.

and other vegetables, but this practice seems now to be confined to the really barren areas. It is also dried in conical piles above high-water mark and sold to small factories at Kilrush (Co. Clare), Maam Cross and Roundstone, where it is dried further, ground, and sold as an additive for animal fodder. The market for this material has been fairly consistent in recent years; of some 8000 tonnes dry weight produced annually in Ireland the greater part comes from Connemara and the Burren. *Chondrus crispus* (Irish moss or *carraigín*) and *Palmaria palmata* (dulse or *creathnach*) are gathered by hand in both areas; the former is used to make a type of blancmange and the latter is eaten uncooked. *Ascophyllum nodosum* is collected by hand in Connemara and transported wet to a factory at Kilkieran, where it is dried, ground, and exported for the extraction of alginic acid, a gel-forming polysaccharide widely used in the textile and food industries. For this purpose, rafts (*climín feamainne*) of *Ascophyllum*, made by surrounding the buoyant weed, after it has been cut, with ropes, are floated to high-water mark on the incoming tide and then transported wet by road to be processed.

More detailed information about the seaweeds of the region may be found in the publications of de Valéra (1958) and de Valéra *et al.* (1979) listed in the bibliography.

FRESHWATER ALGAE

Until recently, the Characeae were given by many field botanists the status, as it were, of honorary flowering plants, and even when their position among the algae was generally conceded they were often collected and recorded by botanists primarily concerned with vascular plants. Thanks to the kindness of Mrs J. A. Moore, who has assembled, and where necessary redetermined, records from the literature and the leading herbaria, we can present a fairly complete account of their representation in Connemara and the Burren.

Nineteen species have been recorded to date, as against twenty-eight for the whole of Ireland; the remaining nine are all eastern or midland in their distribution. There are records for all eight districts of the Flora, but those from districts 4 and 7 are much the most numerous. To what extent this reflects intensity of collecting rather than real distribution remains uncertain. In district 7, however, the small lake at Bunowen appears to be exceptionally rich; from it are recorded six species of *Chara* and three of *Nitella*. It lies surrounded by blanket-bog, but at the base of a basaltic hill, and for this reason seems to be able to accommodate both calcicole and calcifuge species. Two of its species of *Nitella* (*N. confervacea* and *N. spanioclema*) are distinctly rare, being confined in the British Isles to west Ireland and west Scotland, while Bunowen is their only known station in Connemara. Three of the species of *Chara* found here are relatively frequent in Ireland but rather rare in Britain; they are *C. aculeolata*, *C. desmacantha* and *C. rudis*; all occur in the Burren as well as in Connemara. Of the other species which are rare or moderately rare in the British Isles, *Chara canescens* has been found in brackish waters in several places around the shores of Galway Bay, *C. connivens*, which is elsewhere submaritime, has been recorded at two inland stations (Templebannagh L. and 'between L. Mask and L. Corrib'), and *Nitella tenuissima* is known from the Burren and near Galway. Finally, *Lamprothamnion papulosum*, discovered in L. Murree in 1973, is known only from one other station in Ireland (Co. Wexford); in Britain it is known only from southern England where it is rare, and perhaps extinct.

Many of the records cited above are old and in need of updating, but although pollution and drainage may have taken their toll in the area around Galway city, most of the lakes in Connemara and the Burren seem to be in much the same state as they were a hundred years ago, and there is no reason for presuming any great change in their flora.

We can say little about the other groups of freshwater algae. From the visit of W. West in 1890 onwards, a considerable number of plankton samples have been collected from the lakes of Connemara – sufficient to show that their flora of diatoms and especially of desmids is rich – but no very striking characteristic of the planktonic flora has emerged, and we consider that detailed discussion of unicellular organisms lies outside the normal scope of a Flora such as this. Some filamentous forms have been recorded in the plankton as well, but the records are fragmentary, and in many cases the determination has not been pursued to the specific level. (References to the principal papers on planktonic algae are given by McDonnell & Fahy, 1978.) No attempt, however, has so far been made to sample filamentous forms from pools, ditches and turloughs, where they must be much more abundant. Indeed, a striking instance of such abundance is given by Scannell (1972), who noticed in one turlough a growth of *Oedogonium* spp. sufficiently exuberant that when the water-level sank

Fig. 25. Limestone pavement near Poulsallagh, with the solution-hollows filled with the blue-green alga *Nostoc*.

and the alga died its remains covered large areas with a white, paper-like felt. Another feature of the Burren in which algae are involved and which is of some ecological importance is the accumulation of large masses of *Nostoc* spp. (together with various unicellular algae) in solution hollows of the limestone pavement. In dry weather the alga dries up to a friable black mass and is often blown away, but sometimes it serves as the starting-point for the growth of mosses or seedlings of small phanerogams, and so to the development of a sod of vegetation filling the hollow.

Mention should be made, in conclusion, of another alga whose habitat is subaerial rather than aquatic. This is *Phycopeltis arundinacea* (*P. expansa*), a member of the Chaetophorales, which is found throughout both the Burren and Connemara and is locally common, growing as an epiphyte mainly on the leaves of ivy in sheltered situations, but occasionally on the leaves of other plants. The genus is widely distributed, mainly in the warmer regions of the world; we can find no certain records from Europe apart from Ireland, where its abundance seems to be quite exceptional (see Scannell, 1965).

FUNGI

The fungi of the Flora region have received very little attention and are very inadequately recorded. The exhaustive catalogue of Irish Hymenomycetes recently published by Muskett & Malone (1980) mentions nearly 1200 species, but they record fewer than 70 for the Flora region, and only two have been added since. The state of affairs in the Gasteromycetes is even worse, for we can trace only one solitary record. The scarcity of woodland in Connemara and the Burren means, doubtless, that the higher Basidiomycetes are less richly represented here than in many other parts of Ireland, but there can be little doubt that the number so far recorded could be at least quadrupled by systematic search. There has never, unfortunately, been a field mycologist resident in the area, and by mid-autumn, the high season for most species, botanical visitors have all gone home.

Of the Hymenomycetes on record one seems to be very rare in Ireland: *Hygrophoropsis (Cantharellus) albida* was found on heaths in two places in Connemara in 1895, and these remain the only Irish records. Others (also from Connemara) which are moderately rare (or perhaps only under-recorded) are *Asterophora lycoperdoides, Boletus testaceoscaber, Coprinus ephemerus, Hymenochaete corrugata, Pholiota erinacea* and *Craterellus cornucopioides*.

Records of rusts, smuts and Ascomycetes are so scanty that they cannot serve as the basis of any generalization. All we can do is to mention a few species from the region which appear to be rare in Ireland (and in some cases in Britain also). These include:

Uredinales
> *Puccinia tanaceti*, on *Artemisia maritima* at Parkmore, and *Hyalospora adianticapilli-veneris* on *Adiantum* on Inishmore.

Ustilaginales
> *Thecaphora trailii*, on *Cirsium dissectum*, and *Urocystis fischeri*, on *Carex flacca*, both near Ballyvaughan: *Melanotaenium endogenum* on *Galium verum* at Carna.

Ascomycetes
> *Mycosphaerella asplenii* (Pseudosphaeriales), on *Asplenium adiantum-nigrum* near Letterfrack. *Hypocreopsis rhododendri* (Hypocreales), on *Corylus* stems in the Burren. *Rutstroemia lindaviana* (Helotiales) on *Phragmites* near Gentian Hill. *Verpatinia calthicola* (Helotiales), on *Iris pseudacorus* east of Galway. *Isothea rhytismoides* (Sphaeriales) on *Dryas* at Gentian Hill. *Chaetosphaerella (Thaxteria) phaeostroma* (Coronophorales), on rhododendron-wood near Cashel. *Gloniella adianti* (Hysteriales) on *Blechnum* at Derreen.

Studies are at present in progress at University College, Galway, on marine saprophytic Fungi of the Galway region.

BIBLIOGRAPHY

A complete botanical bibliography of Connemara and the Burren lies far outside the scope of this Flora. The references listed below are mainly those cited in the text, but some other publications of ecological or general interest have been added. For further publications on the Burren reference may be made to Malloch (1976).

The following abbreviations have been adopted for journals which recur very frequently:

IN   *Irish Naturalist*
INJ   *Irish Naturalist's Journal*
JB   *Journal of Botany, British and Foreign (London)*
PRIA *Proceedings of the Royal Irish Academy*

Allin, T. (1871). Note on Burrin, Clare. *JB*, **9**, 18.

Andrews, W. (1845). Observations on the botany of the Great Arran Island, Galway Bay. *London Journal of Botany*, **4**, 569–70.

Babington, C. C. (1836). Observations made during a visit to Connemara and Joyce's Country in August, 1835. *Magazine of Natural History*, **9**, 119–30.

Baily, K. S. (later Lady Kane) (1833). *The Irish Flora.* 220 pp. Dublin.

Baker, E. G. (1901). Some British violets. *JB*, **39**, 220–7.

Baker, H. G. (1954). The *Limonium binervosum* complex in western and northern Ireland. *Proceedings of the Botanical Society of the British Isles*, **1**, 131–41.

Balfour, J. H. (1853). A botanical trip to Ireland. *Phytologist*, **4**, 1005–7.

—— (1876). Notes of an excursion to Connemara in September, 1874. *Transactions of the Botanical Society of Edinburgh*, **12**, 371–7.

Ball, J. (1839). Botanical notes of a tour in Ireland, with notices of some new British plants. *Annals of Natural History*, **2**, 28–36.

Bangerter, E. B. (1954). Plant records: *Symphytum peregrinum. Proceedings of the Botanical Society of the British Isles*, **1**, 175.

Bannister, P. & McAllister, H. A. (1966). *Asplenium septentrionale* (L.) Hoffm. in W. Galway. *INJ*, **15**, 149.

Barrington, R. M. (1877). Plants of west Ireland collected by J. Reilly. *JB*, **15**, 178–9.

Boatman, D. J. (1956). *Mercurialis perennis* L. in Ireland. *Journal of Ecology*, **44**, 587–96.

—— (1957). An ecological study of two areas of blanket-bog on the Galway–Mayo peninsula, Ireland. *PRIA*, **59B**, 29–42.

—— (1960). The relationship of some bog communities in western Galway. *PRIA*, **61B**, 141–66.

Borrer, W. (1849). *Allium babingtonii.* In Borrer *et al., supplement* 4, t. 2906, to *English Botany*, by J. Sowerby & J. E. Smith. London.

Boyle, P. J. (1968). Preliminary account of *Spartina pectinata* Link new to Ireland. *INJ*, **16**, 74–5.

Braun-Blanquet, J. (1952). Pflanzensoziologische Ueberlegungen als Hilfsmittel zur Erkennung systematischen Einheiten am Beispiel von *Antennaria hibernica* dargelegt. *Vegetatio*, **3**, 298–300.

Braun-Blanquet, J. & Tüxen, R. (1952). Irische Pflanzengesellschaften. *Veröffentlichungen des geobotanischen Institutes Rübel in Zürich*, **25**, 224–415.

Carter, C. (1846). A botanical ramble in Ireland. *Phytologist*, **2**, 512–14.

Colgan, N. (1893). Notes on the flora of the Aran Islands. *IN*, **2**, 75–8, 106–11.

(1897). *Euphrasia salisburgensis* Funck in Ireland. *IN*, **6**, 105–8.

(1900). Botanical notes on the Galway and Mayo highlands. *IN*, **9**, 111–18.

Colgan, N. & Scully, R. W. (1898). *Cybele Hibernica*, edn 2. 538 pp. Dublin.

Connolly, G. (1930). The vegetation of southern Connemara. *PRIA*, **39B**, 203–31.

Corry, T. H. (1880). Notes on a botanical ramble in the county of Clare, Ireland, in 1879. *Proceedings of the Belfast Natural History and Philosophical Society*, 1879–80, 167–207.

Cullen, J. (1976). The *Anthyllis vulneraria* complex: a resumé. *Notes from the Royal Botanic Garden, Edinburgh*, **35**, 1–38.

Dandy, J. E. (1969). Nomenclatural changes in the list of British vascular plants. *Watsonia*, **7**, 157–78.

de Valéra, M. (1958). *A Topographical Guide to the Seaweeds of Galway Bay*. 36 pp. Dublin: Institute for Industrial Research and Standards.

de Valéra, M., Pybus, C., Casley, B. & Webster, A. (1979). Marine algae of the northern shore of the Burren, Co. Clare. *PRIA*, **79B**, 259–69.

Drabble, E. (1930). Notes on Irish pansies. *JB*, **68**, 141–3.

Druce, G. C. (1909). Notes on Irish plants. *IN*, **18**, 209–13.

(1911). The International Phytogeographical Excursion to the British Isles. III. The floristic results. *New Phytologist*, **10**, 306–28.

(1931*a*). Additions to the Irish flora. *INJ*, **3**, 218.

(1931*b*). New county and other records: *Carex vesicaria*. *Reports of the Botanical Exchange Club*, **9**, 36.

Duncan, U. K. (1956). Plant records: *Euphrasia confusa*. *Proceedings of the Botanical Society of the British Isles*, **2**, 36.

Eager, A. R. & Scannell, M. J. P. (1981). William McCalla (*c*. 1814–1849), phycologist; his published papers of 1846. *Journal of Life Sciences, Royal Dublin Society*, **2**, 109–36.

Evans, E. & Turner, B. (1977). *Ireland's Eye*. 195 pp. Belfast.

Folan, A. C. M. & Mitchell, M. E. (1970). The lichens and lichen-parasites of Derryclare wood, Connemara. *PRIA*, **70B**, 163–70.

Foot, F. J. (1864). On the distribution of plants in the Burren, County of Clare. *PRIA*, **24**, 143–60.

Godwin, H. (1975). *The History of the British Flora*. 541 pp. Cambridge University Press.

Gordon, V. (1954). Plant notes: *Epilobium linnaeoides*. *Proceedings of the Botanical Society of the British Isles*, **1**, 37.

Gorham, E. (1957). The chemical composition of some Irish fresh waters. *PRIA*, **58B**, 237–43.

Gough, K. (1952). *Orchis traunsteinerioides* Pugsl. in Co. Clare. *INJ*, **10**, 273.

Graham, R. (1840). An account of an excursion to the west of Ireland. *Annual Reports of the Botanical Society of Edinburgh*, **3**, 54–7.

Guiry, M. D. (1978). A Consensus and Bibliography of Irish Seaweeds (*Bibliotheca phycologica*, **44**). 287 pp. Vaduz.

Haglund, G. E. (1935). Some *Taraxacum*-species from Ireland and Wales cultivated in the Botanic Gardens, Lund. *Botaniska Notiser*, **1935**, 429–38.

Hall, P. M. & Simpson, N. D. (1936). Plant records: *Viola lactea*. *Reports of the Botanical Exchange Club*, **11**, 241.

Halliday, G., Argent, G. C. & Hawksworth, D. L. (1967). Some Irish plant records. *INJ*, **15**, 313–18.

Hart, H. C. (1875). *A List of Plants found in the Islands of Aran, Galway Bay.* 32 pp. Dublin.

— (1883). Report on the flora of the Mayo and Galway mountains. *PRIA* (*ser. 2; Science*), **3**, 694–768.

Heslop-Harrison, J. (1949). *Pinguicula grandiflora* Lam. in N. Clare. *INJ*, **9**, 311.

Heslop-Harrison, J., Wilkins, D. A. & Greene, S. W. (1961). *Arenaria norvegica* Gunn in Co. Clare. *INJ*, **13**, 267–8.

Hooker, W. J. (1835). Another heath found in Ireland. *Companion to the Botanical Magazine*, **1**, 158.

How, W. (1650). *Phytologia Britannica.* 133 pp. London.

Howard, H. W. & Lyon, A. G. (1950). The identification and distribution of the British watercresses. *Watsonia*, **1**, 228–33.

Hurst, C. P. (1902). Notes on Connemara botany. *IN*, **11**, 45.

Ivimey-Cook, R. B. & Proctor, M. C. F. (1966). The plant communities of the Burren, Co. Clare. *PRIA*, **64B**, 211–301.

Jermy, A. C., Arnold, H. R., Farrell, L. & Perring, F. H. (1978). *Atlas of Ferns of the British Isles.* 101 pp. London.

Jessen, K. (1949). Studies in the late Quaternary deposits and flora-history of Ireland. *PRIA*, **52B**, 85–290.

Jonsell, B. (1968). Studies on the north-west European species of *Rorippa*, s. str. *Symbolae botanicae Upsalienses* **19(2)**, 1–221.

Kelly, D. L. & Kirby, E. N. (1982). Irish native woodlands on limestone. *Journal of Life Sciences, Royal Dublin Society*, (in press).

K'Eogh, J. (1735). *Botanologia universalis Hibernica.* Cork.

Lambe, E., Mitchell, M. E. & O'Connell, M. (1978). A reassessment of the status of *Mercurialis perennis* L. in the Burren, Co. Clare. *INJ*, **19**, 157–8.

Lambert, D. S. (1980). Field records – plants: *Taraxacum. INJ*, **20**, 169.

Lambert, D. S. & Hackney, P. (1976). *Taraxacum* records from Ireland. *INJ*, **18**, 310–12.

— (1979). Field records – plants: *Taraxacum. INJ*, **19**, 370–1.

Levinge, H. C. (1892). *Neotinea intacta* in County Clare. *JB*, **30**, 194–5.

— (1893). *Limosella aquatica* in Ireland. *JB*, **31**, 309–10.

Lhwyd, E. (1712). Some further observations relating to the antiquities and natural history of Ireland. *Philosophical Transactions of the Royal Society of London*, **27**, 524–6.

Linton, E. F. & Linton, W. R. (1886). Notes of a botanical tour in west Ireland. *JB*, **24**, 18–21.

McClintock, D. C. (1960). Plant records: *Polygonum cuspidatum. Proceedings of the Botanical Society of the British Isles*, **3**, 404.

McDonnell, J. K. & Fahy, E. (1978). An annotated list of the algae comprising a lotic community in western Ireland. *INJ*, **19**, 184–7.

MacGowran, B. (1979). *Rorippa islandica* (Oeder *ex* Murray) Borbás in turloughs of south-east Galway (H15). *INJ*, **19**, 326–7.

Mackay, J. T. (1806). A systematic catalogue of rare plants found in Ireland. *Transactions of the Dublin Society*, **5**, 121–84.

— (1830). Notice of a new indigenous heath, found in Cunnamara. *Transactions of the Royal Irish Academy*, **16**, 127–8.

— (1836). *Flora Hibernica.* 354 + 279 pp. Dublin.

— (1859). Additions to the plants of Ireland since the publication of 'Flora Hibernica'.

Proceedings of the Dublin University Zoological and Botanical Association, **1**, 250–3.

Mackie, G. (1970). Plant records from W. Galway and Westmeath. *INJ*, **16**, 319.

Malloch, A. J. C. (1976). An annotated bibliography of the Burren. *Journal of Ecology*, **64**, 1093–105.

Markgraf-Dannenberg, I. (1952). Studien an irischen *Festuca*-rassen. *Veröffentlichungen des geobotanischen Institutes Rübel in Zürich*, **35**, 114–42.

Marshall, E. S. (1896). Irish plants collected in June, 1896. *JB*, **34**, 496–500.

— (1899). Remarks on the 'Cybele Hibernica', edn 2. *JB*, **37**, 269–72.

Marshall, E. S. & Shoolbred, W. A. (1896). Irish plants observed in 1895. *JB*, **34**, 250–8.

Moon, J. McK. (1955). *Epilobium linnaeoides* in Ireland. *INJ*, **11**, 292.

Moore, D. (1854). Notes on some rare plants, including *Ajuga pyramidalis*, in Arran. *Phytologist*, **5**, 189–91.

Moore, D. & More, A. G. (1866). *Contributions towards a Cybele Hibernica*. 399 pp. Dublin.

More, A. G. (1855). Notes on the flora of the neighbourhood of Castle Taylor, in the County of Galway. *Proceedings of the Botanical Society of Edinburgh*, 1855, 26–30.

— (1860). Localities for some plants observed in Ireland, with remarks on the geographical distribution of others. *Proceedings of the Dublin University Zoological and Botanical Association*, **2**, 54–65.

— (1869). Discovery of *Aira uliginosa*, Weihe, at Roundstone, Co. Galway. *JB*, **7**, 265–6.

— (1872). On recent additions to the flora of Ireland. *PRIA* (*ser. 2*; *Science*), **1**, 256–93.

— (1873). On recent additions to the flora of Ireland. *JB*, **11**, 115–19.

— (1876). Report on the flora of Inish-Bofin, Galway. *PRIA* (*ser. 2*; *Science*), **2**, 533–78.

Murray, R. P. (1887). *Arabis ciliata* R. Br. in Clare. *JB*, **25**, 183.

Muskett, A. E. & Malone, J. P. (1980). Catalogue of Irish fungi – II. Hymenomycetes. *PRIA*, **80B**, 197–276.

Nelmes, E. (1948). Plant records: *Carex lepidocarpa*. *Reports of the Botanical Exchange Club*, **13**, 316.

Nelson, E. C. (1981a). Studies in *Erica mackaiana* Bab. I. Distribution in Connemara, Ireland. *INJ*, **20**, 199–202.

— (1981b). William McCalla – a second 'panegyric' for an Irish phycologist. *INJ*, **20**, 275–83.

[Newman, E.] (1846). Occurrence of *Erica ciliaris* in the county Galway, west of Ireland. *Phytologist*, **2**, 683.

Newman, E. (1854). *A History of British Ferns and Allied Plants*, 3rd edn. 344 pp. London.

Nowers, J. E. & Wells, J. G. (1892). The plants of the Aran Islands, Galway Bay. *JB*, **30**, 180–3.

O'Flaherty, J. T. (1824). A sketch of the history and antiquities of the southern islands of Aran. *Transactions of the Royal Irish Academy*, **14**, 79–140.

Ogilby, L. (1845). Notes of a botanical ramble in Connemara and Aran. *Phytologist*, **2**, 345–51.

O'Kelly, P. B. (1903). A botanical ramble last June at Gortbweehan Lake. *Journal of the Limerick Field Club*, **2**, 190.

Oliver, D. (1851). Notes of a botanical ramble in Ireland last autumn. *Phytologist*, **4**, 125–8.

— (1852). Botanical notes of a week in Ireland during the present month (August, 1852). *Phytologist*, **4**, 676–9.

O'Mahony, T. (1860). Notes of a botanical excursion in Clare. *Proceedings of the Dublin Natural History Society*, **1**, 30–4.

Osvald, H. (1949). Notes on the vegetation of British and Irish mosses. *Acta phytogeographica suecica*, **26**, 1–62.

Pearsall, W. H. (1936). *Hydrilla verticillata* Presl, a plant new to Ireland. *INJ*, **6**, 20–1.

Pearsall, W. H. & Lind, E. M. (1941). A note on a Connemara bog type. *Journal of Ecology*, **29**, 62–8.

(1942). The distribution of phytoplankton in some north-west Irish loughs. *PRIA*, **48B**, 1–24.

Perring, F. H. & Walters, S. M. (1962). *Atlas of the British Flora*. 432 pp. London.

Perring, F. H. & Sell, P. D. (1968). *Critical Supplement to the Atlas of the British Flora*. 159 pp. London.

Petiver, J. (1703). Gazophylactii naturae et artis decas tertia. London.

Phillips, R. A. (1900). *Viola lactea* Smith, in Ireland. *IN*, **9**, 244.

(1924). Some new county and vice-county records for Irish plants. *IN*, **33**, 33–8.

Praeger, R. L. (1895a). *Pilularia* in Connemara. *IN*, **4**, 165.

(1895b). Notes on the flora of Aranmore. *IN*, **4**, 249–52.

(1895c). Aran Island brambles. *IN*, **4**, 318.

(1897). Irish plants, collected chiefly in the province of Leinster in 1896. *IN*, **6**, 89–100.

(1899). *Juncus obtusiflorus* in the Aran Islands. *IN*, **8**, 25.

(1900). Botanical exploration in 1899. *IN*, **9**, 135–49.

(1901a). *Habenaria intacta* in Clare. *IN*, **10**, 143.

(1901b). *Irish Topographical Botany*. clxxxviii+410 pp. Dublin. (Published as Vol. 7 of *PRIA* (ser. 3), but usually treated as an independent publication.)

(1902a). Additions to 'Irish Topographical Botany' in 1901. *IN*, **11**, 1–8.

(1902b). Gleanings in Irish Topographical Botany. *PRIA*, **24B**, 61–94.

(1903a). Additions to 'Irish Topographical Botany' in 1902. *IN*, **12**, 23–40.

(1903b). *Pinguicula grandiflora* in Clare. *IN*, **12**, 269.

(1905). Notes on the botany of central Clare. *IN*, **14**, 188–94.

(1906). Notes of a western ramble. *IN*, **15**, 257–66.

(1907). On recent extensions of the range of some rare western plants. *IN*, **16**, 241–3.

(1909a). Botanical notes, chiefly from Lough Mask and Kilkee. *IN*, **18**, 32–40.

(1909b). *Bartsia viscosa* in Connemara. *IN*, **18**, 253.

(1909c). *A Tourist's Flora of the West of Ireland*. 243 pp. Dublin.

(1911). Notes on the flora of Inishbofin. *IN*, **20**, 165–72.

(1912). Plants from western Ireland. *IN*, **21**, 26.

(1932a). The flora of the turloughs: a preliminary note. *PRIA*, **41B**, 37–45.

(1932b). Some noteworthy plants found in or reported from Ireland. *PRIA*, **41B**, 95–124.

(1933). Stations for some uncommon Irish plants. *INJ*, **4**, 193–6.

(1934a). New stations for Irish plants. *INJ*, **5**, 33–6.

(1934b). The *Sorbus aria* group in Ireland. *INJ*, **5**, 50–2.

(1934c). A contribution to the flora of Ireland. *PRIA*, **42B**, 55–86.

(1934d). *The Botanist in Ireland*. 587 pp. Dublin.

(1939). A further contribution to the flora of Ireland. *PRIA*, **45B**, 231–54.

(1946). Additions to the knowledge of the Irish flora, 1939–1945. *PRIA*, **51B**, 27–51.

(1951). Hybrids in the Irish flora: a tentative list. *PRIA*, **54B**, 1–14.

Praeger, R. L. & Carr, J. W. (1895). [Report of the Field Club Union Conference at Galway]: Phanerogams, ferns, &c. *IN*, **14**, 246–52.

Pritchard, N. M. (1959). *Gentianella* in Britain. *Watsonia*, **4**, 169–92.

Pugh, J. P. (1953). The distribution of *Dryopteris borreri* Newm. in the British Isles. *Watsonia* 3, 57–65.

Pugsley, H. W. (1934). *Sorbus porrigens* Hedlund in Ireland. *JB*, 72, 58.

(1935). On some marsh orchids. *Journal of the Linnean Society (Botany)*, 46, 553–92.

(1948). A prodromus of the British *Hieracia*. *Journal of the Linnean Society (Botany)*, 54, 1–356.

Rechinger, K. H. (1961). Notes on *Rumex acetosa* L. in the British Isles. *Watsonia*, 5, 64–6.

Richards, A. J. (1972). The *Taraxacum* flora of the British Isles. *Watsonia*, 9 (*Suppl.*), 1–141.

Rob, C. M. (1948). Plant records: *Carex punctata*. *Reports of the Botanical Exchange Club*, 13, 215.

Roden, C. (1978). Plant records from the mountains around Lough Nafooey. *INJ*, 19, 208.

(1979). The vascular flora and vegetation of some islands of Lough Corrib. *PRIA*, 79B, 223–32.

Rose, F. (1967). *Eriophorum gracile* Roth new to Ireland. *INJ*, 15, 361–2.

Ross, R. (1967). On some Linnean species of *Erica*. *Journal of the Linnean Society (Botany)*, 60, 61–73.

Scannell, M. J. P. (1965). *Phycopeltis*, a genus of alga not previously recorded for the British Isles. *INJ*, 15, 75.

(1972). 'Algal paper' of *Oedogonium* sp., its occurrence in the Burren, Co. Clare. *INJ*, 17, 147–52.

(1973). *Juncus planifolius* R. Br. in Ireland. *INJ*, 17, 308–9.

(1975). *Ruppia cirrhosa* (Petagne) Grande, an addition to the flora of West Galway. *INJ*, 18, 220–1.

(1978). *Asplenium cuneifolium* Viv. in West Galway, Ireland. *INJ*, 19, 245.

Scannell, M. J. P. & McClintock, D. C. (1974). *Erica mackaiana* Bab., in Irish localities and other plants of interest. *INJ*, 18, 81–5.

Scannell, M. J. P. & Synnott, D. M. (1972). *Census Catalogue of the Flora of Ireland*. 127 pp. Dublin.

Scannell, M. J. P. & Webb, D. A. (1976). The identity of the Renvyle *Hydrilla*. *INJ*, 18, 327–31.

Scannell, M. J. P. & White, J. (1976). *Cryptogramma crispa* in West Galway. *INJ*, 18, 336.

Seebohm, H. (1851). List of ferns found in Connemara. *Naturalist*, 1, 220–2.

Shivas, M. G. (1962). The *Polypodium vulgare* complex. *British Fern Gazette*, 9, 65–70.

Sim, J. (1859). Remarks on the flora of Ireland. *Phytologist* (*new series*), 3, 353–7.

Simpson, N. D. (1936). Plant records: *Rubus chlorothyrsus* and *Polygonum polystachion*. *Reports of the Botanical Exchange Club*, 11, 27, 40.

Sledge, W. A. (1942). Plant records: *Callitriche autumnalis*, *Ligusticum scoticum*, *Potamogeton praelongus*. *Reports of the Botanical Exchange Club*, 12, 278, 280, 298.

Sleep, A. (1980). On the reported occurrences of *Asplenium cuneifolium* and *A. adiantum-nigrum* in the British Isles. *Fern Gazette*, 12, 103–7.

Stelfox, A. W. (1946). An erroneous record for *Saxifraga geum* auctt. *INJ*, 8, 437.

(1947). C. C. Babington's record for *Saxifraga geum* in Connemara. *INJ*, 9, 29.

Stirling, A. McG. & Beckett, K. A. (1966). Plant records: *Asplenium septentrionale*. *Proceedings of the Botanical Society of the British Isles*, 6, 236.

Sweeting, M. E. (1953). The enclosed depression of Carran, County Clare. *Irish Geography*, 2, 218–24.

[Tansley, A. G.] (1908). The British Vegetation Committee's visit to the West of Ireland. *New Phytologist*, **7**, 253–60.

Teunissen, D. & Teunissen-van Oorschot, H. G. C. M. (1980). The history of the vegetation in S.W. Connemara (Ireland). *Acta botanica neerlandica*, **29**, 285–306.

Tomlinson, W. J. C. (1909). *Habenaria intacta* in West Galway. *IN*, **18**, 156.

Townsend, F. (1897). A monograph of the British species of *Euphrasia*, parts 2 and 4. *JB*, **35**, 395–406, 465–77.

Tutin, T. G. (1980). *Umbellifers of the British Isles*. 197 pp. London.

Tutin, T. G. & Chater, A. O. (1974). *Asperula occidentalis* Rouy in the British Isles. *Watsonia*, **10**, 170–1.

Vines, J. H. & Druce, G. C. (1914). *An Account of the Morisonian Herbarium*. 350 pp. Oxford.

Wade, W. (1802). Catalogus plantarum rariorum in comitatu Gallovidiae, praecipue Cunnamara inventarum. *Transactions of the Dublin Society*, **2**, 103–27.

(1804). Plantae rariores in Hibernia inventae. *Transactions of the Dublin Society*, **4**, 1–214.

Walters, S. M. (1949*a*). *Alchemilla vulgaris* L. agg. in Britain. *Watsonia*, **1**, 6–18.

(1949*b*). *Aphanes microcarpa* (Boiss. & Reut.) Rothm. in Britain. *Watsonia*, **1**, 163–9.

Watson, W. C. R. (1958). *Handbook of the Rubi of Great Britain and Ireland*. 274 pp. Cambridge.

Watts, W. A. (1963). Late-glacial pollen zones in western Ireland. *Irish Geography*, **4**, 367–76.

(1977). The late Devensian vegetation of Ireland. *Philosophical Transactions of the Royal Society* (*B*), **280**, 273–93.

(1979). Regional variation in the response of vegetation to late-glacial climatic events in Europe. In *Studies in the Late-Glacial of North-West Europe*, ed. J. J. Lowe, J. M. Gray & J. E. Robinson, pp. 1–21. Oxford.

Webb, D. A. (1947). Notes on the acidity, chloride content and other chemical features of some Irish fresh waters. *Scientific Proceedings of the Royal Dublin Society*, **24**, 215–28.

(1948). The Robertsonian saxifrages of Galway and Mayo. *INJ*, **9**, 105–7.

(1950*a*). Hybridization and variation in the Robertsonian saxifrages. *PRIA*, **53B**, 85–97.

(1950*b*). A revision of the dactyloid saxifrages of north-western Europe. *PRIA*, **53B**, 207–40.

(1952). *Alchemilla vulgaris* agg. in Ireland: a preliminary report. *INJ*, **10**, 298–300.

(1954). Notes on four Irish heaths, part 1. *INJ*, **11**, 187–92.

(1955). Woodland relics in the Burren. *INJ*, **11**, 251–2.

(1956). *Spergularia rubra* in Connemara. *INJ*, **12**, 111.

(1957). *Hypericum canadense* L., a new American plant in western Ireland. *INJ*, **12**, 113–16.

(1959). New records for *Spiranthes romanzoffiana*, *Stachys officinalis* and *Eriophorum latifolium*. *INJ*, **13**, 20–1.

(1962). Noteworthy plants of the Burren: a catalogue raisonné. *PRIA*, **62B**, 117–34.

(1963*a*). Notulae systematicae ad Floram Europaeam spectantes: Saxifraga stellaris. *Feddes Repertorium specierum novarum*, **68**, 198–9.

(1963*b*). *Hymenophyllum wilsonii* in the Burren. *INJ*, **14**, 155.

(1965). *Spiranthes romanzoffiana* on Lough Corrib. *INJ*, **15**, 21.

(1966). *Erica ciliaris* in Ireland. *Proceedings of the Botanical Society of the British Isles*, **6**, 221–5.

(1970). Plant records: *Trifolium fragiferum* and *Minulus guttatus × luteus*. *Watsonia*, **8**, 165.

(1977). *An Irish Flora*. 6th edn. 277 pp. Dundalk.

(1978). The status of *Mercurialis perennis* in the Burren. *INJ*, **19**, 200.

(1979). Three trees naturalized in Ireland. *INJ*, **19**, 369.

(1980*a*). The flora of the Aran Islands. *Journal of Life Sciences, Royal Dublin Society*, **2**, 51–83.

(1980*b*). The biological vice-counties of Ireland. *PRIA*, **80B**, 179–96.

(1982). Plant records from Connemara and the Burren, *INJ*, **20**, 466–71.

Webb, D. A. & Glanville, E. V. (1962). The vegetation and flora of some islands in the Connemara lakes. *PRIA*, **62B**, 31–54.

Webb, D. A. & Halliday, G. (1973). The distribution, habitat and status of *Hypericum canadense* L., in Ireland. *Watsonia*, **9**, 333–56.

Webb, D. A. & Hodgson, J. (1968). The flora of Inishbofin and Inishshark. *Proceedings of the Botanical Society of the British Isles*, **7**, 345–400.

White, J. & Doyle, G. (1978). *Neotinea maculata* (*N. intacta*) in woodland. *INJ*, **19**, 187–9.

Williams, P. W. (1966). Limestone pavements with special reference to western Ireland. *Transactions of the Institute of British Geographers*, **40**, 155–72.

Willmot, A. (1979). An ecological survey of the ferns of the Burren, Co. Clare, Eire. *Fern Gazette*, **12**, 9–28.

Wright, E. P. (1867). Notes of a botanical tour to the islands of Arran, Co. Galway. *JB*, **5**, 50–1.

(1871). Notes on the flora of the islands of Arran. *Proceedings of the Dublin Natural History Society*, **5**, 96–105.

MAPS

INDEXES

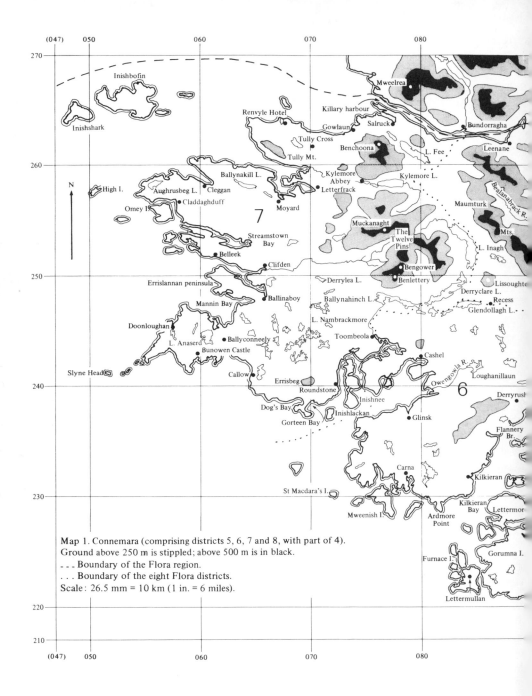

Map 1. Connemara (comprising districts 5, 6, 7 and 8, with part of 4).
Ground above 250 m is stippled; above 500 m is in black.
_ _ _ Boundary of the Flora region.
. . . Boundary of the eight Flora districts.
Scale: 26.5 mm = 10 km (1 in. = 6 miles).

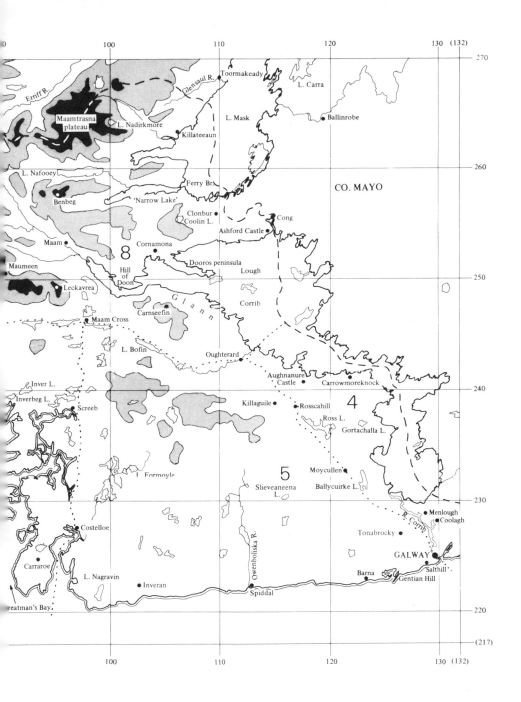

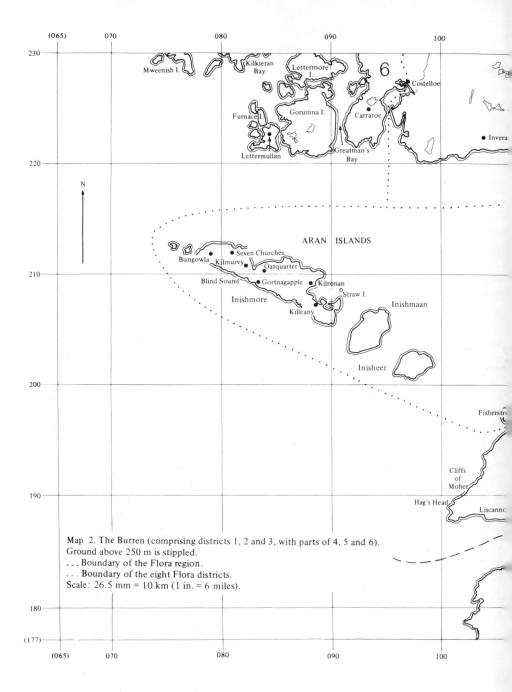

Map 2. The Burren (comprising districts 1, 2 and 3, with parts of 4, 5 and 6).
Ground above 250 m is stippled.
- - - Boundary of the Flora region.
. . . Boundary of the eight Flora districts.
Scale: 26.5 mm = 10 km (1 in. = 6 miles).

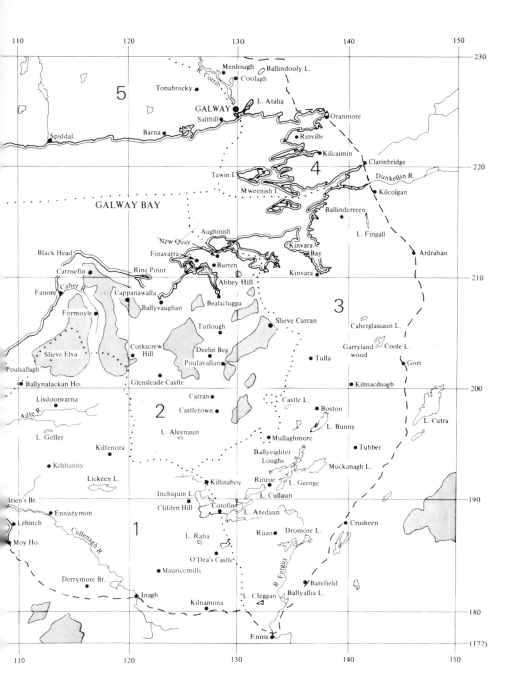

110 120 130 140 150

230

5

Menlough
Ballindooly L.
R. Corrib
Coolagh
Tonabrocky

L. Atalia
GALWAY
Salthill
Oranmore

Spiddal
Barna
Rinville

Kilcaimin
4
Clarinbridge
220

Tawin I.
Dunkellin R.
Kilcolgan

Mweenish I.
Ballinderreen

GALWAY BAY
Ballindereen

Aughinish
L. Fingall

New Quay
Kinvara
Finavarra
Bay
Ardrahan

Black Head
Rine Point
Burren
Kinvara

Carnsefin
Abbey Hill
210

Fanore
Caher R.
Cappanawalla
Caherglassaun L.

Formoyle
Ballyvaughan
Bealaclugga
Slieve Carran
3

Garryland
Coole L.
wood

Turlough
Gort

Poulsallagh
Corkscrew
Hill
Deelin Beg
Tulla

Slieve Elva
Poulavallan

Ballynalackan Ho.
Glensleade Castle
Kilmacduagh
200

Lisdoonvarna
Carran
Castle L.
L. Cutra

Aille R.
2
Castletown
Boston

L. Aleenaun
L. Bunny

L. Goller
Mullaghmore

Kilfenora
Tubber

Kilshanny
Ballyeighter
Loughs
Muckanagh L.

Lickeen L.
Killinaboy
Rinroe
L. George

Inchiquin L.
L. Cullaun

Brien's Br.
Clifden Hill
Corofin
L. Atedaun
190

Ennistymon
Ruan
Dromore L.
Crusheen

Lehinch
L. Raha

Moy Ho.
Cullenagh R.
O'Dea's Castle

R. Fergus

Derrymore Br.
Mauricemills
Barefield

Inagh
Ballyallia L.
Kilnamona
L. Cleggan

180

Ennis

(177)

110 120 130 140 150

307

TOPOGRAPHICAL INDEX

The figure in parentheses following each name indicates the district in which it is located. In the case of names marked on the maps in this book, there follows an indication of the 10-km square in which it is to be found. This is done by citing the grid-lines numbered in the margin of the map: thus 130/210 for Abbey Hill indicates that it is in the square bounded on the west by grid-line 130 and on the south by grid-line 210. In the case of names not marked on our maps a conventional grid-reference is given, by which they can be easily located on the ½-inch Ordnance Survey maps, but to save readers the necessity of referring to a ½-inch map we also give an indication of compass-bearing and distance from a place which *is* given on the maps in this book.

Nearly all the place-names used are to be found on the ½-inch maps; those few that are not are indicated by an asterisk.

Abbey Hill (2) 120–130/210.
Aille R. (1) 100–110/190.
Aillebrack L. (7) L 58 43. 1½ km S.S.W. of Doonloughan.
Aleenaun, L. (2) 120/190.
Altnagaighera (7) L 75 60. Mountain 2 km N. of Kylemore Abbey.
Anaserd, L. (7) 060/240.
*Anessaundoo, L. (7) L 73 46. 3 km N.W. of Toombeola.
Annaghvaan I. (6) L 90 30. N.E. of Lettermore I.
Aran Islands (2) 070–090/200–210.
Ard Bay (6) L 76 31. N. of Mweenish I.
Ardbear Bay (7) L 64 48. 2 km S. of Clifden.
Ardderry L. (5) L 99 45. 2 km E. of Maam Cross.
Ardfry Ho. (4) M 34 21. 3 km W. of Kilcaimin.
Ardmore (7) L 59 52. 2 km W.N.W. of Belleek.
Ardmore Point (6) 080/220.
Ardnasillagh (4) M 15 41. 3½ km E.S.E. of Oughterard.
Ardrahan (3) 140/210.
Ashford Castle (8) 110/250.
Atalia, L. (4) 130/220. The name is also used in earlier literature for the turlough-like marsh on Inishmore, now usually called Turloughmore.
Atedaun, L. (3) 120–130/180.
Athola, L. (7) L 62 48. On the S. side of the Errislannan peninsula.
Athry L. (6) L 81 46. 1 km S. of the S. end of Derryclare L.
Aughawoolia, L. (6) L 97 41. 3 km N. of Screeb.
Aughinish (3) 120/210.
Aughnanure Castle (4) 110/240.

Aughrusbeg L. (7) 050/250.
Aughrusmore (7) L 55 57. 1 km S.E. of Aughrusbeg L.

Ballinaboy (7) 060/240.
Ballinderreen (3) 130/210.
Ballindooly L. (4) 130/220.
Ballyallia Ho. and L. (3) 130/180.
Ballyclery (3) M 39 12. 2½ km N.E. of Kinvara.
Ballyconneely (7) 060/240. Ballyconneely Bay lies to the S. of the village.
Ballycotteen, Lower (1) R 04 92. 3½ km N.N.W. of Liscannor.
Ballycuirke L. (4) 120/230.
Ballycullinan L. (1) R 28 85. 3 km S. of Corofin.
Ballydoo L. (8) M 03 53. 1 km N.N.W. of Cornamona.
Ballyeighter Loughs (3) 130/190.
*Ballyloughan (4) M 31 24. A south-eastern suburb of Galway.
Ballynacloghy (4) M 35 19. 2½ km S.W. of Kilcaimin.
Ballynahinch Castle (7) L 77 46. On the S. side of Ballynahinch L.
Ballynahinch L. (7) 070/240.
Ballynakill L. (7) 060/250.
Ballynakill harbour (7) L 67 58. Between Moyard and Tully Mt.
Ballynalackan Castle and Ho. (1) 110/200.
Ballyogan Loughs (3) R 37 91. 1 km S. of Muckanagh L.
*Ballyportry L. (3) R 29 90. 2 km N.E. of Corofin.
Ballyvaughan (2) 120/200.
Barefield (3) 130/180.
Barna (5) 120/220.
Beaghcauneen L. (7) L 67 46. 2 km E.S.E. of Ballinaboy.

Bealaclugga (2) 120/200. Sometimes called Bell harbour.

Bealanabrack R. (8) 080–090/250.

Belleek (7) 060/250.

Benbaun (7) L 78 53. The highest summit of the Twelve Pins, 2 km E.S.E. of Muckanaght.

Benbeg (8) 090/250.

Benbreen (7) L 78 51. One of the Twelve Pins, 1 km N. of Bengower.

Benchoona (7) 070/260.

Bencorr (7) L 81 52. The highest peak on the eastern side of the Twelve Pins.

Bencorragh (8) L 98 57. 3 km E.N.E. of Benbeg.

Bengower (7) 070/250.

Benlettery (7) 070/240.

*Benwee (8) L 87 59. 2½ km S. of Leenane.

*Bilberry I. (8) M 04 51. 2 km S. of Cornamona.

Black Head (2) 110/210.

*Bleanoran (4) M 24 40. On the shore of L. Corrib, 3 km E.S.E. of Carrowmoreknock.

Bofin, L. (5) 100/240.

Bofin, L. (7) L 52 65. On Inishbofin, W. of the centre.

Bohaun (8) M 02 57. On the S. shore of the 'Narrow Lake'.

Bollard, L. (7) L 69 42. 2 km N. of the summit of Errisbeg.

Boston (3) 130/190.

Bovraughaun Hill (5) M 01 28. 5 km E.N.E. of Costelloe.

Briskeen, L. (3) M 41 00. 1½ km E.N.E. of Kilmacduagh.

*Brontcen I. (8) M 14 45. 3½ km N.E. of Oughterard.

Bungowla (2) 070/210.

Bunnacunneen (8) L 93 57. Summit 2 km W. of Benbeg.

Bunnagippaun (5) M 11 39. 4 km S.S.W. of Oughterard.

Bunnahown (6) L 80 40. 2½ km S.S.E. of Cashel.

Bunny, L. (3) 130/190.

Bunowen (7) 050/240.

Burnthouse (4) M 24 41. On the shore of L. Corrib 2 km E. of Carrowmoreknock.

Burren (village) (2) 120/210.

Caher R. (2) 110/200.

*Chercloghane (1) M 15 01 3½ km N.E. of Lisdoonvarna.

Caherconnell (2) R 22 99. 4 km S.E. of the Corkscrew Hill.

Caherglassaun L. (3) 140/200.

Callaherick L. (6) L 80 38. 2 km N.N.E. of Glinsk.

Callow (7) 060/240.

Canal Br. (7, 8) L 80 47. Between Ballynahinch and Derryclare Loughs.

Cancregga (1) R 00 87. 2 km S. of Hag's Head.

Cannaver I. (8) M 09 48. In L. Corrib, 1¼ km S. of the tip of the Dooros peninsula. In the older literature usually spelt 'Canova'.

Canower (6) L 77 41. 3½ km S.W. of Cashel.

Cappanawalla (2) 110/200.

Carna (6) 070/230.

Carnseefin (8) 100/240.

Carnsefin (2) 110/210.

Carran (2) 120/190.

Carraroe (6) 090/220.

Carrick (8) M 06 52. 2 km E. of Cornamona.

Carrowgarriff (3) M 41 11. 3 km E.N.E. of Kinvara.

Carrowmore (4) M 36 18. 5½ km W.S.W. of Clarinbridge.

Carrowmoreknock (4) 120/240.

Cashel (6) 070–080/240.

Cashla Bay (5, 6) L 95 24. E. of Carraroe; Costelloe is at its head.

Cashleen (7) L 66 62. 2 km S.W. of Renvyle Hotel.

Castle L. (3) 130/190.

Castledodge R. (3) R 39 94. Flows past Tubber into the E. end of Muckanagh L.

Castlekirk I. (8) L 99 50. In the N.W. arm of L. Corrib.

Castletown (2) 120/190.

*Church L. (7) L 55 65. On Inishbofin, near the S.E. corner.

Clab, Glen of (2) M 29 02. Immediately E. of Poulavallan.

Claddagh (5) M 29 24. A district in the S.W. part of the city of Galway.

Claddaghduff (7) 050/250.

Clarinbridge (4) 140/220.

Cleggan (7) 060/250.

Cleggan, L. (3) 130/180.

Clifden (7) 060/250.

Clifden Hill (1) 120/180.

Cloghanulk (2) M 11 04. On the coast 3 km N.E. of Poulsallagh.

Clonbur (8) 100/250.

Cloonagat L. (7) L 69 47. 4½ km S.E. of Clifden.

Clooncoose (2) M 27 95. 3 km S.W. of Castletown.

Cong (8) 110/250.

*Conga, L. (7) L 73 46. 2 km N. of L. Nambrackmore.

Coolagh (4)　120/220.
Coole L. (3)　140/200.
Coolin L. (8)　100/250.
*Corcomroe Abbey (2)　M 29 08. 1 km N.W. of Bealaclugga.
Corkscrew Hill (2)　120/200.
Cornamona (8)　100/250.
Corofin (1, 3)　120/180.
Corranroo Bay (3)　M 32 11. 2 km N.E. of Abbey Hill.
Corrib, L. (4, 8)　090–130/230–250.
Corrib, R. (4)　120/220.
Costelloe (5, 6)　090/220.
Costelloe Br. (5)　L 97 26. 1 km S.E. of Costelloe village.
Courhoor L. (7)　L 59 57. 1½ km S.W. of Cleggan.
*Craiggamore (7)　L 71 45. The name often used in the literature for L. Nambrackmore and the adjacent hill.
*Cregduff L. (7)　L 71 39. 1 km S.W. of Roundstone.
Cregg Mt. (7)　L 71 52. A westerly spur of the Twelve Pins, 5½ km E.N.E. of Cliffden.
Crusheen (3)　130/180.
*Culfin R. (7)　L 75 63. Drains L. Muck and L. Fee, running into the sea N.E. of Gowlaun.
Cullaun, L. (3).　130/190.
Currarevagh Ho. (8)　M 10 46. On the shore of L. Corrib, 5 km N.W. of Oughterard.

Dawros R. (7)　L 72 58. 1 km N. of Letterfrack. Dawros Br. is at its mouth.
Dealagh R. (1)　R 11 91. Flows from L. Goller through Kilshanny to the sea at O'Brien's Br.
Deelin Beg (2)　120/200.
Derreen　(7)　L 63 52.　3 km　N.W.　of Clifden.
Derroogh　(5)　L 99 25.　3 km　S.E.　of Costelloe.
Derryclare L. (7)　080/240.
Derryerglinna (8)　M 06 42. 6 km W. of Oughterard.
Derrylea L. (7)　070/240.
Derryloughaun (5)　M 16 21. 3 km E. of Spiddal.
Derrymore Br. (1)　110/180.
Derryrush (6)　080/230.
Devilsmother (8)　L 91 62. Mountain 3½ km E.N.E. of Leenane. Its northern face lies outside the Flora region.
Diamond Hill (7)　L 73 58. A north-westerly spur of the Twelve Pins, 2 km S.E. of Letterfrack.

Dirkbeg L.　(8)　L 99 65.　1 km N. of L. Nadirkmore.
Dog's Bay (7)　060/230. The name is often used so as to include also Gorteen Bay.
Doolin　(2)　R 07 97.　1 km　N.E.　of Fisherstreet.
Doon, Hill of (8)　100/240. Doon Hill, near Bunowen in district 7, is referred to as Bunowen Hill to avoid confusion.
Doonagore L. (1)　R 07 94. 2 km S.E. of Fisherstreet.
Doonloughan (7)　050/240.
Dooros peninsula (8)　100/250.
Doughruagh (7)　L 75 59. Mountain 1 km N. of Kylemore Abbey.
Drimneen R. (5)　M 18 40. Flows northwards from Killaguile Ho. to L. Corrib.
Dromore L. and Castle (3)　130/180.
Drumconora (3)　R 35 83. 2 km N.W. of Barefield.
Drumcullaun L. (1)　R 19 82. 2 km N.W. of Inagh.
Dunkellin R. and estuary (3, 4)　130–140/210.

Ebor Hall (8)　M 08 53. 3 km S.S.W. of Clonbur.
Eenagh, L. (1)　R 26 82. 2½ km N.N.W. of Kilnamona.
Elmvale Ho. (1)　R 25 92. 2 km W. of Killinaboy.
Ennis (3)　130/170. Only the northern suburbs are in the Flora region.
Ennistymon (1)　110/180.
Errisbeg (7)　060/240.
Errislannan peninsula (7)　060/240–250.

Fadda, L. (7)　L 66 45. 2 km S.S.E. of Ballinaboy.
Fanore (2)　110/200.
Fee, L. (7)　070–080/260.
Feenagh (2)　M 19 06. 2 km E. of Formoyle.
Fergus, R. (1, 3)　120–130/170–190. It rises E. of Killinaboy and flows through Inchiquin, Atedaun and Dromore Loughs to Ennis.
Ferry Br. (8)　100/250.
Finavarra (2)　120/210.
Fingall, L. (3)　140/210.
Finny R. (8)　M 00 59. Runs from L. Nafooey into the 'Narrow Lake'.
Fisherstreet (1, 2)　100/190.
Flannery Br. (6)　080/230.
Formoyle (2)　110/200.
Formoyle, L.　100/230.
Fox Hill (8)　M 06 60. 2½ km N. of Ferry Br.
Freaghillaun (6)　L 73 35. 2½ km S.S.E. of Inishlackan.

Funshinmore (3) M 33 08. Road junction 3½ km S.E. of Abbey Hill.
Furnace I. (6) 080/220.

*Gable, Mt. (8) A name used by Marshall, not to be found in any gazetteer or map. The context suggests that it *might* indicate Benlevy, the hill to the W. of Clonbur.
Galway (4) 120/220. Most of the city lies in district 4, but the Cathedral and the area south and west of it is in district 5.
Garraun (7) L 76 61. 2½ km N. of Kylemore L.
Garraunbaun L. (7) L 66 57. 1 km N.W. of Moyard.
Garryland wood (3) M 42 03. Immediately S.W. of Coole L.
Garrynagry (4) M 19 38. 2 km E. of Rosscahill.
Garvillaun L. (1) R 25 82. 2 km E.S.E. of Mauricesmills.
*Gentian Hill (5) 120/220. A local name, long applied to the complex of promontories of calcareous glacial drift, marked on the ½-inch map as 'Blake's Hill' and 'Seaweed Point', lying 2–3 km E. of Barna.
George, L. (3) 130/190.
Glann (8) 100/240.
Glasgeivnagh Hill (2) R 29 96. 2 km S.E. of Castletown.
Glenbrickeen L. (7) L 66 53. 2½ km N.N.E. of Clifden.
Glencolumbkille Ho. (2) R 31 99. 3 km N.W. of Castle L.
Glendollagh L., Ho. and wood (6) 080/240.
Gleninagh Mt. (2) M 17 09. 3½ km S.E. of Black Head.
Glenquin (2) M 30 96. 2 km W. of Mullaghmore.
Glensleade Castle (2) 120/200.
Glentrasna (5) L 99 41. 3 km N.E. of Screeb.
Glinsk (6) 070/230.
Goller, L. (1) 110/190.
Gort (3) 140/200.
Gortachalla L. (4) 120/230.
Gortaclare Mt. (2) M 30 04. 3 km S.W. of Slieve Carran.
Gortdrishagh (8) M 11 46. Estate by L. Corrib, 4½ km N.W. of Oughterard.
Gorteen Bay (7) 060–070/230.
Gortmore (6) L 92 38. 4 km W.N.W. of Screeb.
Gortnagapple (2) 080/200.
Gorumna I. (6) 080/220.
Gowlanagower, L. (7) L 54 66. On Inishbofin, N.E. of the centre.

Gowlaun (7) 070/260.
Gragan Castle (2) M 20 03. At the foot of the Corkscrew Hill.
Greatman's Bay (6) 090/220.

Hag's Head (1) 100/180.
Hibbert, L. (6) L 88 22. Near the middle of Gorumna I.
High I. (7) 050/250.

*Ilaunmahon (4) M 25 37. 3 km E. of Gortachalla L.
Ilion West (8) L 86 52. 2 km E. of the centre of L. Inagh.
Inagh (1) 120/180.
Inagh, L. (7) 080/250.
Inchagoill (8) M 12 49. Island in L. Corrib, 6 km N. of Oughterard.
Inchiquin L. (1) 120/180–190.
Inishbofin (7) 050/260.
Inishdawros (7) L 63 40. Tidal island in the S.E. part of Ballyconneely Bay.
Inishdoorus (8) M 09 49. Adjacent to the tip of the Dooros peninsula.
Inisheer (2) 090/200.
Inishlackan (7) 070/230.
Inishmaan (2) 090/200.
Inishmore (2) 070–090/200 210.
Inishnee (7) 070/230–240.
Inishshark (7) 040/260.
Inver L. (6) 090/230–240.
Inveran (5) 100/220.

Keeraun, L. (6) L 78 31. Immediately S. of Carna.
Keeraunnagark (5) L 97 23. 2 km N.W. of L. Nagravin.
Keeraunnageeragh (8) M 05 47. Alternative name for Carnseefin.
*Kilbeg ferry (4) M 23 41. 2 km E.N.E. of Carrowmoreknock.
Kilbrickan (6) L 91 35. 5½ km S.W. of Screeb.
Kilcaimin (4) 130/220.
Kilcolgan (3, 4) 140/210.
Kilfenora (1) 110/190.
Kilkieran (6) 080/230.
Killaguile (5) 110/230.
Killateeaun (8) 100/260.
Killeany (2) 080/200.
Killinaboy (1, 3) 120/190.
Kilmacduagh (3) 140/200.
Kilmurvy (2) 080/210.
Kilnamona (1) 120/180.
Kilroghter (4) M 30 30. 2 km N.E. of Menlough.

Kilronan (2)   080/200.
Kilshanny (1)   110/190.
Kiltartan L.   (3)   M 45 05.   3 km N.E.   of Coole.
Kingstown Bay (7)   L 59 53.   3 km N.W.   of Belleek.
Kinvara (3)   130/210.
Knocka L.   (5)   M 15 28.   6 km N.N.E.   of Spiddal.
Knockboy   (6)   L 77 35.   2 km   S.W.   of Glinsk.
Knocknagreena   (5)   M 20 22.   2 km   W.   of Barna.
Knocknahillion (8)   L 87 53.   3 km E.N.E.   of the middle of L. Inagh.
Knocknalarabana (1)   R 07 94.   2 km S.S.E.   of Fisherstreet.
Kylemore Abbey and L. (7)   070/250.
Kylesa (6)   L 86 34.   1 km S.S.W.   of Flannery Br.

Leamaneh Castle (2)   R 23 93.   5 km E.   of Kilfenora.
Leckavrea Mt. (8)   090/240.
Leenane (8)   080/260.
Lehanagh Loughs (8)   L 87 51.   2 km N.E.   of Lissoughter.
Lehinch (1)   100/180.
Letterbreckaun (8)   L 85 55.   Peak in the Maumturk range 2½ km E. of the N. end of L. Inagh.
Lettercraffroe L. (5)   M 05 38.   9 km E.   of Screeb.
Letterdife   (7)   L 72 41.   1 km   N.   of Roundstone.
Letterfrack (7)   070/250.
Lettermullan (6)   080/220.
Lickeen L. (1)   110/190.
Liscannor (1)   100/180.
Lisdoonvarna (1)   110/190.
Lissoughter (8)   080/240.
Loughauneirin (8)   M 00 45.   3½ km   E.   of Maam Cross.
Loughanillaun (6)   080/240. The name is used also for several other lakes in Connemara, but in this work refers only to the lake so designated on our map.
Lugnabrick   (8)   L 96 55.   2 km   N.   of Maam.
Luick, L. (2)   M 28 07.   Immediately S.   of Bealaclugga.
Lusteenmore I. (8)   M 06 61.   In L. Mask, 2 km S.W. of Killateeaun.

Maam (8)   090/250.
Maam Cross (5, 6, 8)   090/240.

Maamtrasna (8)   M 00 60.   1 km N.E. of the E. end of L. Nafooey.
Maamtrasna plateau   090/260. The large, flat-topped mountain mass N. of L. Nafooey, of which only the S.E. half lies within the Flora region.
Mace Head (6)   L 73 32.   5½ km W. of Carna.
Mannin Bay (7)   060/240.
Mask, L. (8)   100–110/250–260.
Maumeen (8)   090/250.
Maumturk Mts. (8)   080/250.
Mauricesmills (1)   120/180.
Menlough (4)   120/220.
Merlin Park (4)   M 33 25. 4 km E. of Galway.
Moher, Cliffs of (1)   100/190.
Moneen Mt.   (2)   M 26 07.   3 km   E.   of Ballyvaughan.
Mooghna L.   (1)   R 13 85.   4 km   S.   of Ennistymon.
Moy Ho. (1)   100/180.
Moyard (7)   060/250.
Moycullen (4, 5)   120/230.
Moyrus (6)   L 75 34.   3½ km N.W. of Carna.
Muck, L. (7)   L 77 62.   Immediately N.W. of L. Fee.
Muckanagh L. (3)   130/190.
Muckanaght (7)   070/250.
Muckinish Bay (2)   M 27 10.   Extends northwards   and   north-westwards   from Bealaclugga.
Mullaghmore (2)   130/190.
Murree, L.   (2)   M 25 12.   1 km   W.   of Finavarra.
Mweenish I. (4)   130/210.
Mweenish I. (6)   070/220–230.

Nacarrigeen, L. (8)   L 80 59.   1 km S. of the S.E. end of L. Fee.
Nacorrussaun, L. (7)   L 64 45.   2½ km E. of Ballyconneely.
Nacreeva, L. (5)   L 97 32.   6 km S. of Screeb.
Nadirkmore, L. (8)   100/260.
Nafooey, L. (8)   090/250.
Nafurnace, L.   (5)   L 97 37.   1 km   S.E.   of Screeb.
Nagravin, L. (5)   090/220.
*Nagrooaun L. (7)   L 55 66.   On Inishbofin, near the N.E. corner.
Nahasleam, L. (6)   L 97 43.   2 km S.S.W. of Maam Cross.
Nalawney, L. (7)   L 69 41.   On the N. flank of Errisbeg.
Nambrackmore, L. (7)   070/240.
Naneevin, L.   (5)   M 16 38.   1 km   N.W.   of Rosscahill.
'Narrow Lake'   100/250.

New Quay (2)   120/210.
*Newtown Castle (2)   M 22 06. 2 km S.W. of Ballyvaughan.

Oatquarter (2)   080/210.
O'Brien's Br. (1)   100/180.
O'Dea's Castle (1)   120/180.
Omey I. (7)   050/250.
Oorid L. (6)   L 93 46. 4 km W. of Maam Cross.
Oranmore (4)   130/220.
Oughterard (5, 8)   110/240.
Owenboliska R. (5)   110/220–230.
Owenbrin R. (8)   M 02 65. Flows from the Maamtrasna plateau into L. Mask, S. of Killateeaun.
Owenbristy (3)   M 43 11.   3 km W. of Ardrahan.
Owengowla R. (6)   080/230–240.
Owenree R. (8)   M 01 47. Flows from Loughauneirin to enter L. Corrib due S. of the Hill of Doon.

Parkmore (4)   M 34 20.   3 km S.W. of Kilcaimin.
Pollacappul L. (7)   L 76 58. Immediately S.W. of Kylemore L.
Porridgetown (4)   M 16 40. 5 km S.E. of Oughterard.
Portcowrugh (2)   L 85 10.   1 km N.E. of Oatquarter.
Portmurvy (2)   L 82 10. On the coast, N.E. of Kilmurvy.
Poulavallan (2)   120/200.
Poulsallagh (2)   100/200.

Raha, L. (1)   120/180.
Recess (6, 8)   080/240.
*Renmore barracks (4)   M 32 25. On the coast 2½ km E. of Galway.
Renvyle (7)   060/260. The name is here used to indicate the neighbourhood of the hotel, not the village of 'Rinvyle', which lies 2 km to the S.E.
Rine Point (2)   120/210.
Rinroe (3)   130/190.
Rinville (4)   130/220.
Roadford (1, 2)   R 07 97.   2 km N.E. of Fisherstreet.
Rock I. (2)   L 75 12. The more westerly of the two islets off the W. end of Inishmore.
Roscaun Point (4)   M 35 24. Half-way between Oranmore and Galway.
Rosmuck (6)   L 92 33. 5½ km S.W. of Screeb.
Ross L. (4)   110–120/230.
Rosscahill (4, 5)   110/230.
Roundstone (7)   070/230–240.

Ruan (3)   130/180.
*Rusheenduff, L. (7)   L 67 63. Beside Renvyle Hotel.

St Macdara's I. (6)   070/230.
Salruck (7)   070/260.
Salthill (5)   120/220.
Screeb (6)   090/230.
Seven Churches (2)   080/210.
Shanakeever L. (7)   L 66 52. 2 km N.E. of Clifden.
Shandangan L. (3)   R 30 91. 2½ km E.S.E. of Killinaboy.
Shindilla, L. (8)   L 96 46. 1½ km W. of Maam Cross.
Skannive, L. (6)   L 81 32. 2 km E. of Carna.
Slieveaneena L. (5)   110/230.
Slievebeg (1)   R 17 89.   4 km E.N.E. of Ennistymon.
Slieve Carran (2)   130/200.
Slieve Elva (1)   110/200.
Slyne Head (7)   The term 'Slyne Head peninsula' is used to denote the peninsula extending westwards from Ballyconneely (050–060/240). Slyne Head itself is an island off its western tip.
Spiddal (5)   110/220.
Srahnalong (8)   M 00 61. 2 km N.E. of the N.E. corner of L. Nafooey.
Srue (4)   M 18 40. 3 km W. of Carrowmoreknock.
Streamstown Bay (7)   060/250.

Tanny, L. (7)   L 64 55.   2½ km S.W. of Moyard.
Tawin I. (4)   130/210.
Teernakill Br. (8)   L 96 52. 1 km S.S.W of Maam.
Templebannagh L.   (3)   R 38 94.   2 km W.N.W. of Tubber.
Terryland (4)   M 29 26. 2½ km N. of Galway.
Tirneevin (3)   M 40 01. 4 km W. of Gort.
Tonabrocky (5)   120/220.
Toombeola (6, 7)   070/240.
Townross (2)   M 24 07.   1½ km S.E. of Ballyvaughan.
Travaun L. (3)   R 35 97. 2 km W. of L. Bunny.
Truska, L. (7)   L 59 45.   2 km E. of Doonloughan.
Truskan, L. (6)   L 81 30. 3 km S.E. of Carna.
Tubber (3)   140/190.
Tulla (3)   130/200.
Tullaghnafrankagh L. (3)   M 43 15.   4 km N.N.W. of Ardrahan.
Tully Cross (7)   070/260.
Tully Mt. (7)   060/260.

Tullywee Br. (6, 8)   L 87 47. 1½ km E. of Recess.

Turlough village (2)   120/200.

Twelve Pins (7)   070–080/240–250. The mountain group bounded by Kylemore L. on the north, Ballynahinch L. on the south and L. Inagh on the east. There is no clear western boundary.

White I. (8)   M 09 57. 1½ km N. of Clonbur.

INDEX OF ENGLISH AND IRISH
NAMES

Adder's tongue 263
Agrimony 67
Airgead luachra 60
Aiteann 52
Alder 189
Alexanders 91
Allseed 44
American ladies'-tresses 196
Anemone 6
Angelica 97
Annual rock-rose 26
Apple 71
Ash 140
Aspen 194
Avens 62 63

Bainne muice 121
Balsam 48
Barberry 11
Barren strawberry 65
Bearberry 128
Bearnán lachan 142
Bedstraw 101
Bee orchid 201
Beech 191
Beech fern 273
Beet 175
Beith 188
Bent 245
Bilberry 128
Bindweed 146
Biolar 16
Biolar griagáin 17
Biolar trá 21
Birch 188
Bird's-foot trefoil 56
Bird's-nest orchid 195
Bishop's-weed 93
Blackberry 61
Black bryony 207
Black currant 74
Blackheads 118
Black medick 53
Blackthorn 72
Bladder campion 32
Bladderwort 162

Bloody cranesbill 45
Bluebell 205
Bog asphodel 206
Bogbean 142
Bog-cotton 228
Bog myrtle 188
Bog pimpernel 139
Bog thistle 117
Bracken 265
Bramble 61
Brioslán 64
Brooklime 155
Brookweed 139
Broom 53
Broomrape 161
Bryony 207
Buachalán buí 115
Buckthorn 50
Bugle 171
Buidhe mór 26
Bulbous buttercup 9
Bullace 73
Bulrush 212
Burdock 116
Bur-marigold 110
Burnet rose 68
Bur-reed 213
Burren rock-rose 27
Bush vetch 58
Butterbur 114
Buttercup 6
Butterfly orchid 203
Butterwort 163

Cabáiste Phádhraic 75
Caithleach dearg 12
Calamint 167
Campion 32
Canadian water-weed 217
Caorthann 70
Carline thistle 118
Carnán caisil 78
Carrot 98
Cat's ear 120
Cat's foot 109
Ceannabhán 228

Celandine 10 13
Celery 91
Centaury 141
Chaffweed 139
Chamomile 111
Charlock 21
Cherry 73
Cherry laurel 73
Chervil 94
Chickweed 34
Chicory 119
Cinquefoil 64 65
Cleavers 103
Clover 54
Cock's-foot 251
Coigeal na mban sidhe 212
Coll 189
Colt's-foot 114
Columbine 10
Comfrey 143
Copóg 182
Corncockle 33
Corn marigold 112
Corn spurrey 38
Cowslip 137
Cow-wheat 159
Crab apple 71
Cranberry 128
Cranesbill 45
Crann bán 51
Crann silín 73
Créachtach 85
Creeping buttercup 9
Creeping cinquefoil 64
Creeping Jenny 138
Creeping red fescue 253
Creeping thistle 118
Crowberry 135
Cuckoo-flower 17
Cudweed 109
Cúig-mhéarach 64
Cuileann 49
Cuillean trá 90
Currant 74

Dair 190
Daisy 107 113
Dame's violet 19
Damson 73
Dandelion 123
Danesweed 99
Dead-nettle 169
Devil's bit 106
Dewberry 61
Dock 182
Dodder 148
Dog-daisy 113
Dog rose 69
Dog's mercury 185
Dogwood 99
Dorset heath 130
Draighean 72
Dréimire gorm 149
Dréimire Mhuire 141
Dris 61
Druichtín móna 80
Dubh chosach 265
Duckweed 214
Duilleóg bháite 11
Duilleóg mhaith 120

Earball caitín 85
Earball rí 47
Early purple orchid 199
Eidhneán 99
Elder 100
Elecampane 110
Elm 185
Enchanter's nightshade 88
Eyebright 155

Fairy flax 44
False oat 247
Fea 191
Fearnóg 189
Feithleóg 100
Fennel 95
Feochadán 117
Feoras 50
Fescue 252
Field madder 104
Figwort 151
Filmy fern 264
Fiorin-grass 244
Fleabane 110
Flea sedge 231
Fliodh 34
Flowering rush 216
Fly orchid 201
Fool's parsley 96
Forget-me-not 144
Fothram 151

Foxglove 152
Foxtail 243
Fragrant orchid 201
Fraoch 129
Fraoch gallda 133
Fraoch na h-aon choise 133
Fraochán 128
Fraochóg 128
Frochan 128
Frogbit 218
Frog orchid 203
Fuchsia 88
Fuinnseóg 140
Fumitory 14
Furze 52

Gairleóg 206
Gallán mór 114
Garbhlus 103
Garlic 207
Garlic mustard 19
Gentian 141
Giant hogweed 98
Giant rhubarb 82
Giolcach 249
Giolcach sléibhe 53
Gipsywort 166
Giuis 260
Glasswort 177
Gliogán 274
Glúineach 31
Glúineach dhearg 179
Glúineach the 179
Goat's beard 121
Golden rod 107
Golden saxifrage 77
Goldilocks 9
Gooseberry 74
Goosefoot 174
Goose-grass 103
Gorse 52
Grass of Parnassus 77
Green-veined orchid 198
Grey poplar 194
Gronnlus 115
Ground ivy 167
Groundsel 115
Guelder-rose 100

Harebell 127
Hare's-foot clover 54
Hart's-tongue fern 266
Hawthorn 72
Hazel 189
Heartsease 29
Heather 129
Hedge mustard 19

Hedge-parsley 94
Helleborine 197
Hemlock 90
Hemp-agrimony 106
Hempnettle 169
Henbane 149
Herb-Robert 47
Hogweed 97
Holly 49
Holly fern 271
Honeysuckle 100
Hop 186
Hop clover 55
Horned poppy 13
Hornwort 11
Horse-chestnut 51
Horse-radish 21
Horsetail 274

Irish whitebeam 70
Italian rye-grass 252
Iúr 262
Ivy 99
Ivy-leaved crowfoot 8

Juniper 260

Kidney vetch 56
King-cup 10
Knapweed 118

Ladies' bedstraw 102
Ladies' tresses 196
Lady-fern 269
Lady's mantle 66
Lady's smock 17
Lamb's lettuce 105
Laurel 73
Leamhán 185
Leathín 62
Léine 17
Liath uisce 163
Liathán 112
Ling 129
Lobelia 127
London pride 76
Loosestrife 85 137
Lords and ladies 214
Lousewort 160
Luachair 207
Lucerne 53
Luibh Eoin Bhaiste 40
Lus na gcnamh mbriste 143

Machall coille 63
Machall uisce 63
Mackay's heath 130

Madder 101
Maidenhair fern 265
Maidenhair spleenwort 267
Male fern 270
Mallow 43
Mare's-tail 83
Marjoram 166
Marram 245
Marsh cinquefoil 65
Marsh cudweed 109
Marsh fern 273
Marsh mallow 43
Marsh marigold 10
Marsh pennywort 90
Marsh thistle 117
Meadow buttercup 9
Meadow-rue 6
Meadowsweet 60
Meadow vetchling 59
Méaracán gorm 127
Méaracán phuca 152
Medick 53
Mediterranean heath 132
Michaelmas daisy 107
Milkwort 31
Mint 164
Monkey-flower 151
Montbretia 204
Moonwort 263
Mountain ash 70
Mountain avens 62
Mountain everlasting 109
Mountain fern 272
Mountain pansy 30
Mouse-ear chickweed 34
Mugwort 113
Mullach dubh 118
Mullein 149
Mustard 19 21

Nead coille 6
Nealfheartach 64
Neanntóg 187
Nettle 187
New Zealand flax 205
Nightshade 149
Nipplewort 120
Nóinín 107
Nóinín mór 113

Oak 190
Ox-eye daisy 113

Pansy 30
Parsley 92
Parsley fern 265
Parsley-piert 66

Parsnip 97
Pearlwort 38
Pellitory 187
Pennywort 78
Peppermint 165
Perennial rye-grass 252
Periwinkle 143
Pignut 94
Pimpernel 138 139
Pine 260
Pipewort 224
Plantain 173
Polypody 273
Pondweed 218
Poplar 194
Poppy 12 13
Praiseach bhuí 21
Primrose 137
Privet 140
Purple loosestrife 85
Pyramidal orchid 198

Quaking-grass 251
Queen Anne's lace 94

Radish 25
Ragged robin 32
Ragwort 115
Railleóg 188
Raithneach 265
Raspberry 60
Red campion 32
Red dead-nettle 169
Red rattle 160
Redshank 179
Red valerian 105
Reed 249
Rest-harrow 53
Rock-rose 26 27
Roisnín radhairc 155
Rós 68
Rose 68
Rose-bay 86
Roseroot 78
Rowan 70
Royal fern 264
Rush 207
Rusty-back 269
Rye-grass 252

Sail chuach 27
Saileach 191
St Dabeoc's heath 133
St John's wort 40
St Patrick's cabbage 75
Salad burnet 67
Sally 191

Saltwort 178
Samhadh 183
Samhadh caorach 183
Samphire 95
Sandhill pansy 30
Sanicle 90
Scabious 106
Scarlet pimpernel 138
Sceach geal 72
Scotch rose 68
Scots pine 260
Scurvy-grass 21
Scutch-grass 258
Sea beet 175
Sea holly 90
Seakale 25
Sea lavender 135
Sea milkwort 138
Sea-pink 136
Sea purslane 175
Sea rocket 25
Sea spleenwort 266
Sea stock 14
Sea wormwood 114
Seamar chré 152
Seamróg 54
Seileastram 204
Self-heal 168
Shamrock 55
Sheep's bit 127
Sheep's fescue 252
Sheep's sorrel 183
Shepherd's needle 94
Shepherd's purse 23
Silverweed 64
Skull-cap 168
Slánlus 173
Sneezewort 111
Snowberry 100
Soapwort 31
Sorrel 183
Sow-thistle 121
Spanish bluebell 206
Spearmint 165
Spear thistle 117
Spearwort 8
Speedwell 152
Spindle-tree 50
Spleenwort 266
Spotted persicaria 179
Spurge 184
Squinancy-wort 103
Sreang bogha 53
Stitchwort 34
Stock 14
Stone bramble 60
Stonecrop 78

Storksbill 47
Strawberry 63
Strawberry clover 55
Subh craobh 60
Subh talún 63
Sundew 80
Sweet violet 27
Swine's cress 24
Sycamore 51

Taith-fheithleann 100
Tansy 113
Teasel 105
Thale-cress 20
Thistle 117 118
Thrift 136
Thyme 167
Timothy 243
Toothwort 161
Torachas biadhain 8
Tormentil 64
Traveller's joy 5
Tree mallow 43
Trom 100
Tufted vetch 57
Turlough violet 29
Turnip 20
Tutsan 40
Twayblade 195

Ubhall fhiadhain 71

Urach bhallach 106

Valerian 104 105
Vervain 164
Vetch 57
Violet 27

Wall lettuce 123
Water avens 63
Water chickweed 39
Watercress 16
Water-lily 11
Water lobelia 127
Water milfoil 81
Water mint 165
Water-pepper 179
Water plantain 215
Water-starwort 84
Water-thyme 217
Welsh poppy 13
Weld 26
Whitebeam 70
White clover 55
White dead-nettle 170
White mustard 21
Whitethorn 72
White water-lily 11
Whitlow-grass 23
Willow 191
Willow-herb 85
Winter-cress 18

Wintergreen 134
Winter heliotrope 114
Wood anemone 6
Wood avens 63
Woodbine 100
Woodruff 103
Wood-rush 212
Wood sage 170
Wood sorrel 48
Wood violet 28
Woody nightshade 149
Wormwood 113
Woundwort 168

Yarrow 111
Yellow bird's-nest 135
Yellow clover 55
Yellow flag 204
Yellow fleabane 110
Yellow loosestrife 137
Yellow pimpernel 138
Yellow rattle 161
Yellow water-lily 11
Yellow-weed 26
Yellow-wort 140
Yew 262
Yorkshire fog 246

Zig-zag clover 54

INDEX OF LATIN NAMES

Acaena 68
Acer 51
Aceraceae 51
Achillea 111
Adiantum 265
Aegopodium 93
Aesculus 51
Aethusa 96
Agrimonia 67
Agropyron 258
Agrostemma 33
Agrostis 244
Aira 248
Ajuga 171
Alchemilla 66
Alisma 215
Alismataceae 215
Alliaria 19
Allium 206
Alnus 189
Alopecurus 243
Althaea 43
Amaryllidaceae 205
Ammophila 245
Anacamptis 198
Anagallis 138
Anchusa 144
Anemone 6
Angelica 97
Antennaria 109
Anthemis 111
Anthoxanthum 245
Anthriscus 94
Anthyllis 56
Aphanes 66
Apium 91
Apocynaceae 143
Aquifoliaceae 49
Aquilegia 10
Arabis 18
Araceae 214
Arabidopsis 20
Araliaceae 99
Arctium 116
Arctostaphylos 128
Arenaria 35 36 37

Armeria 136
Armoracia 21
Arrhenatherum 247
Artemisia 113
Arum 214
Asperula 103
Asplenium 266
Aster 107
Astragalus 57
Athyrium 269
Atriplex 175
Avenula 246

Baldellia 215
Balsaminaceae 48
Barbarea 18
Bartsia 160
Bellis 107
Berberidaceae 11
Berberis 11
Berula 93
Beta 175
Betula 188
Betulaceae 188
Bidens 110
Bilderykia 181
Blackstonia 140
Blechnum 266
Blysmus 228
Boraginaceae 143
Botrychium 263
Brachypodium 258
Brassica 20
Briza 251
Bromus 254 257
Buddleja 143
Buddlejaceae 143
Butomaceae 216
Butomus 216

Cakile 25
Calamagrostis 245
Calamintha 167
Callitrichaceae 84
Callitriche 84
Calluna 129

Caltha 10
Calystegia 146
Campanula 127
Campanulaceae 127
Cannabaceae 186
Caprifoliaceae 99
Capsella 23
Cardamine 17
Carduus 117
Carex 231
Carlina 118
Caryophyllaceae 31
Castanea 190
Catabrosa 251
Catapodium 254
Celastraceae 50
Centaurea 118
Centaurium 141
Centranthus 105
Centunculus 139
Cephalanthera 198
Cerastium 33
Ceratophyllaceae 11
Ceratophyllum 11
Ceterach 269
Chaenorhinum 150
Chaerophyllum 94
Chamaemelum 111
Chamaenerion 86
Chamomilla 112
Cheiranthus 15
Chelidonium 13
Chenopodiaceae 174
Chenopodium 174
Chrysanthemum 113
Chrysosplenium 77
Cicerbita 122
Cichorium 119
Circaea 88
Cirsium 117
Cistaceae 26
Cladium 230
Clematis 5
Clinopodium 164
Cochlearia 21
Coeloglossum 203

Comarum 65
Compositae 106
Conium 90
Conopodium 94
Convolvulaceae 146
Convolvulus 147
Cornaceae 99
Cornus 99
Coronopus 24
Corylaceae 189
Corylus 189
Cotoneaster 72
Crambe 25
Crassulaceae 78
Crataegus 72
Crepis 125
Crithmum 95
Crocosmia 204
Cruciferae 14
Cryptogramma 265
Cupressaceae 260
Cuscuta 148
Cymbalaria 150
Cynosurus 252
Cyperaceae 225
Cystopteris 270
Cytisus 53

Daboecia 133
Dactylis 251
Dactylorchis 199
Dactylorhiza 199
Danthonia 249
Daucus 98
Deschampsia 247
Desmazeria 254
Digitalis 152
Dioscoreaceae 207
Diphasiastrum 276
Diphasium 276
Diplotaxis 21
Dipsacaceae 105
Dipsacus 105
Draba 23
Drosera 80
Droseraceae 80
Dryas 62
Dryopteris 270

Elatinaceae 40
Elatine 40
Eleocharis 225
Eleogiton 227
Elodea 217
Elymus 258
Empetraceae 135
Empetrum 135

Endymion 205
Epilobium 85
Epipactis 197
Equisetaceae 274
Equisetum 274
Erica 129
Ericaceae 128
Erigeron 108
Erinus 150
Eriocaulaceae 224
Eriocaulon 224
Eriophorum 228
Erodium 47
Erophila 23
Eryngium 90
Erysimum 19
Escallonia 74
Euonymus 50
Eupatorium 106
Euphorbia 184
Euphorbiaceae 184
Euphrasia 155

Fagaceae 190
Fagus 191
Fallopia 181
Festuca 252 255
Festulolium 252
Filaginella 109
Filago 108
Filices 263
Filipendula 60
Foeniculum 95
Fragaria 63
Frangula 50
Fraxinus 140
Fuchsia 88
Fumaria 14

Galeopsis 169
Galium 101
Gentiana 141
Gentianaceae 140
Gentianella 141
Geraniaceae 45
Geranium 45
Geum 63
Glaucium 13
Glaux 138
Glechoma 167
Glyceria 256
Gnaphalium 108 109
Gramineae 242
Gunnera 82
Guttiferae 40
Gymnadenia 201

Halimione 175
Haloragaceae 81
Hammarbya 195
Hebe 155
Hedera 99
Helianthemum 26 27
Helictotrichon 246
Helxine 186
Heracleum 97
Hesperis 19
Hieracium 125
Hippocastanaceae 51
Hippuridaceae 83
Hippuris 83
Holcus 246
Honkenya 37
Humulus 186
Huperzia 276
Hyacinthoides 205
Hydrilla 217
Hydrocharis 218
Hydrocharitaceae 217
Hydrocotyle 90
Hymenophyllum 264
Hyoscyamus 149
Hypericum 40
Hypochoeris 120

Ilex 49
Impatiens 48
Inula 110
Iridaceae 204
Iris 204
Isoetaceae 278
Isoetes 278
Isolepis 227

Jasione 127
Juncaceae 207
Juncaginaceae 218
Juncus 207
Juniperus 260

Knautia 106
Koeleria 246

Labiatae 164
Lactuca 122 123
Lamium 169
Lapsana 120
Lathraea 161
Lathyrus 59
Lavatera 43
Leguminosae 52
Lemna 214
Lemnaceae 214
Lentibulariaceae 162

Leontodon 120
Lepidium 24
Lepidotis 277
Lepturus 250
Leucanthemum 113
Leucorchis 202
Leycesteria 101
Ligusticum 97
Ligustrum 140
Liliaceae 205
Limonium 135
Limosella 155
Linaceae 44
Linaria 150
Linum 44
Listera 195
Lithospermum 145
Littorella 173
Lobelia 127
Logfia 108
Lolium 252
Lonicera 100
Lotus 56
Luzula 212
Lychnis 32 33
Lycopodiaceae 276
Lycopodiella 277
Lycopodium 276 277
Lycopus 166
Lysimachia 137
Lythraceae 85
Lythrum 85

Malaxis 195
Malus 71
Malva 43
Malvaceae 43
Marrubium 164
Matricaria 111
Matthiola 14
Meconopsis 13
Medicago 53
Melampyrum 159
Melandrium 32
Melica 250
Mentha 164
Menyanthaceae 142
Menyanthes 142
Mercurialis 185
Milium 243
Mimulus 151
Minuartia 36
Moehringia 36
Molinia 249
Monotropa 135
Montia 39
Mycelis 123

Myosotis 144
Myrica 188
Myricaceae 188
Myriophyllum 81

Najadaceae 224
Najas 224
Narcissus 205
Nardus 250
Narthecium 206
Nasturtium 16
Neotinea 202
Neottia 195
Nepeta 164
Nuphar 11
Nymphaea 11
Nymphaeaceae 11

Odontites 159
Oenanthe 95
Oleaceae 140
Omalotheca 108
Onagraceae 85
Ononis 53
Ophioglossum 263
Ophrys 201
Orchidaceae 195
Orchis 198
Origanum 166
Orobanchaceae 161
Orobanche 161
Osmunda 264
Oxalidaceae 48
Oxalis 48
Oxycoccus 128
Oxyria 182

Papaver 12
Papaveraceae 12
Parapholis 250
Parentucellia 160
Parietaria 187
Parnassia 77
Parnassiaceae 77
Pastinaca 97
Pedicularis 160
Pentaglottis 144
Peplis 85
Pernettya 134
Petasites 114
Petroselinum 92
Phalaris 242
Phegopteris 273
Phleum 243
Phormium 205
Phragmites 249
Phyllitis 266

Pilosella 126
Pilularia 265
Pimpinella 93
Pinaceae 260
Pinguicula 163
Pinus 260
Plantaginaceae 173
Plantago 173
Platanthera 203
Plumbaginaceae 135
Poa 255
Polygala 31
Polygalaceae 31
Polygonaceae 178
Polygonum 178 181
Polypodium 273
Polystichum 271
Populus 194
Portulacaceae 39
Potamogeton 218
Potamogetonaceae 218
Potentilla 64
Poterium 67
Primula 137
Primulaceae 137
Prunella 168
Prunus 72
Pseudorchis 202
Pteridium 265
Puccinellia 256
Pulicaria 110
Pyrola 134
Pyrolaceae 134

Quercus 190

Radiola 44
Ranunculaceae 5
Ranunculus 6
Raphanus 25
Reseda 26
Resedaceae 26
Reynoutria 181
Rhamnaceae 50
Rhamnus 50
Rhinanthus 161
Rhodiola 78
Rhododendron 133
Rhynchospora 229
Ribes 74
Rorippa 15 16
Rosa 68
Rosaceae 59
Rubia 101
Rubiaceae 101
Rubus 60
Rumex 182

Ruppia 222
Ruppiaceae 222

Sagina 37
Salicaceae 191
Salicornia 177
Salix 191
Salsola 178
Sambucus 99
Samolus 139
Sanguisorba 67
Sanicula 90
Saponaria 31
Sarothamnus 53
Saussurea 118
Saxifraga 74
Saxifragaceae 74
Scandix 94
Schoenoplectus 227
Schoenus 230
Scirpus 227
Scrophularia 151
Scrophulariaceae 149
Scutellaria 168
Sedum 78
Selaginella 277
Selaginellaceae 277
Senecio 115
Sesleria 249
Sherardia 104
Sieglingia 249
Silene 32
Silybum 118
Sinapis 21
Sisymbrium 19
Sisyrinchium 204
Sium 92

Smyrnium 91
Solanaceae 149
Solanum 149
Soleirolia 186
Solidago 107
Sonchus 121
Sorbus 69
Sparganium 213
Spergula 38
Spergularia 38
Spiraea 59 60
Spiranthes 196
Stachys 168
Stellaria 34
Suaeda 177
Subularia 24
Succisa 106
Symphoricarpos 100
Symphytum 143

Tamus 207
Tanacetum 113
Taraxacum 123
Taxaceae 262
Taxus 262
Teucrium 170
Thalictrum 5
Thelycrania 99
Thelypteris 272
Thlaspi 23
Thymus 167
Torilis 98
Tragopogon 121
Trichophorum 227
Trifolium 54
Triglochin 218
Triodia 249

Tripleurospermum 111
Trisetum 247
Tritonia 204
Tuberaria 26
Tussilago 114
Typha 212
Typhaceae 212

Ulex 52
Ulmaceae 185
Ulmus 185
Umbelliferae 89
Umbilicus 78
Urtica 187
Urticaceae 186
Utricularia 162

Vaccinium 128
Valeriana 104
Valerianaceae 104
Valerianella 105
Verbascum 149
Verbena 164
Verbenaceae 164
Veronica 152
Viburnum 100
Vicia 57
Vinca 143
Viola 27
Violaceae 27
Vulpia 254

Zannichellia 223
Zannichelliaceae 223
Zostera 223
Zosteraceae 223